AF901329

The Roles of Remote Sensing in Nature Conservation

Ricardo Díaz-Delgado • Richard Lucas
Clive Hurford
Editors

The Roles of Remote Sensing in Nature Conservation

A Practical Guide and Case Studies

Springer

Editors
Ricardo Díaz-Delgado
Estación Biológica de Doñana (EBD),
Consejo Superior de Investigaciones
Científicas (CSIC)
Sevilla, Spain

Clive Hurford
Natural Resources Wales
Cardiff, Wales, UK

Richard Lucas
Centre for Ecosystem Science, School of
Biological, Earth and Environmental
Sciences
The University of New South Wales
Kensington, NSW, Australia

ISBN 978-3-319-64330-4 ISBN 978-3-319-64332-8 (eBook)
DOI 10.1007/978-3-319-64332-8

Library of Congress Control Number: 2017953259

© Springer International Publishing AG 2017
This work is subject to copyright. All rights are reserved by the Publisher, whether the whole or part of the material is concerned, specifically the rights of translation, reprinting, reuse of illustrations, recitation, broadcasting, reproduction on microfilms or in any other physical way, and transmission or information storage and retrieval, electronic adaptation, computer software, or by similar or dissimilar methodology now known or hereafter developed.
The use of general descriptive names, registered names, trademarks, service marks, etc. in this publication does not imply, even in the absence of a specific statement, that such names are exempt from the relevant protective laws and regulations and therefore free for general use.
The publisher, the authors and the editors are safe to assume that the advice and information in this book are believed to be true and accurate at the date of publication. Neither the publisher nor the authors or the editors give a warranty, express or implied, with respect to the material contained herein or for any errors or omissions that may have been made. The publisher remains neutral with regard to jurisdictional claims in published maps and institutional affiliations.

Printed on acid-free paper

This Springer imprint is published by Springer Nature
The registered company is Springer International Publishing AG
The registered company address is: Gewerbestrasse 11, 6330 Cham, Switzerland

Many indigenous artists imagine themselves hovering over the land (country) observing both the natural and metaphysical forms and markings of the landscape. These bird's-eye views are characteristic of a hunter and gatherer society. They read the earth surface closely for signs of life, for tracking animals and for recognizing recent events.[1]

[1] Extracted from https://www.aboriginal-art-australia.com/aboriginal-art-library/the-story-of-aboriginal-art/

Acknowledgements

Books are usually the final expression of an idea, as was the case with this one. From the moment of conception to its delivery and publication, many people have emerged to silently push the editors in the right direction. This book has been possible only because of the cooperation and support of very many friends and colleagues over recent years. Most of them are devoted to conservation manage at different scales and in different subject areas but with a common requirement: for technology to enhance their day-to-day work. Thanks to these we decided to stop the race and walk slowly to share our experiences in using remote sensing tools, applications and products.

Initially our thanks must go to the corresponding authors who have worked closely with us to provide the chapters for the book, namely, Jeroen Vanden Borre, Pete Bunting, Anna Allard, Gwawr Jones, Katie Medcalf, Marcos Jiménez, Abigail Sanders, Madhura Niphadkar, Jens Oldeland, Michael Schneider and Sonali Ghosh. We are also grateful to the many associate authors involved in producing the chapters of the book: these appear in the list of contributors.

Our gratitude is also extended to the publishing editors and technicians at Springer NL, particularly Nel van der Werf, for their faith, encouragement, patience and support throughout the writing and collation phases of the publication process.

Last but not least, we would like to thank our families, particularly Yolanda, Jara and Chris, for their endless patience, particularly during the period of collating this book. Furthermore, we will be forever grateful to those who decide to participate in our living survey "Remote Sensing Awareness by Nature Conservation Staff and Its Use", available at https://goo.gl/GqC63f.

Contents

Contributors

Editors

Ricardo Díaz-Delgado Laboratory of Remote Sensing and GIS and Natural Processes Monitoring Team, ICTS-RBD, Estación Biológica de Doñana, CSIC, Sevilla, Spain

Richard Lucas Centre for Ecosystem Science, School of Biological, Earth and Environmental Sciences, The University of New South Wales, Kensington, NSW, Australia

Clive Hurford Natural Resources Wales, Cardiff, Wales, UK

Authors

Anna Allard Swedish University for Agricultural Sciences, SLU, Umeå, Sweden

David Aragonés Laboratory of Remote Sensing and GIS, Estación Biológica de Doñana, CSIC, Sevilla, Spain

Gemma Bell Environment Systems Ltd, Aberystwyth, UK

Johanna Breyer Environment Systems Ltd, Aberystwyth, UK

Peter Bunting Earth Observation and Ecosystem Dynamics Group, Department of Geography and Earth Sciences, Aberystwyth University, Aberystwyth, UK

David Canal Estación Biológica de Doñana, CSIC, Sevilla, Spain

Holger Dettki Umeå Center for Wireless Remote Animal Monitoring (UC-WRAM), Department of Wildlife, Fish, and Environmental Studies, Swedish University of Agricultural Sciences, Umeå, Sweden

Miguel Ferrer Estación Biológica de Doñana, CSIC, Sevilla, Spain

Sonali Ghosh Wildlife Institute of India, Dehradun, Uttarakhand, India

A. Große-Stoltenberg Applied Landscape Ecology and Ecological Planning Institute of Landscape Ecology, Münster, Germany

Birgen Haest Remote Sensing Department, Flemish Institute for Technological Research (VITO), Mol, Belgium

Bird Migration Department, Institute of Avian Research, Wilhelmshaven, Germany

Martin Horlock Norfolk Biodiversity Information Service, Norwich, UK

Marcos Jiménez Remote Sensing Area, National Institute of Aerospace Technology (INTA), Madrid, Spain

Gwawr Jones Joint Nature Conservation Committee (JNCC), Peterborough, UK

Aberystwyth University, Aberystwyth, UK

Manuel Mañez Natural Processes Monitoring Team, ICTS-RBD, Estación Biológica de Doñana, CSIC, Sevilla, Spain

Antonio Martínez Natural Processes Monitoring Team, ICTS-RBD, Estación Biológica de Doñana, CSIC, Sevilla, Spain

Katie Medcalf Environment Systems Ltd, Aberystwyth, UK

Anthea Mitchell Centre for Ecosystem Science, School of Biological, Earth and Environmental Sciences, The University of New South Wales, Kensington, NSW, Australia

L. Naftal Faculty of Natural Resources and Spatial Sciences, Namibia University of Science and Technology, Windhoek, Namibia

Samuel Neal Norfolk Biodiversity Information Service, Norwich, UK

J. Oldeland Biodiversity, Ecology and Evolution of Plants, Biocentre Klein Flottbek and Botanical Garden, University of Hamburg, Hamburg, Germany

Jacqueline Parker Environment Systems Ltd, Aberystwyth, UK

Paul Robinson Joint Nature Conservation Committee (JNCC), Peterborough, UK

Abigail Sanders Computing, Engineering and Science Department, Glyntaff, Trefforest Campus, University of South Wales, South Glamorgan, Wales

Michael Schneider Västerbotten County Administration, Umeå, Sweden

Toon Spanhove Research Institute for Nature and Forest (INBO), Brussels, Belgium

B.J. Strohbach Faculty of Natural Resources and Spatial Sciences, Namibia University of Science and Technology, Windhoek, Namibia

Jeroen Vanden Borre Research Institute for Nature and Forest (INBO), Brussels, Belgium

Part I
An Introduction to Remote Sensing Tools for Habitat Mapping and Monitoring

Introducing the Book "The Roles of Remote Sensing in Nature Conservation"

Ricardo Díaz-Delgado, Clive Hurford, and Richard Lucas

Abstract Although many books describe recent advances in remote sensing, there is a gap in the market for a book dedicated to the application of remote sensing to aid nature conservation. Our activities in training workshops and seminars with conservation managers and practitioners have clarified their requirements, particularly in relation to the low transferability of remote sensing tools in their day to day work. Here, we outline each chapter and present the results of a survey carried out in the context of a workshop on remote sensing tools for conservation management. We provide the original questions and answers which support the common narrative found in the different conservation forums: a need for examples of good practice in the application of remote sensing tools.

Keywords Remote sensing applications • Remote sensing tools • Nature conservation • Nature conservation management • Workshop • Survey

The Need for a Book on RS and Conservation: Filling the gap

If we 'Google' for "Remote Sensing", we find almost 30 million entries, and if we then filter to "books", we discover half a million entries. So why are we writing a new remote sensing book? Well, the discipline of remote sensing offers the tools for countless and varied applications, so much so that if we then search Google for books on remote sensing applied in nature conservation, then the answer supplied by Google Inc. is just one book (Katsch and Vogt 1999; Nagendra et al. 2013; Spanhove et al. 2012). The reader will convene with us there is a gap to fill.

R. Díaz-Delgado (✉)
Estación Biológica de Doñana (EBD), Consejo Superior de Investigaciones Científicas (CSIC), Sevilla, Spain
e-mail: rdiaz@ebd.csic.es

C. Hurford
Natural Resources Wales, Cardiff, Wales, UK

R. Lucas
Centre for Ecosystem Science, School of Biological, Earth and Environmental Sciences, The University of New South Wales, High Street, Kensington (Sydney), NSW 2052, Australia

© Springer International Publishing AG 2017

R. Díaz-Delgado et al. (eds.), *The Roles of Remote Sensing in Nature Conservation*,
DOI 10.1007/978-3-319-64332-8_1

Apart from statistics, we, the editors, had the common feeling that much work had to be done in order to effectively transfer remote sensing technology, findings and tools to managers in protected areas. This idea comes mainly from several European meetings on habitat mapping under Natura 2000 and the implementation of EU Habitats Directive. In one such workshop held in 2016, we proposed to the delegates a survey on their awareness and usability of remote sensing applications. The survey is still accessible and shown in the next section. One of the preliminary findings was the urgent need for a real transference from scientific methods to applicable tools for managers.

In Europe, one of the major challenges of nature conservation is related to the need to provide updated spatial information on habitat status and sustaining long-term monitoring activities (Vanden Borre et al. 2011). The managers of protected areas are aware of such needs and of the potential applications. However, they suddenly face difficult technical issues and background limitations to locally implement remote sensing or to interpret European-wide remote sensing products. So, our main reason for this book was to raise awareness of very basic remote sensing applications and to draw attention to its limitations; but also to highlight and convey the opportunities. To some extent, the book is intended for conservation site managers rather than scientists or technicians. Obviously, we may fail to fulfill these aims, but is up to you, the manager with the book in your hands, who has to decide this.

Book Structure

The book is composed of contributions by several authors. Most of these authors are at the interface between managers in nature conservation and researchers. They have largely contributed to spread the word of remote sensing applications to audiences of managers at international, national and local scales. Their chapters usually provide guidelines to overcome the shortcomings in implementing a technique or taking decisions according to method accuracy and easiness.

The book is therefore structured in four sections: A first introductory section on habitat mapping and monitoring, focusing on the needs of ecologists and managers and providing several adopted solutions. Following this, the second section provides a group of case studies focusing on detailed vegetation mapping, which can provide a baseline for habitat and biodiversity mapping and monitoring; The third section presents case studies that are devoted to remote sensing applications and new technologies which are used to identify, map or track individual or groups of species. We showcase several studies that have used remote sensing tools for faunal surveys, nowadays clearly enhanced by the use of drones. The final section looks ahead through very recent methods that are relevant to nature conservation and a final chapter briefly summarizes the future tangible options to be transferred to managers from the remote sensing community.

Thus, under the first section, Jeroen Vande Borre and colleagues introduce the challenge of continuous habitat mapping by managers. This is a pending request by ecologists to the remote sensing community and raises the critical issue of the transferability of remote sensing methods, which is the motivation behind the whole book.

After posing the needs from ecologists and managers, Peter Bunting shares with us a very valuable manual on the basic remote sensing techniques to elucidate the data source to use and the processing procedures to apply.

In order to enrich the vision of the many utilities of remote sensing in long-term ecological monitoring, Ricardo Díaz-Delgado briefly describes the different monitoring protocols in the protected area of Doñana National Park in Spain that are carried out at the landscape scale by means of remote sensing.

The second section starts with a nationwide case study of habitat mapping. Anna Allard provides a detailed chapter on the setup of the Swedish programme to monitor landscape changes. This chapter is a very good example on how to implement a wise design enhancing monitoring efficiency with the help of remote sensing data sources. Afterwards, Gwawr Jones and colleagues share their experiences in the use of very high resolution (VHR) remote sensing for mapping and assessing te condition of terrestrial coastal habitats listed as Annex I by the EC's Habitats Directive.

With a continued focus on vegetation, Katie Medcalf and colleagues introduce a biodiversity monitoring system that has been implemented in Norfolk, UK, and uses earth observation data. In the final chapter of this second section, Marcos Jiménez and Díaz-Delgado provide a good example on plant species mapping by means of subpixel classification through application of spectral unmixing methods to airborne hyperspectral images.

The third section, which offers case studies focusing on single species, begins with two chapters on the successful use of drones for mapping plant species. Abigail Sanders describes how drones can be used with standard cameras to retrieve the understorey distribution of the invasive. Jens Oldeland and colleagues then provide evidence on the use of proximal sensing with drones equipped with visible cameras to identify different tree species in Namibia.

Still under the section, several chapters converge on the use of new technologies including drones and trap cameras, and innovative digital image analysis to inventory fauna species. The chapter by Michael Schneider and Holger Dettki opens the floor for less traditional remote sensing methods (e.g., camera traps and biotelemetry with Global Positioning Systems or GPS) in the surveillance and monitoring of large carnivores in Sweden. Then, Sonali Gosh and Richard Lucas introduce the integration of camera trap surveys with remote sensing information on land cover, fire and human activities to better understand the distribution, movement and behavior of tigers and other large cats in Manas National Park in northeast India. Clive Hurford in his chapter reviews different digital image analysis techniques and open source software to automatically count bird species in pictures of bird flocks. Finally, Ricardo Díaz-Delgado and colleagues share with us an essay where drones, commonly referred to as unmanned airborne systems or vehicles (UAS or UAVs), are contributing to improve the efficiency of bird breeding monitoring in the Doñana protected area.

In the final section, Richard Lucas and Anthea Mitchell presents an overview of the Earth Observation Data for Ecosystem Monitoring (EODESM) system which was developed through two European-funded projects (BIO_SOS and ECOPTENTIAL). This system facilitates the integration of a diverse range of remote sensing observations and derived products, ancillary spatial datasets and field data to classify land covers and habitats using consistent taxonomies and provide evidence-based assessments of change. The book is closed with a short epilogue on what are the most likely innovations in remote sensing that will be available in the foreseeable future for nature conservation and management.

The Feeling of Managers: A Survey

We, the authors, are engaged in several activities involving information exchange with conservation managers, either at local, regional, national or international level. These include workshops, seminars, training lectures and participatory meetings and so on. In the preparation of the 2016 workshop on the application and potential of remote sensing in nature conservation management, a survey was prepared to review the awareness of conservation managers of remote sensing applications and their experiences of working with data and derived information. Twenty of the workshop participants provided a response to the survey, with these being either managers or technicians working in protected areas although several few scientists involved in research projects within protected areas were also present. This sample size was too small and hence we would like to keep the exercise open to invite views from others involved in remote sensing for nature conservation. The survey can therefore be accessed at https://goo.gl/GqC63f and we will endeavour to write up the outcomes as a journal article in the near future.

Below, we have listed the questions together with the answers and the percent assigned to each one by the respondents, although these are available on the survey website:

1. Have you ever used remote sensing in your day-to-day work? (exclusive answer)
   (a) Yes (90%)
   (b) No
   (c) I don't know (10%)
2. In such a case, for what purpose? (not exclusive answer)
   (a) Data source for management planning (63%)
   (b) Data source for decision-making (54%)
   (c) Data source for monitoring of management actions (36%)
   (d) Data source for measuring the extent of disturbances or extreme events such as wildfires, invasive species spread, pollution, etc. (45%)

3. What remote sensing applications do you know in nature conservation and management of protected areas? (not exclusive answer)

   (a) Use of either historical or recent aerial pictures, (90%)
   (b) Use of land use cover, vegetation or habitat maps (100%)
   (c) Use of multispectral indices (NDVI, EVI, GEMI, etc.) (72%)
   (d) Others, please specify (18%)

4. Are you aware about remote sensing research projects carried out in your protected area? (exclusive answer)

   (a) Yes, many (46%)
   (b) Yes, some (46%)
   (c) Not at all (8%)

5. In such a case, please briefly describe one example:

   Some answers were: monitoring of natural systems, LiDAR flights, mapping of invasive species, among others.

6. To what extent do you think remote sensing and its products play a relevant role in the day-to-day management of a protected area? (exclusive answer)

   (a) It is essential (54%)
   (b) It is a complementary information (46%)
   (c) It is not important (0%)

7. What advantages do you believe remote sensing provides to nature conservation? (not exclusive answer)

   (a) Synoptic view of the protected area (81%)
   (b) Historical background (91%)
   (c) Multispectral information (wavelengths other than visible) (72%)
   (d) I don't know (0%)
   (e) Others, please specify, (0%)

8. Is there a staff member in your management headquarters skilled in the use of remote sensing? (exclusive answer)

   (a) Yes, he/she's an expert (9%)
   (b) Yes, he/she's a technician with some background (36%)
   (c) No (55%)

9. In the case you could incorporate a new technician in your working team, how relevant would it be that he/she has a background in remote sensing applications to nature conservation? (exclusive answer)

   (a) Of critical importance (46%)
   (b) Of relative importance (54%)
   (c) Not relevant (0%)

10. Please, indicate which of the following remote sensing applications you are aware of (not exclusive answer)

    (a) Interpretation of aerial pictures (82%)
    (b) Digital classification of multi and hyperspectral images (for land use cover maps) (55%)
    (c) Thermal imagery (36%)
    (d) Radar imagery or data (27%)
    (e) LiDAR data or derived information such as Digital Elevation Models (72%)
    (f) Field spectroscopy and spectral signatures (27%)
    (g) Vegetation indices (NDVI, EVI, GEMI, etc.) (73%)
    (h) Portable ground radar (9%)
    (i) Terrestrial LiDAR (54%)
    (j) Airborne images (by airplane, drones or other platform) (73%)
    (k) Bathimetric mapping with sonar (36%)
    (l) Others, please specify

11. Do you know which remote sensing applications have been used in the protected area you work for? (not exclusive answer)

    (a) Iinterpretation of aerial pictures (90%)
    (b) Digital classification of multi and hyperspectral images (for land use cover maps) (55%)
    (c) Thermal imagery (0%)
    (d) Radar imagery or data (0%)
    (e) LiDAR data or derived information such as Digital Elevation Models (55%)
    (f) Field spectroscopy and spectral signatures (9%)
    (g) Vegetation indices (NDVI, EVI, GEMI, etc) (46%)
    (h) Portable ground radar (0%)
    (i) Terrestrial LiDAR (36%)
    (j) Airborne images (by airplane, drones or other platform) (46%)
    (k) Bathymetric mapping with sonar (18%)
    (l) Others, please specify

12. Which ecosystem/s is/are the most representative/s of the protected area where you work? (not exclusive answer)

    (a) Forest and shrubs (82%)
    (b) Wetlands (27%)
    (c) Alpine grasslands and meadows (45%)
    (d) Coastal and marine ecosystems (55%)

13. Did you ever hear about Copernicus European programme (formerly known as GMES)? (exclusive answer)

    (a) Yes indeed (45%)
    (b) Not at all (36%)
    (c) It sounds familiar to me (18%)

14. Do you know about Remote Sensing agencies and associations in your country? (exclusive answer)
    (a) Yes, I do (36%)
    (b) I am afraid I don't (55%)
    (c) I am aware of something similar (9%)
15. Overall and from your point of view, the main remote sensing applications are: (not exclusive answer)
    (a) Mapping of vegetation, habitats, species etc. (81%)
    (b) Valuable images to visualize the territory (64%)
    (c) Information to characterize land use cover change through time (100%)
    (d) Very detailed images to detect, identify and discriminate specific features such as wildfires, forest decay, etc. (91%)
    (e) Landscape analysis (connectivity, fragmentation, etc.) (73%)
    (f) Quantitative information of the territory (biomass, primary production, etc.) (73%)
    (g) Others, please specify
16. Which of the following do you think are the constraints of using remote sensing applications in your day-to-day work: (not exclusive answer)
    (a) A too high degree of knowledge required to deal with the information provided by remote sensing sources (73%)
    (b) Very high prices of the very high resolution images (both for satellite, airborne and drone) (55%)
    (c) Overall ignorance of the operational applications (46%)
    (d) Low temporal resolution of the available remote sensing images (9%)
    (e) Low spatial resolution of the available remote sensing images (18%)
    (f) Others, please specify: *lack of time and low interest level of the staff on such techniques*
17. Do you think there are enough lectures on remote sensing in the University degrees related to nature conservation? (such as Biological or Environmental Sciences, Forestry, Geography, etc.) (exclusive answer)
    (a) Yes (0%)
    (b) No (54%)
    (c) I don't know (46%)

Although the size of the sampled population is low, the survey provides a preliminary overview of the knowledge that this group of conservation managers had on the topic. They mostly knew about different and specific remote sensing applications such as LiDAR or vegetation indices, while they miss more skilled staff at the management agencies.

We kindly encourage the readers to read the book and fill the survey available online in order to enhance the knowledge on the success in transferring remote sensing applications to nature conservation and management.

Acknowledgments This book is the result of discussions by the authors during several meetings and workshops of the ENCA, Eurosite and other agencies where a common ground was identified to target the challenge of transferring remote sensing applications to the nature conservation managers. The authors are therefore very grateful to these agencies and the participants to these workshops who kindly provided their experience and feedback on our requests. We also thank to Springer its kind support and patience in editing and typesetting the book.

References

Katsch, C., Vogt, H.: Remote sensing from space – present and future applications in forestry, nature conservation and landscape management. South. Afr. For. J. **185**, 14–27 (1999)

Nagendra, H., Lucas, R., Honrado, J.P., Jongman, R.H.G., Tarantino, C., Adamo, M., Mairota, P.: Remote sensing for conservation monitoring: assessing protected areas, habitat extent, habitat condition, species diversity, and threats. Ecol. Indic. **33**, 45–59 (2013). doi:10.1016/j.ecolind.2012.09.014

Spanhove, T., Vanden Borre, J., Delalieux, S., Haest, B., Paelinckx, D.: Can remote sensing estimate fine-scale quality indicators of natural habitats? Ecol. Indic. **18**, 403–412 (2012). doi:10.1016/j.ecolind.2012.01.025

Vanden Borre, J., Paelinckx, D., Mücher, C.A., Kooistra, L., Haest, B., De Blust, G., Schmidt, A.M.: Integrating remote sensing in Natura 2000 habitat monitoring: prospects on the way forward. J. Nat. Conserv. **19**, 116–125 (2011). doi:10.1016/j.jnc.2010.07.003

Towards a Mature Age of Remote Sensing for Natura 2000 Habitat Conservation: Poor Method Transferability as a Prime Obstacle

Jeroen Vanden Borre, Toon Spanhove, and Birgen Haest

Abstract Over the past decades, remote sensing has been repeatedly identified as a promising and powerful tool to aid nature conservation. Many methods and applications of remote sensing to monitor biodiversity have indeed been published, and continue to be at an increasing rate. As such, remote sensing is seemingly living up to its expectations; yet, its actual use in nature conservation (or rather the lack thereof) contradicts this. We argue that, at least for the practical implementation of regular vegetation monitoring, including within protected areas (e.g., European Natura 2000 sites), a lack of transferability of remote sensing methods is an overlooked factor that hinders its effective operational use for nature conservation. Among the causes of poor method transferability is the large variation in objects of interest, user requirements, ground reference data, and image properties, but also the lack of consideration of transferability during method development. To stimulate the adoption of remote sensing based techniques in vegetation monitoring and conservation, we recommend that a number of actions are taken. We call upon remote sensing scientists and nature monitoring experts to specifically consider and demonstrate method transferability by using widely available image data, limiting ground reference data dependence, and making their preferably open-source programming code publicly available. Furthermore, we recommend that nature conservation specialists are open and realistic about potential outcomes by not expecting the replacement of current in-place methodologies, and actively contributing to the thought process of generating transferable and repeatable methods.

We believe a new focus on method repeatability instead of novelty, would herald a mature era for remote sensing in nature conservation, in which remote sensing, through its operational use, would truly live up to its potential for nature conservation.

J. Vanden Borre (✉) • T. Spanhove
Research Institute for Nature and Forest (INBO), Kliniekstraat 25, 1070 Brussels, Belgium
e-mail: Jeroen.VandenBorre@inbo.be; Toon.Spanhove@inbo.be

B. Haest
Remote Sensing Department, Flemish Institute for Technological Research (VITO), Boeretang 200, 2040 Mol, Belgium

Bird Migration Department, Institute of Avian Research, An der Vogelwarte 21, 26386 Wilhelmshaven, Germany
e-mail: birgen.haest@gmail.com

© Springer International Publishing AG 2017

R. Díaz-Delgado et al. (eds.), *The Roles of Remote Sensing in Nature Conservation*,
DOI 10.1007/978-3-319-64332-8_2

Keywords Earth observation • EU Habitats Directive • Annex I habitats • Habitat mapping • Habitat monitoring • Conservation status • User requirements

Introduction

Remote sensing and, in particular, earth observation (i.e., the gathering of information about the earth's surface and events on it from the air or from space; Jones and Vaughan 2010), has been around as a tool and a science for several decades. Over the years, technological advancements in sensors (increasing spatial and spectral resolution, active sensors), platforms (satellites, airborne, and more recently unmanned systems), image analysis techniques (e.g., object-based image analysis; Blaschke et al. 2014) and computer processing capacity have made the tool ever more powerful. Not surprisingly, remote sensing has repeatedly been identified as a very promising and powerful tool to aid biodiversity mapping and monitoring (e.g., Stoms and Estes 1993; Innes and Koch 1998; Nagendra 2001; Kerr and Ostrovsky 2003; Turner et al. 2003; Aplin 2005; Xie et al. 2008; Gross et al. 2009; Wang et al. 2010; Nagendra et al. 2013; Corbane et al. 2015; Pettorelli et al. 2016), and many studies have set out to develop practical applications of remote sensing to specific needs of the ecological and conservation communities (e.g., Feilhauer et al. 2014; Franke et al. 2012; Kopeć et al. 2016; Neumann et al. 2015; Riedler et al. 2015; Thoonen et al. 2013; Zlinszky et al. 2014). Indeed, the use of remote sensing offers a number of distinct advantages over field-based data collection, such as the provision of a spatially explicit and consistent view over larger areas, that is free of *a priori* interpretation at the time of data collection. It also allows for easier updates, and even retrospective evaluation of changes (when archive images are available) (Vanden Borre et al. 2011a).

Natura 2000 is a European Union-wide policy aimed at the restoration and long-term maintenance of the most typical and most threatened habitat types and species in Europe. As part of its implementation, EU member states have selected outstanding natural areas within their territories to become part of the Natura 2000 network (Evans 2012), and are monitoring and reporting the status of species and habitat types on their territory in six-yearly intervals. Vanden Borre et al. (2011b) analysed the use of remote sensing in this process and found that its application was mostly limited to research contexts. Despite the potential advantages and continuous scientific and technological progress, the step towards operational monitoring using remote sensing appeared to be too big a hurdle to take. It has been suggested that ecologists have lagged behind in adopting the new technological opportunities provided by remote sensing (Newton et al. 2009; Horning et al. 2010), and one could easily think that this is because remote sensing requires technical skills that are too far from general ecologists' skills. But is that really the case? And if so, is it the only reason? Ecologists and nature conservationists have not been particularly slow in taking up other useful technologies when they emerged, such as GPS and radio tracking, population genetic analysis, aerial photography and GIS, computer-intensive statistical procedures, or citizen science, each of which requiring specialized skills and (often expensive) equipment.

In this paper, we focus on another critical factor that, in our view, may form a major barrier to the operationalization of remote sensing in nature conservation applications, but is all too often overlooked. That factor is method transferability, or more precisely, the lack thereof. We argue that poor transferability of remote sensing methods may be the cause of many disappointments among potential users, and therefore hinder further attempts to their operational implementation in biodiversity conservation. We discuss different causes of poor method transferability, and illustrate this with some examples based on Natura 2000 habitat monitoring, an application field where operationalization of remote sensing monitoring could have a large impact. Finally, we discuss pathways to overcome the current mismatch, and make remote sensing applications more widely useful and transferable.

Transferability as a Critical Success Factor: The Case of Natura 2000 Habitat Monitoring

Vanden Borre et al. (2011b) noticed a low uptake of operational remote sensing in Natura 2000 habitat monitoring. Several possible causes were put forward, both on the technical and on the practical side. Technical challenges include, for instance, the fact that habitat types strongly vary in scale, show high variation across Europe, and sometimes can only be reliably identified by the presence of a selected set of indicator species (the latter mostly occurring in low numbers) (Vanden Borre et al. 2011a). Challenges of a more practical kind are *inter alia* the difficulty of obtaining suitable image data at a suitable time, and the potentially high cost associated with that. Also, many potential user organizations lack the profound GIS and remote sensing skills needed to effectively make use of this type of information (Kennedy et al. 2009).

While each of these causes may be valid, another aspect has, until now, been largely ignored but could prove to be a critical factor, namely: a lack of method transferability. With the right kind of data and properly fine-tuned methods, remote sensing can deliver highly detailed spatial and thematic information. Many biodiversity monitoring schemes, such as Natura 2000 habitat monitoring, require, or at least benefit from, high spatial and thematic detail. However, when details are important, and the potential to extract such detail is there, approaches not surprisingly tend to be specifically geared to the problem at hand. After all, remote sensing scientists, like other scientists, need to publish to persist. And failures usually do not make it to publication (Fanelli 2012; Matosin et al. 2014). Additionally, in a method-focused field like remote sensing, coming up with novel methods further increases the chances of getting the work published. In recent years, the number of published methods for detailed remote sensing of Natura 2000 habitats has soared. Many of these include extensive fine-tuning to tackle complex problems. As such, the methods are often adjusted to one particular location, to such an extent that it may jeopardize their applicability elsewhere.

Possible Causes of Poor Transferability

Objects of Interest Differ (Even if They Carry the Same Name)

Biodiversity is extremely diverse, as are the stakeholders in biodiversity conservation and the (living) objects they are studying, monitoring and/or conserving. Although designated by the same name, objects may actually be extremely variable across Europe or even globally, and therefore hard to define. Forest, for instance, is a widely used and intuitively understood term, but the variation in national forest definitions is huge, with different thresholds applied for crown closure, tree height, minimum area and minimum width (Lawrence et al. 2010). As a result, what is considered a forest in one country, may not classify as a forest in a neighbouring country, or *vice versa.*

Natura 2000 habitat types provide another example of different definitions under the same name. Although this typology is embedded in European legislation, guidance from the EU on the identification of the different habitat types has been rather limited and biased towards regions with pre-existing vegetation typologies (UK, Germany, Nordic countries) at the time when the list was drawn up (Evans 2006). As a result, most member states produced their own national interpretation handbooks for the habitat types on their territory, with varying levels of agreement between states. To illustrate this variation, we screened a selection of these handbooks (see Annex) for the national definitions of two rather widespread habitat types; the 'European dry heaths' (Natura 2000 code: 4030) (45 descriptive definitions of the type or subtypes found, in 18 different sources, from 10 member states) and '*Luzulo-Fagetum* beech forests' (Natura 2000 code: 9110) (19 descriptions, 16 sources, 8 member states), We noted the plant species mentioned as 'characteristic', 'typical', 'indicative', or similar wordings. Looking at these species lists, the extent of the variation (or the difference in interpretation) becomes immediately obvious (Fig. 1). The vast majority (68%) of species were mentioned only once or twice. Of the 219 vascular plant species and 37 lichens and mosses typical of 'European dry heaths', only eleven vascular plant species (i.e., 5%) and one moss were mentioned in more than a quarter of the descriptions. For the '*Luzulo-Fagetum* beech forests', the species list is somewhat more uniform, with 23 out of 90 vascular plants (26%) and three out of nine mosses mentioned in more than a quarter of the habitat descriptions. The low agreement in component species is partly due to different ecological characteristics of the habitat across Europe. However, different interpretations between member states, and even between regions within a member state, undoubtedly also play a role (Evans, 2010).

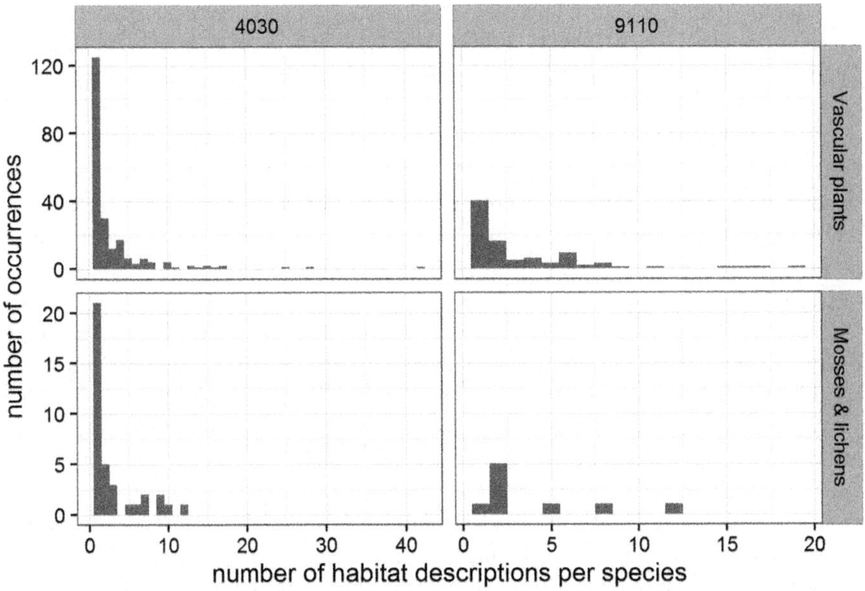

Fig. 1 Histogram of the number of national habitat definitions in which a plant species is mentioned, for the two Natura 2000 habitat types European dry heaths (4030) and *Luzulo-Fagetum* beech forests (9110) and their subtypes. The majority of species are only mentioned for a few (sub) types and in a few member states. In contrast, only a few species, mainly vascular plants, are mentioned in the bulk of the descriptions

User Requirements Abound and Differ (Even if They Appear Similar)

Understanding user requirements is an integral part of the design of any kind of product, and is critical to its eventual success. Discovering the real requirements, however, is often difficult. Different people will define a specific set of needs in different ways, mostly depending on their personal experience, interests and expectations. Furthermore, most users typically think along the lines of their current framework and existing workflows, rather than being innovative and focusing on the real needs (both current and future) and exploiting technical developments. Similarly, users are not always aware of the specifications of current products and of the potential of future products. As a result, what is a similar requirement at first sight (e.g., conservation status of a habitat) may actually reveal very different underlying data needs. In heathlands, for instance, desiccation and atmospheric nitrogen deposition both lead to the dominance of Purple moorgrass (*Molinia caerulea*). From a remote sensing perspective, the observation (i.e., the grass cover) is the same, but from a management perspective, different actions are required.

Conservation status reporting under Natura 2000 requires four aspects for each habitat to be monitored and assessed against thresholds or reference values: habitat area, habitat range, specific structures and functions (incl. typical species), and future prospects (ETC/BD 2011; see also Vanden Borre et al. 2011b). To assess the specific structures and functions, several EU member states have taken an approach to evaluate local habitat quality at (all or a sample of) individual habitat occurrences, and identified relevant indicators to grade habitat quality in the field (e.g., North Rhine-Westphalia, Germany: Verbücheln et al. 2002; Austria: Ellmauer 2005; Flanders, Belgium: T'jollyn et al. 2009). Such an approach benefits local conservation managers, by providing data on where habitats are in good or poor condition. This may be directly used to prioritize the areas where conservation actions are needed most.

Generally, the indicators cover floristic and/or fauna composition (e.g., number of key species present), vegetation structure (e.g., height, cover, proportion of dead wood), disturbances (e.g., invasive species), and landscape configuration (e.g., connectivity and isolation). All these indicators relate to specific properties of the habitat (Bock et al. 2005), and hence contain useful information for managers and policy-makers (Spanhove et al. 2012; T'jollyn et al. 2009). However, their relative importance in the eventual conservation status assessment strongly depends on the context (geographically, socio-economically, etc.). Adding to that the large variation between and within habitat types, it becomes clear that 'default' remote sensing methods will only be able to address a small number of indicators. A degree of adaptation of the method will generally be unavoidable.

As an example, we evaluated the indicators that are used for the assessment of natural habitats in Flanders (T'jollyn et al. 2009) and Germany (PAN & ILÖK 2010). In Flanders, 120 indicators have been defined for 43 habitat types (69 subtypes). Frequently used indicators are the number and spatial coverage of key species, and some widespread threats such as forest, grass and tall herb encroachment and alien invasive species. The majority of the indicators, however, are only relevant for a few habitats, and almost half of them (48%) are used for only one habitat type or subtype (Fig. 2). Figure 3 shows the results of a 'rarefaction'-like analysis, where the number of conservation status indicators or key species is modelled as a function of the number of habitat (sub) types evaluated, for Flanders (T'jollyn et al. 2009) and Germany (PAN & ILÖK 2010). None of the curves shows signs of nearing a plateau value, indicating that any habitat type added will further raise the total number of indicators/species on which data need to be gathered.

Available Ground Reference Data Differ

Ground reference data, if available, come in many forms, both sampling-wise and content-wise (Stehman and Czaplewski 1998). Depending on the applied sampling design, samples can be provided as points (but referring to a spatial sample unit on the ground of all possible sizes and shapes) or as objects (polygons). Their selection can have followed random or non-random sampling principles, and they can have been collected in the field, or deduced from other (usually older) map information.

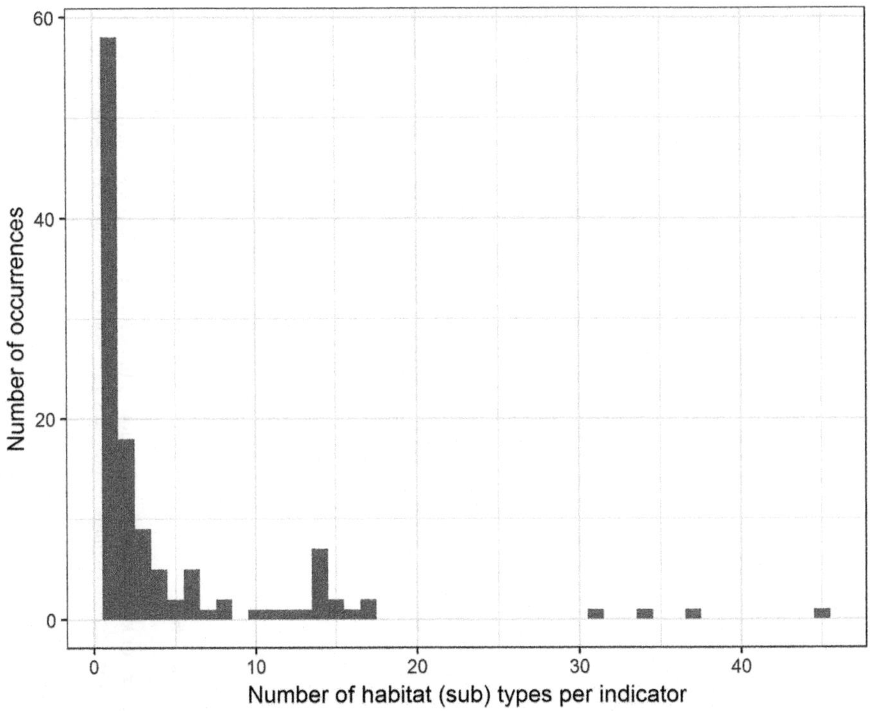

Fig. 2 Histogram of the number of habitat (sub) types in which a given indicator is used in the Flemish manual for conservation status assessments (T'jollyn et al. 2009). The majority of the indicators are used for one or a few habitat (sub) types only

With respect to content, the dataset can contain all classes in the area, or only target classes. Likewise, it can have mixtures and gradients between classes included, or (un-)intentionally excluded from the dataset. Sample sizes can be more or less evenly spread over the classes, or highly skewed. Also, the data can be topical or (partially) outdated. In some cases, reference data can even be completely absent.

All these aspects affect the potential outcome and application of specific remote sensing methods to a certain degree. Given the currently limited sharing of reference data (Pettorelli et al. 2014), which further impedes the availability of similar reference data over different areas, the impact of ground reference data availability on remote sensing method transferability cannot be disregarded.

Image Properties Differ

In common with ground reference data, image data also come in a variety of forms. Not only do specifications for each sensor differ (number and location of bands, band widths and spectral sensitivities, spatial and radiometric resolutions, known

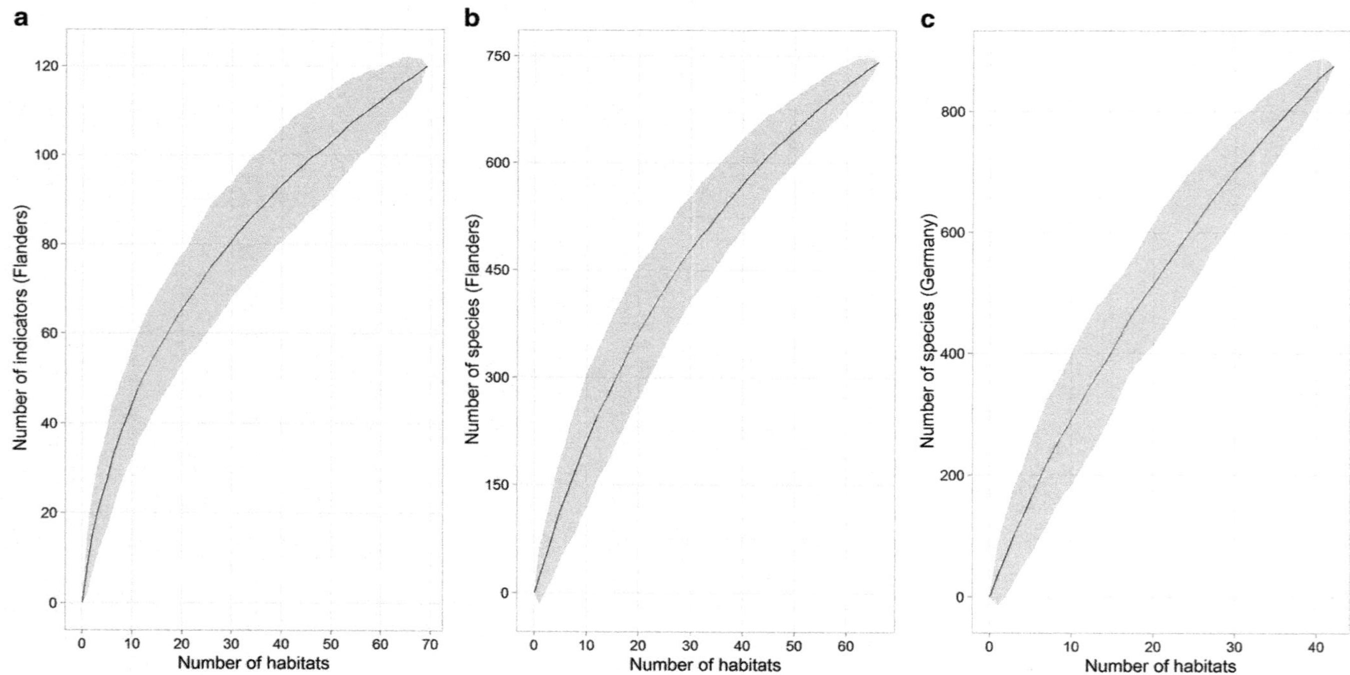

Fig. 3 'Rarefaction'-like curves of the total number of conservation status indicators or key species as a function of the number of habitats evaluated. (**a**) and (**b**) Conservation status indicators resp. key species for Flanders. (**c**) Key species for Germany. Shading indicates 95% confidence intervals, based on Monte Carlo bootstrap method. For none of these cases, a plateau value is reached, indicating that additional habitat types would require information on a mostly new set of indicators or species

and unknown flaws and anomalies,...), commonly applied pre-processing algorithms may further inflate differences, even between two images of the same sensor (through resampling and reductions in spatial, spectral and/or radiometric resolution) (Jones and Vaughan 2010). For vegetation monitoring however, constraints in applications most often arise from difficulties in obtaining suitable high-resolution imagery. A limited number of data providers, cloudy weather, and yearly variations in vegetation phenology, make it very difficult to obtain truly comparable pairs of image data from two different timestamps.

Real-World Examples

Next, we present some examples of transfer cases from our own experience, illustrating the type of problems that occurred, and the solutions that were (sometimes) found.

The method we tested for transferability was originally developed in the framework of a Belgian research project called 'HABISTAT'. Its aim was to map Natura 2000 habitat types and their conservation status in Natura 2000 sites, with a focus on Atlantic heathland in Flanders. The method is described in Haest et al. (2010) and Haest et al. (2017). In short, it consists of a supervised classification (Linear Discriminant Analysis) of vegetation using classes geared towards remote identification, followed by a rule-based reclassification into occurrences ('patches') of Natura 2000 habitat.

In the frame of a follow-up project 'MS.MONINA' (EU 7th Framework Programme), three transfer cases were tested, with the method being applied to: (a) the original study area, but at later dates, (b) a Continental heathland ecosystem in Germany, and (c) a coastal dune ecosystem in Flanders. In the course of the tests, several method adjustments had to be made:

- In the original context of the HABISTAT project, the study area of Atlantic heathland was intensively visited, with many hundreds of field reference plots recorded in either carefully selected homogeneous vegetation patches or in random locations. These served as training and validation data respectively. In a more operational remote sensing context, such large datasets are rarely available, and therefore, the method had to be extended to accommodate different types of training data. In particular, the method was adapted to work with extracted training data from existing (field-driven) vegetation maps, rather than the point-based field plots in the original version. We considered two options, using either single or multiple points per polygon of the field map.
- For the new study sites in the transfer cases, a new remote-sensing oriented classification scheme had to be developed, based on vegetation descriptions or maps of the new study sites. It was not always straightforward to 'translate' existing field data into the new classification scheme, either because the existing classification was too detailed or not detailed enough, or because the existing field information was biased towards a few specific vegetation types only.

- The reclassification algorithms to convert a vegetation map into a habitat map also had to be further developed. Reclassification rules were added or adapted to deal with the known habitats that occurred on the new study sites. Furthermore, the rules for delineating 'relevant' habitat patches (based on the BIOHAB method; Bunce et al. 2008) were extended to work not only in a heathland context, but (at least in theory) in all European ecosystems.

It is important to notice that in the tests, we did not strive for success at all cost. It was our intention to evaluate what could be done with the method in – what we considered – a realistic timeframe. We therefore limited the available time to approximately 2 weeks of work per case. If, after that time, we did not feel yet we were on track to a result in the very near future, we would stop the test.

Transfer Test Case 1: To Later Dates

Study Site

The heathland of Kalmthoutse Heide is part of a cross-border Natura 2000 site in Belgium and the Netherlands, located some 20 km northeast of the city of Antwerp (Belgium), in the Atlantic Biogeographical Region of Europe. It consists of a core area of over 1000 ha of open heathland and inland dunes, surrounded by broad-leaved forests (oak, birch) and pine plantations. Prevailing Natura 2000 habitat types are wet (4010) and dry heathland (4030) and open habitats on inland dunes (2310 and 2330). The area is also home to a wide range of heathland-associated birds, reptiles, amphibians and invertebrates.

The site's location, amidst an intensively used agricultural area (especially fertilizer-demanding maize culture) and in the vicinity of the port of Antwerp's industry, makes it extremely prone to nutrient enrichment from atmospheric depositions. This results in the encroachment of Purple moorgrass (*Molinia caerulea*) at the expense of the ericoid dwarf shrub vegetation (*Calluna vulgaris* and *Erica tetralix*) that is typical of heathlands. Other pressures include dune fixation by an invasive exotic moss species (*Campylopus introflexus*), desiccation from drinking water extraction, recreational use, and uncontrolled wildfires. The management primarily aims at counteracting the negative effects of nutrient accumulation, through grazing, mowing and sod-cutting, and restoring the natural dynamics of the inland dunes.

Method Application

The original application of our method was on a Airborne Hyperspectral Scanner (AHS) image acquired on 2nd June 2007 by the Spanish National Institute of Aeronautics (INTA). The same method was then applied (the transfer case) to several hyperspectral images of the same area acquired by the German Aerospace Centre's (DLR) Airborne Prism Experiment (APEX) in 2010, 2011 and 2012. The same hierarchical classification scheme (Table 1) and the same reference dataset

Table 1 Four-level classification scheme for the Grenspark De Zoom-Kalmthoutse Heide

Level 1		Level 2		Level 3		Level 4	
H	Heath-land	Hd	Dry heathland	Hdc	*Calluna vulgaris*-dominated heathland	Hdcy	*Calluna*-stand of predominantly young age
						Hdca	*Calluna*-stand of predominantly adult age
						Hdco	*Calluna*-stand of predominantly old age
						Hdcm	*Calluna*-stand of mixed age classes
		Hw	Wet heathland	Hwe	*Erica tetralix*-dominated heathland	Hwe-	*Erica tetralix*-dominated heathland
		Hg	Grass-encroached heathland	Hgm	*Molinia caerulea*-dominated heathland	Hgmd	*Molinia*-stand on dry soil
						Hgmw	*Molinia*-stand on moist soil
				Hgd	*Deschampsia flexuosa*-dominated heathland	Hgd-	*Deschampsia flexuosa*-dominated heathland
		Hs	Shrub/tree-encroached heathland	Hst	Tree-encroached heathland	Hst-	Tree-encroached heathland
				Hsr	*Rubus*-encroached heathland	Hsr-	*Rubus*-encroached heathland
G	Grass-land	Gt	Temporary grassland	Gt-	Temporary grassland	Gt--	Temporary grassland
		Gp	Permanent grassland	Gpa	Permanent grassland in intensive agricultural use	Gpap	Species-poor permanent agricultural grassland
						Gpar	Species-rich permanent agricultural grassland
				Gpn	Permanent grassland with semi-natural vegetation	Gpnd	Dry semi-natural permanent grassland
				Gpj	*Juncus effusus*-dominated grassland	Gpj-	*Juncus effusus*-dominated grassland

(continued)

Table 1 (continued)

Level 1		Level 2		Level 3		Level 4	
F	Forest	Fc	Coniferous forest	Fcp	Pine forest	Fcpc	Corsican pine
						Fcps	Scots pine
		Fd	Deciduous forest	Fdb	Birch forest	Fdb-	Birch forest
				Fdq	Oak forest	Fdqz	Pedunculate oak
S	Sand dune	Sb	Bare sand	Sb-	Bare sand	Sb--	Bare sand
		Sf	Fixed sand dune	Sfg	Sand dune with grasses as important fixators	Sfgm	Sand dune fixed by grasses and mosses
				Sfm	Sand dune with mosses as dominating fixators	Sfmc	Fixed sand dune with predominantly *Campylopus introflexus*
						Sfmp	Fixed sand dune with predominantly *Polytrichum piliferum*
W	Water body	Wo	Oligotrophic water body	Wov	Shallow, vegetated oligotrophic water body	Wov-	Shallow, vegetated oligotrophic water body
				Wou	Unvegetated oligotrophic water body	Wou-	Unvegetated oligotrophic water body
A	Arable fields	Ac	Arable field with crop	Acm	Arable field – Maize	Acm-	Arable field – Maize
				Aco	Arable field – other crops	Aco-	Arable field – other crops

(containing data from 2007–2009) were re-used on each occasion. However, all field data were checked on aerial photos to account for major events that would render them invalid for further use (e.g., sod cutting, tree removal and wildfires). Therefore, the effective sample size for training and validation gradually decreased from 2010 to 2012.

An overview of accuracies obtained after transfer can be found in Table 2. Not surprisingly, accuracy levels decreased from level 1 (6 classes) to level 4 (24 classes) because of the increasing level of complexity and fewer training samples per class. It was also apparent that accuracy levels dropped from 2007 to 2012. This can be explained by the increasing time gap between the acquisitions of the training data (2007–2009) and the imagery.

Figure 4 shows the dynamics of an excerpt of the Kalmthoutse Heide as revealed by the time series of image classifications. The sudden and large increase of *Sfm* (bare sand with some mosses) in June 2011 was a result of a wildfire in May 2011. In the following months and years, this bare sand was gradually recolonized, mainly by *Molinia caerulea* (*Hgm*).

Table 2 Overview of obtained accuracies of method transfer to later dates, for the Grenspark De Zoom-Kalmthoutse Heide

Image date	Image type	Training sample size	Level	OA (%)	Avg. PA (%)	Avg. UA (%)	Kappa
2007–June–02	AHS	938	1	93.82	93.19	94.00	0.93
			2	90.19	89.63	90.09	0.89
			3	87.10	79.83	84.85	0.86
			4	81.24	69.87	73.64	0.80
2010–June–28	APEX	1375	1	89.09	86.40	88.03	0.86
			2	83.93	83.76	84.52	0.82
			3	81.16	74.96	78.57	0.79
			4	70.84	63.65	66.43	0.69
2011–June–27	APEX	687	1	87.48	87.59	87.86	0.85
			2	83.99	83.10	84.38	0.82
			3	77.87	71.32	74.49	0.76
			4	69.87	62.67	64.45	0.68
2011–Sept–24	APEX	687	1	85.15	85.34	85.09	0.82
			2	80.49	77.77	80.31	0.78
			3	75.84	66.22	72.85	0.74
			4	68.70	59.62	63.53	0.67
2012–July–02	APEX	686	1	87.32	87.70	87.71	0.85
			2	81.49	78.91	79.16	0.79
			3	74.20	66.13	68.97	0.72
			4	67.64	58.55	59.59	0.65

AHS Airborne Hyperspectral Scanner, *APEX* Airborne Prism Experiment. Level: see Table 1, *OA* overall accuracy, *Avg. PA* average producer's accuracy, *Avg. UA* average user's accuracy

Transfer Test Case 2: To a Different Site – From Atlantic to Continental Heathland

Study Site

The Wahner Heide is a heathland in between the cities of Cologne and Bonn in Germany, in the Continental Biogeographical Region of Europe. In terms of vegetation (and Natura 2000 habitats), the area is similar to the Kalmthoutse Heide: a complex mixture of dry and wet heathlands, inland dunes, open grasslands and forested areas. Its history, however, is different. Whereas Kalmthoutse Heide has been a nature reserve attracting visitors for almost a century, Wahner Heide was used as military training grounds up until the 1990s. During that time, exercises with heavy military tanks kept the area open, while at the same time inaccessible to the public. Once the military use was discontinued, the area underwent a transition into successional forests and the use as a local recreation area increased. A management plan is currently in place to preserve the open habitats.

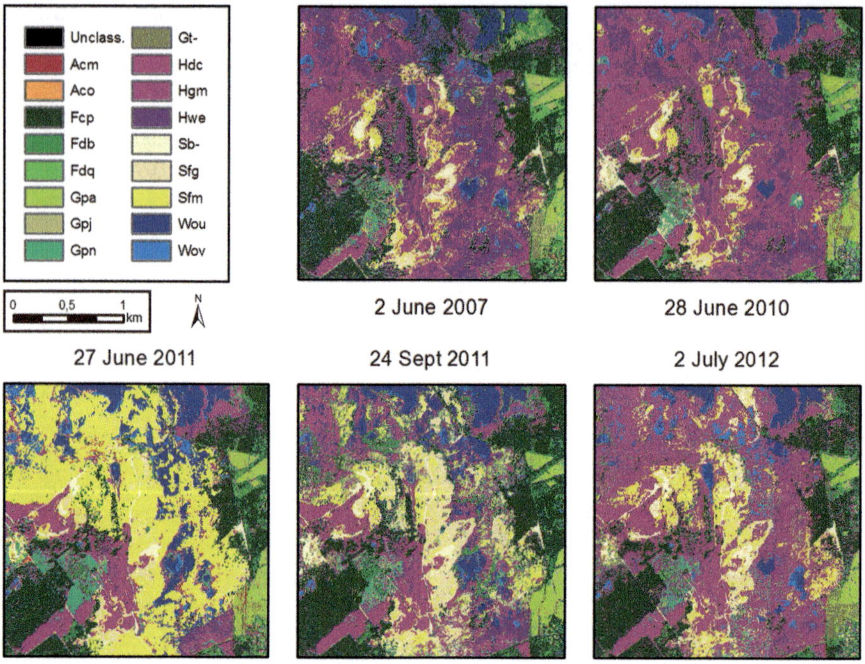

Fig. 4 Excerpts of the Kalmthoutse Heide time series, 2007–2012. Classification at level 3 of the classification scheme (Legend: see Table 1)

Method Application

Good field reference data for training is crucial for a successful application of the method. For Wahner Heide, no tailor-made dataset (as in the case of Kalmthout) was available. Instead, data from three different sources were compiled: two sets of vegetation relevés from 2011 and an old vegetation map of 1990. However, despite this apparent wealth of data, we did not succeed in drawing up a useful classification scheme for the area. The datasets turned out to be too different in typology and detail, and lacked almost all spatial overlap. No other information on vegetation types or Natura 2000 habitats was available, and we had no personal acquaintance with the area. As a result, the transfer attempt had to be ceased before obtaining any result.

Transfer Test Case 3: To a Different Ecosystem – from Heathlands to Coastal Dunes

Study Site

The Flemish nature reserve De Westhoek is one of the oldest nature reserves in Belgium and the largest remaining area of coastal dunes in the country, bordering France at its westernmost tip. It hosts a variety of Natura 2000 habitats, such as Marram (*Ammophila arenaria*) dunes (2120), grey dunes (2130), Sea-buckthorn (*Hippophae rhamnoides*) shrub (2160), Creeping willow (*Salix repens*) shrub (2170), wooded dunes (2180), and humid dune slacks (2190). Although this array of habitats is completely different from Kalmthoutse Heide, the landscape is structurally very similar, consisting mainly of dry sandy dunes, either bare or covered with grasses, mosses or low shrubs, interspersed with humid depressions and occasional snippets of woodland. Moreover, there are also parallels in threats and disturbances to both sites: dune fixation, invasive exotic species, groundwater extractions and recreational use are affecting the functioning of the ecosystem. Therefore, we considered it appropriate to test the transferability of our method from heathland ecosystems to coastal dune ecosystems.

Method Application

Hyperspectral APEX images of the area were acquired on 14 June 2011. The first task was to draw up a remote sensing oriented classification scheme for this site. This was especially crucial since it was the first time the method was to be applied outside heathlands. For the Westhoek, a detailed (almost to the level of dominant species) and recent (2010) vegetation map was available, which proved very valuable for this task. The resulting classification scheme is shown in Table 3.

Training data extraction proved less straightforward. Since the source data were provided as polygons, selecting random pixels within these polygons for training could not exclude the possibility of including impure pixels. Furthermore, the training dataset was highly imbalanced, with a high amount of Hawthorn (*Crataegus monogyna*). The first trial classification runs indeed showed this imbalance to cause problems, but after several adaptations (see introduction to section "Real-world examples"), the accuracy levels could be brought to an acceptable level (see Table 4; validation based on a separate dataset extracted from the same source map of 2010). However, the next step, the translation into a habitat map by the rule-based reclassification algorithm (Haest et al. 2017), resulted in clear errors: the 'No habitat type' class dominated overall although, in reality, the area is almost entirely covered by Natura 2000 habitats. It seems that the higher species diversity in coastal dunes, compared to heathlands, proved difficult to accommodate in simple rules. Substantial additional fine-tuning of the rule-base would be needed to make this step work satisfactorily, which we did not pursue, as it would have taken more time and raised the risk of over-fitting to one site. The classification result at Level 3 and an excerpt of the erroneous habitat map is shown in Fig. 5.

Table 3 Four-level classification scheme for the Flemish nature reserve De Westhoek

Level 1		Level 2		Level 3		Level 4	
P	Pioneer vegetations on sand	PS	Pioneer vegetations on sand	PS-a	Pioneer vegetations of dynamic dunes	Ammopare	Marram (*Ammophila arenaria*)
						Festujun	Rush-leaved fescue (*Festuca juncifolia*)
						Pion_mix	Mixed vegetation or other dominant species on sand
L	Low non-woody vege-tations	LM	Moss dunes	LM-t	Moss dunes	Tortorur	*Syntrichia (Tortula) ruraliformis*
		LH	Herb-dominated vegetations of fixed dunes	LH-g	Herb-dominated vegetations on dry sand	Grassveg	Low grassland vegetation on dry sand
				LH-j	Herb-dominated vegetations on wet sand	Slackveg	Low dune slack vegetation
H	High non-woody vege-tations	HT	Tall grasses and herbs	HT-c	Tall grasses	Calamepi	Wood small-reed (*Calamagrostis epigejos*)
				HT-u	Tall herbs (nettles, thistles)	Tallherb	Tall herbs (nettles, thistles)
		HF	Helophyte vegetations	HF-f	Helophyte vegetations	Phragaus	Common reed (*Phragmites australis*)
S	Shrubs and bushes	SL	Low shrubs	SL-s	Creeping willow	Salixrep	Creeping willow (*Salix repens*)
				SL-i	Burnet rose	Rosa_pim	Burnet rose (*Rosa pimpinellifolia*)
				SL-r	*Rubus*	Rubusfru	Bramble (*Rubus fruticosus* agg.)

		SH	High shrubs	SH-h	Sea-buckthorn	Hipporha	Sea-buckthorn (*Hippophae rhamnoides*)
				SH-p	Other high shrubs	Cratamon	Hawthorn (*Crataegus monogyna*)
						Ligusvul	Wild privet (*Ligustrum vulgare*)
						Prunuspi	Blackthorn (*Prunus spinosa*)
						Salixcin	Grey willow (*Salix cinerea*)
						Clemavit	Traveller's joy (*Clematis vitalba*)
						Sambunig	Elder (*Sambucus nigra*)
				SHex	High shrubs – exotic	Prunuser	Rum cherry (*Prunus serotina*)
T	Trees	TD	Deciduous trees	TDIb	Indigenous deciduous trees	Acer_pse	Sycamore (*Acer pseudoplatanus*)
						Betulspp	Birch (*Betula* spp.)
						Fraxiexc	Ash (*Fraxinus excelsior*)
						Quercrob	Common oak (*Quercus robur*)
						Salixalb	White willow (*Salix alba*)
						Alnusglu	Alder (*Alnus glutinosa*)
				TDNb	Non-indigenous deciduous trees	Populxca	Hybrid black poplar (*Populus x canadensis*)
						Populalb	White/Grey poplar ('abele'; *Populus alba/canescens*)
U	Unvegetated	US	Bare sand	US-o	Bare sand	Baresand	Bare sand
W	Water	WW	Water	WWVw	Water – with aquatic vegetation	Waterveg	Water – with aquatic vegetation – no Charaphyceae
						Watercha	Water – with aquatic vegetation incl Charaphyceae

Table 4 Accuracy of vegetation classification for De Westhoek based on a separate validation dataset

Level	No of classes	No of samples	OA (%)	Avg. PA (%)	Avg. UA (%)	Kappa
1	7	1619	83.8	65.4	75.2	0.75
2	10	1618	79.5	50.9	63.6	0.71
3	17	1618	76.1	48.6	57.9	0.69
4	31	1388	72.2	43.3	56.4	0.68

Level: see Table 3; *OA* overall accuracy, *Avg. PA* average producer's accuracy, *Avg. UA* average user's accuracy

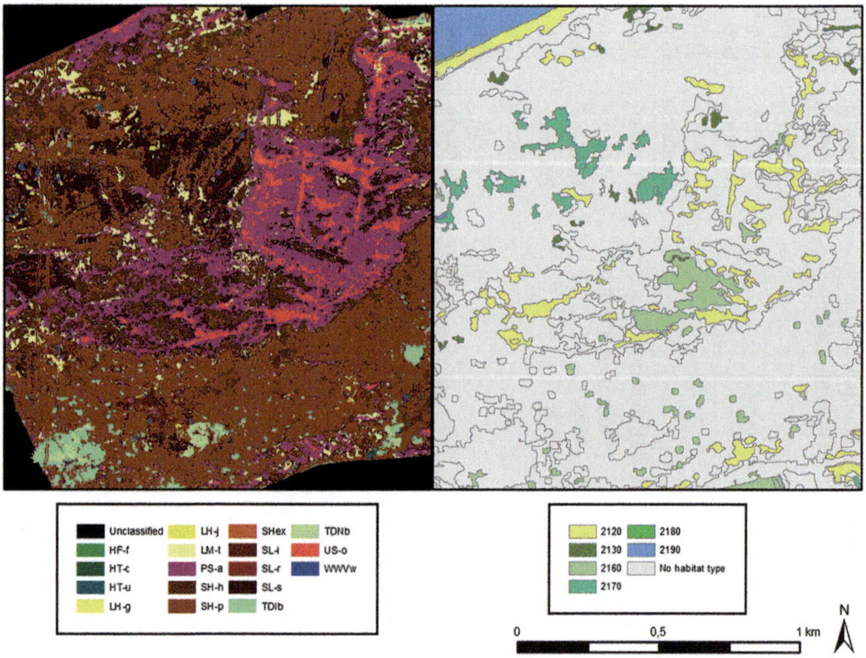

Fig. 5 Classification excerpt at level 3 (*left*) and resulting (erroneous) habitat map (*right*) for De Westhoek. (Classification legend: see Table 3; habitat code legend: see main text section "Study site")

Towards Better Transferability

In the previous sections, we identified the lack of transferability as a potential major impediment for the use of remote sensing methods, discussed possible causes, and illustrated the problem with some examples from our own work. In the following, we want to explore some possible solutions, and make recommendations to overcome the problem and stimulate the uptake of remote sensing in Natura 2000 habitat conservation.

Use More Widely Available Image Data

User requirements in vegetation remote sensing can be diverse and demanding. As a result, there is a tendency to use state-of-the-art image data (hyperspectral and/or very high spatial resolution) to fulfil requirements as much as possible. Although this has its scientific merit, obtaining good quality data of such types can be quite cumbersome: data providers are few, demand is high and data are costly. Less expensive images may be purchased from the archive, but this gives less control over timing of imagery. Moreover, clouds and year-to-year phenological variations may make it virtually impossible to obtain truly comparable image pairs of different years suitable for change detection analysis. At the same time, there is a range of cheaper or even free data products whose potential has not been fully tested, at least not for detailed Natura 2000 habitat conservation: Landsat, Aster and, since 2015, Sentinel-2 all provide multispectral data of a somewhat lower spatial resolution (10–60 m) than is mostly desired but it is hard to imagine that this excludes any application on Natura 2000 habitats, even at a local level (e.g., Feilhauer et al. 2014). Moreover, many countries routinely acquire aerial orthophotos of very high spatial resolution (<2 m pixel resolution), which are sometimes available for free to government agencies and NGOs, and which could be combined with satellite data. Even with these free data sources, acquiring the perfect data set (e.g., in terms of timing, cloud cover etc.) may still be a challenge but at least these data have the advantage of not consuming substantive components of the available budget before the work even starts.

Limit the Dependence on Ground Reference Data

Obtaining good reference data (i.e., the right variables, collected at the right time, in the right way and the right format, and in sufficient amount) is often a problem, both for training and for validation. Existing data may turn out to be (wholly or partly) of limited use for various reasons. For example, they may be too old, do not cover the core/entire area or are inconsistent in content and coverage. Collecting new data may be too expensive or impractical because of, for example, inaccessible terrain and sub-optimal seasons. Generally, it is advisable to rely on easily accessible reference data (e.g., from online map services like Google Earth) or limit the reference data dependency altogether to enhance transferability. The best way to achieve this is probably to reduce the complexity of the reference dataset as much as possible, to a level where it balances fitness for purpose with technical feasibility. Ontology-based methods (e.g., Nieland et al. 2015) may also boost the chances of successfully re-using existing, non-tailor-made datasets. However, more research is needed to ensure these methods are applicable for non-experts.

Exploit the Strengths of Remote Sensing

Vegetation types are complex phenomena. Over the course of decades, ecologists have developed ways of studying and describing this complexity, mostly involving the classification of a continuous phenomenon into vegetation types (i.e., categories characterized by plant species and their abundances). Through careful study, temporal changes in vegetation types have also been linked to changes in environmental conditions and, further, to pressures and driving forces causing these changes. This provided a framework for monitoring vegetation status, which was eagerly adopted in Natura 2000 monitoring.

With the advent of remote sensing, attempts were (and are) made to apply this new technology as an alternative data source for established monitoring methods. However, this is not necessarily the best way to exploit remote sensing. Vanden Borre et al. (2011b) already pointed out that remote sensing may be far more suited for studying vegetation, and unravelling linkages, in ways that were previously unimaginable. However, this requires active involvement and forward-thinking creativity of both users and method developers to come up with methods that do not solely replace existing monitoring schemes but instead result in new approaches to biodiversity monitoring in which remote sensing methods are integrated with other methods, such as field surveys.

In recent years, a number of such approaches have emerged. For instance, Buck et al. (2015) identified features of relevance for Natura 2000 grasslands, and mapped these as raster information layers over larger areas from various remote sensing or other sources. These information layers constitute a type of primitives that can be flexibly used by ecologists to infer conclusions about grassland habitat types, such as probability of occurrence, intensity of use, threats, etc. Since these information layers (e.g., patch size and shape, spectral homogeneity, line structures, temporal profile of biomass) are usually derived from established remote sensing methods, they are less error-prone than a direct grassland habitat classification, and more flexible in accommodating changed relationships (e.g., over space or time) between these information layers and the habitat types. Hence, they are expected to be more easily transferable.

Lucas et al. (2015) used the principle of data primitives in various stages of their EODHaM system, a comprehensive method aimed at consistent mapping of land cover and Natura 2000 habitat types in and around Natura 2000 sites. Whereas obtaining the data primitives from remote sensing data is more or less straightforward (e.g., spectral indices, but also: size, shape and density of small objects within larger objects, e.g., tree crowns within a forest versus an orchard), their translation to Natura 2000 habitat types is achieved through rulesets based on expert input, which allows for a more flexible adaptation to different geographical settings.

More Considerations for Repeatability and Transferability

Transferability is undervalued, and rarely explicitly assessed in publications (one exception is Keramitsoglou et al. 2015). Nevertheless, we are convinced that more consideration of, and transparency about, transferability will have beneficial effects on the reputation of remote sensing, and eventually its uptake, in the Natura 2000 monitoring community. It will help potential users to gain an idea of what is achievable with remote sensing (avoiding unrealistic expectations). It will also help remote sensing scientists to detect flaws and dependencies in their methods, and fix them. Therefore, we call upon the scientific community to be more considerate towards the transferability of remote sensing methods, by testing these themselves, or by making it easy for others to do so, e.g., by releasing their (preferably open-source) programming code into the public domain.

Conclusions

Remote sensing has been around as a tool for several decades, but its uptake in operational monitoring for nature conservation remains rather limited. In this paper, we argued that poor method transferability may be a hitherto overlooked cause of that. Until now, the complexity of biodiversity has been captured using mostly tailor-made remote sensing approaches, which were then promoted among other members of the nature conservation community with presumably the same requirements. However, transfer of the approach often resulted in unsatisfactory results, leading to disappointment in the potential of remote sensing among intended users. Although poor method transferability is not the only cause for poor uptake of remote sensing products in nature conservation, we call upon both the remote sensing and the nature conservation communities to consider wider applicability at an early stage when new methods are devised. Doing so might be exactly what is needed to unlock the full potential of remote sensing for biodiversity monitoring, and induce its regular operational use.

Acknowledgments The research leading to these results received funding from the Belgian Science Policy Office's STEREO II–programme (HABISTAT project – contract SR/00/103), and from the European Community's 7th Framework Programme (FP7/2007–2013) under grant agreement no 263479 (MS.MONINA project). We thank Markus Reinhold and Sebastian Schmidtlein for providing the Wahner Heide data, and Sam Provoost for the Westhoek data. The views expressed in this manuscript are entirely those of the authors, and do not necessarily reflect the views of their employers or the funding sources.

Annex

Habitat manuals and other sources of habitat descriptions used for the analysis in section "Objects of interest differ (even if they carry the same name)", listed per EU member state (and region), with number of subtypes per habitat (4030: total 45; 9110: total 19).

Member state (region)	Habitat manual	No of descriptions 4030	No of descriptions 9110
EU	European Commission – DG Environment – Nature and biodiversity 2007. *Interpretation manual of European Union – Habitats EUR 27.* [in English]	5	1
	http://ec.europa.eu/environment/nature/legislation/habitatsdirective/docs/2007_07_im.pdf		
AT	Ellmauer T. 2005. *Entwicklung von Kriterien, Indikatoren und Schwellenwerten zur Beurteilung des Erhaltungszustandes der Natura 2000-Schutzgüter. Band 3: Lebensraumtypen des Anhangs I der Fauna-Flora-Habitat-Richtlinie.*[in German]	1	1
	http://www.umweltbundesamt.at/fileadmin/site/umweltthemen/naturschutz/Berichte_GEZ/Band_3_FFH-Lebensraumtypen.pdf		
AT	Ellmauer T. & Traxler A. 2000. *Handbuch der FFH-Lebensraumtypen Österreichs.* Umweltbundesamt GmbH, Vienna. [in German]	1	1
	http://www.umweltbundesamt.at/fileadmin/site/publikationen/M130z.pdf		
BE (Flanders)	T'jollyn F., Bosch H., Demolder H., De Saeger S., Leyssen A., Thomaes A., Wouters J., Paelinckx D. & Hoffmann M. 2009. *Ontwikkeling van criteria voor de beoordeling van de lokale staat van instandhouding van de NATURA 2000-habitattypen, versie 2.0.* Instituut voor Natuur- en Bosonderzoek, Brussel. [in Dutch]	1	1
	https://data.inbo.be/pureportal/files/718815/Tjollyn_etal_2009_OntwikkeligCriteriaVoorBeoordelingLokaleStaatInstandhoudingNatura2000Habitattypen.pdf		
BE (Flanders)	Sterckx G., Paelinckx D., Decleer K., De Saeger S., Provoost S., Denys L., Packet J., Wouters J., Demolder H., Thomaes A., Vandekerkhove K. & De Keersmaeker L. 2007. *Habitattypen Bijlage 1 Habitatrichtlijn.*In: Decleer K. (Ed.), *Europees beschermde natuur in Vlaanderen en het Belgisch deel van de Noordzee. Habitattypen/Dier- en Plantensoorten.* pp. 59–359. Instituut voor Natuur- en Bosonderzoek, Brussel. [in Dutch]	1	1
	https://data.inbo.be/pureportal/files/8692848/Decleer_2007_EuropeesBeschermdeNatuurVlaanderenBelgischDeelNoordzee.pdf		

BE (Wallonia)	CRNFB 2006. *Cahiers «Natura 2000». Habitats de l'Annexe I de la Directive Habitats présents en Wallonie. Version 3 provisoire.*Centre de Recherche de la Nature, des Forêt et du Bois, Gembloux. [in French]	3	1
CZ	Chytrý M., Kučera T. & Kočí M. (Eds) 2001. *Habitat catalogue of the Czech Republic. Interpretation manual for the European programmes Natura 2000 and Emerald.*Agency for Nature Conservation and Landscape Protection of the Czech Republic, Prague. [in Czech]	3	1
	http://www.sci.muni.cz/botany/chytry/Katalog.pdf		
DK	Søgaard B., Skov F., Ejrnæs R., Nielsen K.E ., Pihl S., Clausen P., Laursen K., Bregnballe T., Madsen J., Baatrup-Pedersen A., Søndergaard M., Lauridsen T.L., Møller P.F., Riis-Nielsen T., Buttenschøn R.M., Fredshavn J., Aude E. & Nygaard B. 2005. *Kriterier for gunstig bevaringsstatus. Naturtyper og arter omfattet af EF-habitatdirektivet & fugle omfattet af EF-fuglebeskyttelsesdirektivet. 3. udgave.*Danmarks Miljøundersøgelser, Aarhus. [in Danish]	1	1
	http://www2.dmu.dk/1_viden/2_Publikationer/3_fagrapporter/rapporter/FR457.PDF		
DE	PAN & ILÖK 2010. *Bewertung des Erhaltungszustandes der Lebensraumtypen nach Anhang I der Fauna-Flora-Habitat-Richtlinie in Deutschland.*Bundesamt für Naturschutz, Bonn. [in German]	1	1
	http://www.bfn.de/fileadmin/MDB/documents/themen/monitoring/Bewertungsschemata_LRT_Sept_2010.pdf		
DE (Bayern)	LfU & LWF 2010. *Handbuch der Lebensraumtypen nach Anhang I der Fauna-Flora-Habitat-Richtlinie in Bayern.* Bayerisches Landesamt für Umwelt & Bayerische Landesanstalt für Wald und Forstwirtschaft, Augsburg. [in German]	1	1
	http://www.lfu.bayern.de/natur/biotopkartierung_flachland/kartieranleitungen/doc/lrt_handbuch_201003.pdf		
DE (Brandenburg)	LUGV 2012. *Lebensraumtypen nach Anhang I der FFH-Richtlinie in Brandenburg.* Landesamt für Umwelt, Gesundheit und Verbraucherschutz, Potsdam. [in German]	1	1
	http://www.mugv.brandenburg.de/cms/detail.php/lbm1.c.234908.de		

(continued)

Member state (region)	Habitat manual	No of descriptions 4030	9110
DE (Brandenburg)	Beutler H. & Beutler D. 2002. Katalog der natürlichen Lebensräume und Arten der Anhänge I und II der FFH-Richtlinie in Brandenburg. *Naturschutz und Landschaftspflege in Brandenburg* 11 (1, 2): 2-175.	1	1
	http://www.lugv.brandenburg.de/cms/media.php/lbm1.a.3310.de/lebensr_gesamt.pdf		
DE (Nordrhein-Westfalen)	MfUNLV 2004. *Lebensräume und Arten der FFH-Richtlinie in NRW. Beeinträchtigungen, Erhaltungs- und Entwicklungsmaßnahmen, Bewertung des Erhaltungszustandes.* Ministerium für Umwelt und Naturschutz Nordrhein-Westfalen, Düsseldorf. [in German]	1	1
	http://www.naturschutzinformationen-nrw.de/ffh-arten/web/babel/media/ffh_broschuere_akt2005.pdf		
ES	Auct. pl. 2009. *Bases ecológicas preliminares para la conservación de los tipos de hábitat de interés comunitario en España.* Dir. Gral. de Medio Natural. Ministerio de Medio Ambiente, y Medio Rural y Marino, Madrid. [in Spanish]	3	
	http://www.jolube.es/Habitat_Espana/documentos/4030.pdf		
FR	Bensettiti F., Rameau J.-C. & Chevallier H. (coord.) 2001. *«Cahiers d'habitats» Natura 2000. Connaissance et gestion des habitats et des espèces d'intérêt communautaire. Tome 1 – Habitats forestiers.* MATE/MAP/MNHN. Éd. La Documentation française, Paris, 2 volumes : 339 p. et 423 p. + cédérom. [in French]		4
	https://inpn.mnhn.fr/docs/cahab/tome1.pdf		
FR	Bensettiti F., Boullet V., Chavaudret-Laborie C. & Deniaud J. (coord.) 2005. *«Cahiers d'habitats» Natura 2000. Connaissance et gestion des habitats et des espèces d'intérêt communautaire. Tome 4 – Habitats agropastoraux.* MEDD/MAAPAR/MNHN. Éd. La Documentation française, Paris, 2 volumes: 445 p. et 487 p. + cédérom. [in French]	18	
	https://inpn.mnhn.fr/docs/cahab/tome4_1.pdf		
NL	Anon. 2006. *Natura 2000 profielen habitattypes.* Alterra, Wageningen. [in Dutch]	1	1
	https://www.synbiosys.alterra.nl/natura2000/gebiedendatabase.aspx?subj=profielen		
NL	Janssen J.A.M. & Schaminée J.H.J. 2003. *Europese Natuur in Nederland. Habitattypen.* Uitgeverij KNNV, Utrecht. [in Dutch]	1	1
UK	JNCC 2012. *Special Areas of Conservation – Annex I habitat account.* [in English]	1	
	http://jncc.defra.gov.uk/ProtectedSites/SACselection/SAC_habitats.asp		

References

Aplin, P.: Remote sensing: ecology. Prog. Phys. Geogr. **29**, 104–113 (2005). doi:10.1191/030913305pp437pr

Blaschke, T., Hay, G.J., Kelly, M., et al.: Geographic object-based image analysis – towards a new paradigm. ISPRS J. Photogramm. Remote Sens. **87**, 180–191 (2014). doi:10.1016/j.isprsjprs.2013.09.014

Bock, M., Rossner, G., Wissen, M., et al.: Spatial indicators for nature conservation from European to local scale. Ecol. Indic. **5**, 322–338 (2005). doi:10.1016/j.ecolind.2005.03.018

Buck, O., Millán, V.E.G., Klink, A., Pakzad, K.: Using information layers for mapping grassland habitat distribution at local to regional scales. Int. J. Appl. Earth Obs. Geoinf. **37**, 83–89 (2015). doi:10.1016/j.jag.2014.10.012

Bunce, R.G.H., Metzger, M.J., Jongman, R.H.G., et al.: A standardized procedure for surveillance and monitoring European habitats and provision of spatial data. Landsc. Ecol. **23**, 11–25 (2008). doi:10.1007/s10980-007-9173-8

Corbane, C., Lang, S., Pipkins, K., et al.: Remote sensing for mapping natural habitats and their conservation status – new opportunities and challenges. Int. J. Appl. Earth Obs. Geoinf. **37**, 7–16 (2015). doi:10.1016/j.jag.2014.11.005

Ellmauer, T.: Entwicklung von Kriterien, Indikatoren und Schwellenwerten zur Beurteilung des Erhaltungszustandes der Natura 2000- Schutzgüter. Band 3: Lebensraumtypen des Anhangs I der Fauna-Flora-Habitat-Richtlinie. Vienna (2005)

ETC/BD: Assessment and reporting under Article 17 of the habitats directive. Explanatory notes & guidelines for the period 2007–2012. Final version July 2011. Paris (2011)

Evans, D.: The habitats of the European Union habitats directive. Biol. Environ. Proc. R. Irish Acad. **106**, 167–173 (2006). doi:10.3318/BIOE.2006.106.3.167

Evans, D.: Interpreting the habitats of annex I: past, present and future. Acta Bot. Gall. **157**, 677–686 (2010). doi:10.1080/12538078.2010.10516241

Evans, D.: Building the European Union's Natura 2000 network. Nat. Conserv. **1**, 11 (2012). doi:10.3897/natureconservation.1.1808

Fanelli, D.: Negative results are disappearing from most disciplines and countries. Scientometrics. **90**, 891–904 (2012). doi:10.1007/s11192-011-0494-7

Feilhauer, H., Dahlke, C., Doktor, D., et al.: Mapping the local variability of Natura 2000 habitats with remote sensing. Appl. Veg. Sci. **17**, 765–779 (2014). doi:10.1111/avsc.12115

Franke, J., Keuck, V., Siegert, F.: Assessment of grassland use intensity by remote sensing to support conservation schemes. J. Nat. Conserv. **20**, 125–134 (2012). doi:10.1016/j.jnc.2012.02.001

Gross, J.E., Goetz, S.J., Cihlar, J.: Application of remote sensing to parks and protected area monitoring: introduction to the special issue. Remote Sens. Environ. **113**, 1343–1345 (2009). doi:10.1016/j.rse.2008.12.013

Haest, B., Thoonen, G., Vanden Borre, J., et al.: An object-based approach to quantity and quality assessment of heathland habitats in the framework of Natura 2000 using hyperspectral airborne AHS images. In: The International Archives of the Photogrammetry, Remote Sensing and Spatial Information Sciences, vol. XXXVIII-4/C7, pp. 1–6. ISPRS (2010)

Haest, B., Vanden Borre, J., Spanhove, T., et al.: Habitat mapping and quality assessment of NATURA 2000 heathland using airborne imaging spectroscopy. Remote Sens. **9**, 266 (2017). doi:10.3390/rs9030266

Horning, N., Robinson, J.A., Sterling, E.J., et al.: Remote Sensing for Ecology and Conservation – A Handbook of Techniques. Oxford University Press, Oxford (2010)

Innes, J.L., Koch, B.: Forest biodiversity and its assessment by remote sensing. Glob. Ecol. Biogeogr. **7**, 397–419 (1998). doi:10.1046/j.1466-822X.1998.00314.x

Jones, H.G., Vaughan, R.A.: Remote sensing of vegetation: Principles, techniques, and applications, p. 384. Oxford University Press, Oxford/New York (2010)

Kennedy, R.E., Townsend, P.A., Gross, J.E., et al.: Remote sensing change detection tools for natural resource managers: understanding concepts and tradeoffs in the design of landscape monitoring projects. Remote Sens. Environ. **113**, 1382–1396 (2009). doi:10.1016/j.rse.2008.07.018

Keramitsoglou, I., Stratoulias, D., Fitoka, E., et al.: A transferability study of the kernel-based reclassification algorithm for habitat delineation. Int. J. Appl. Earth Obs. Geoinf. **37**, 38–47 (2015). doi:10.1016/j.jag.2014.11.002

Kerr, J., Ostrovsky, M.: From space to species: ecological applications for remote sensing. Trends Ecol. Evol. **18**, 299–305 (2003). doi:10.1016/S0169-5347(03)00071-5

Kopeć, D., Michalska-Hejduk, D., Sławik, Ł., et al.: Application of multisensoral remote sensing data in the mapping of alkaline fens Natura 2000 habitat. Ecol. Indic. **70**, 196–208 (2016). doi:10.1016/j.ecolind.2016.06.001

Lawrence, M., McRoberts, R.E., Tomppo, E., et al.: Comparisons of National Forest Inventories. In: Tomppo, E., Gschwantner, T., Lawrence, M., McRoberts, R.E. (eds.) National Forest Inventories: Pathways for Common Reporting, pp. 19–32. Springer, Dordrecht (2010)

Lucas, R., Blonda, P., Bunting, P., et al.: The earth observation data for habitat monitoring (EODHaM) system. Int. J. Appl. Earth Obs. Geoinf. **37**, 17–28 (2015). doi:10.1016/j.jag.2014.10.011

Matosin, N., Frank, E., Engel, M., et al.: Negativity towards negative results: a discussion of the disconnect between scientific worth and scientific culture. Dis. Model. Mech. **7**, 171–173 (2014). doi:10.1242/dmm.015123

Nagendra, H.: Using remote sensing to assess biodiversity. Int. J. Remote Sens. **22**, 2377–2400 (2001). doi:10.1080/01431160117096

Nagendra, H., Lucas, R., Honrado, J.P., et al.: Remote sensing for conservation monitoring: assessing protected areas, habitat extent, habitat condition, species diversity, and threats. Ecol. Indic. **33**, 45–59 (2013). doi:10.1016/j.ecolind.2012.09.014

Neumann, C., Weiss, G., Schmidtlein, S., et al.: Gradient-based assessment of habitat quality for spectral ecosystem monitoring. Remote Sens. **7**, 2871–2898 (2015). doi:10.3390/rs70302871

Newton, A.C., Hill, R.A., Echeverria, C., et al.: Remote sensing and the future of landscape ecology. Prog. Phys. Geogr. **33**, 528–546 (2009). doi:10.1177/0309133309346882

Nieland, S., Kleinschmit, B., Förster, M.: Using ontological inference and hierarchical matchmaking to overcome semantic heterogeneity in remote sensing-based biodiversity monitoring. Int. J. Appl. Earth Obs. Geoinf. **37**, 133–141 (2015). doi:10.1016/j.jag.2014.09.018

PAN & ILÖK: Bewertung des Erhaltungszustandes der Lebensraumtypen nach Anhang I der Fauna-Flora-Habitat-Richtlinie in Deutschland. Überarbeitete Bewertungsbögen der Bund-Länder-Arbeitkreise als Grundlage für ein bundesweites FFH-Monitoring. Bonn (2010)

Pettorelli, N., Laurance, W.F., O'Brien, T.G., et al.: Satellite remote sensing for applied ecologists: opportunities and challenges. J. Appl. Ecol. **51**, 839–848 (2014). doi:10.1111/1365-2664.12261

Pettorelli, N., Wegmann, M., Skidmore, A., et al.: Framing the concept of satellite remote sensing essential biodiversity variables: challenges and future directions. Remote Sens Ecol Conserv. **2**, 122–131 (2016). doi:10.1002/rse2.15

Riedler, B., Pernkopf, L., Strasser, T., et al.: A composite indicator for assessing habitat quality of riparian forests derived from earth observation data. Int. J. Appl. Earth Obs. Geoinf. **37**, 114–123 (2015). doi:10.1016/j.jag.2014.09.006

Spanhove, T., Vanden Borre, J., Delalieux, S., et al.: Can remote sensing estimate fine-scale quality indicators of natural habitats? Ecol. Indic. **18**, 403–412 (2012). doi:10.1016/j.ecolind.2012.01.025

Stehman, S.V., Czaplewski, R.L.: Design and analysis for thematic map accuracy assessment: fundamental principles. Remote Sens. Environ. **64**, 31–344 (1998)

Stoms, D.M., Estes, J.E.: A remote sensing research agenda for mapping and monitoring biodiversity. Int. J. Remote Sens. **14**, 1839–1860 (1993)

T'jollyn, F., Bosch, H., Demolder, H., et al.: Ontwikkeling van criteria voor de beoordeling van de lokale staat van instandhouding van de Natura 2000 habitattypen. Versie 2.0. Brussels (2009)

Thoonen, G., Spanhove, T., Vanden Borre, J., Scheunders, P.: Classification of heathland vegetation in a hierarchical contextual framework. Int. J. Remote Sens. **34**, 96–111 (2013). doi:10.1080/01431161.2012.708061

Turner, W., Spector, S., Gardiner, N., et al.: Remote sensing for biodiversity science and conservation. Trends Ecol. Evol. **18**, 306–314 (2003). doi:10.1016/S0169-5347(03)00070-3

Vanden Borre, J., Haest, B., Lang, S., et al.: Towards a wider uptake of remote sensing in Natura 2000 monitoring: Streamlining remote sensing products with users' needs and expectations. In: 2011 2nd International Conference on Space Technology. IEEE, pp. 1–4 (2011)

Vanden Borre, J., Paelinckx, D., Mücher, C.A., et al.: Integrating remote sensing in Natura 2000 habitat monitoring: prospects on the way forward. J. Nat. Conserv. **19**, 116–125 (2011). doi:10.1016/j.jnc.2010.07.003

Verbücheln, G., Börth, M., Hinterlang, D., et al.: Anleitung zur Bewertung des Erhaltungszustandes von FFH-Lebensraumtypen. Recklinghausen (2002)

Wang, K., Franklin, S.E., Guo, X., Cattet, M.: Remote sensing of ecology, biodiversity and conservation: A review from the perspective of remote sensing specialists. Sensors. **10**, 9647–9667 (2010). doi:10.3390/s101109647

Xie, Y., Sha, Z., Yu, M.: Remote sensing imagery in vegetation mapping: a review. J. Plant Ecol. **1**, 9–23 (2008). doi:10.1093/jpe/rtm005

Zlinszky, A., Schroiff, A., Kania, A., et al.: Categorizing grassland vegetation with full-waveform airborne laser scanning: a feasibility study for detecting Natura 2000 habitat types. Remote Sens. **6**, 8056–8087 (2014). doi:10.3390/rs6098056

Pre-processing of Remotely Sensed Imagery

Peter Bunting

Abstract A common obstacle to the use of remote sensing data for nature conservation is the difficulty in obtaining or generating data that are pre-processed to a standard that gives confidence in their subsequent use. Such processing is essential in order to facilitate physical measurement (e.g., of temperature, surface reflectance, height) and compare data (e.g., reflectance or radar backscatter) acquired for different dates or areas. For optical and radar data, this pre-processing includes orthorectification, calibration, atmospheric and topographic correction and, in the case of LiDAR, ground return classification and surface height retrieval. This chapter therefore provides an overview of the common pre-processing steps that are undertaken or needed in order to create what has been recently termed an analysis ready data (ARD) product. Increasingly, such products are being provided routinely to minimize the effort of data users but knowledge of how this is achieved is important in determining the integrity and understanding the use of the data. The information provided should help users to identify, select and use data with confidence or to perform their own processing of the raw data.

Keywords Earth observation • Optical • Radar • Lidar • Preprocessing • Atmosphere • Topography • Geometric correction

Introduction

Pre-processing of all remotely sensed imagery, whether airborne or spaceborne, first involves a geometric correction, with this ensuring accurate spatial location of datasets on the Earth's surface (Lillesand et al. 2004). Standardization to a scientific unit is then undertaken such that the data are comparable to that acquired from the same or different sensors (Analysis Ready Data; ARD), with this including calibration of optical data to radiometric units and atmospheric correction to units of reflectance, transformation of Synthetic Aperture Radar (SAR) data to backscatter and other

P. Bunting (✉)
Earth Observation and Ecosystem Dynamics Group, Department of Geography and Earth Sciences, Aberystwyth University, Aberystwyth SY23 3DB, UK
e-mail: pfb@aber.ac.uk

© Springer International Publishing AG 2017

R. Díaz-Delgado et al. (eds.), *The Roles of Remote Sensing in Nature Conservation*,
DOI 10.1007/978-3-319-64332-8_3

Table 1 Standard processing levels and products that could be requested

Sensor	Type	Typical Pre-processing	Derived products
Optical	Spaceborne multispectral	Surface reflectance using a modeled atmosphere.	–
	Airborne multispectral	Surface reflectance using ground targets and/or ground reflectance targets.	–
	Airborne Hyperspectral	Surface reflectance using ground targets and/or ground reflectance targets.	–
	UAV multispectral	Surface reflectance using ground reference targets	Stereo-derived Digital Surface Model (DSM).
LiDAR	Airborne (small footprint)	Ground returns classified and return height above surface defined	Digital Terrain Model (DTM), Digital Surface Model (DSM) and Canopy Height Model (CHM).
SAR	Spaceborne/airborne	Normalised radar cross section (σ^0), commonly displayed in decibels (dB).	–

units or classification of LiDAR ground returns to surface elevation (Table 1). The generation of ARD is often performed by the data providers, reseller or consultant specialists with knowledge of the algorithms and procedures but where in-house expertise and software are available, costs can be reduced. The following sections discuss each of these products and pre-processing routines with specific guidance on the selection of data request specifications and implications of pre-processing decisions.

Geometric Correction of Airborne and Spaceborne Data

In many instances, particularly for modern spaceborne and airborne LiDAR acquisitions, high-quality geometric correction is provided by the data provider (Shan and Toth 2009). However, there are several considerations when undertaking or contracting geometric correction of data.

For airborne datasets, the quality of the geometric correction is defined by the accuracy of the 3-dimensional (3-D) position and orientation of the aircraft during the acquisition (Schlapfer and Richter 2002). An inertial motion unit (IMU) and differential Global Positioning System (dGPS) measure the position and orientation of the aircraft and it is the frequency and accuracy of these measurements that need to be considered when commissioning airborne data acquisitions. For satellite datasets, the location of the satellite and the parameters of the acquisition are key and should be provided with the image data by the data provider in a standard format (format is commonly customised to each data provider) for ingestion to the appropriate processing software (e.g., Schwind et al. 2009).

Once the location of the instrument (whether aircraft or satellite) has been defined and recorded, a model of the acquisition is defined in software. LiDAR directly measures the 3-D component of the environment but for optical (e.g., multispectral and hyperspectral) and SAR, a Digital Elevation Model (DEM) is required to perform an orthorectification. Orthorectification removes the geometric distortion from the image acquisition (i.e., being captured from a single point) such that there is a common viewpoint or datum plane (Fig. 1; Lillesand et al. 2004). However, at extreme viewing angles, full correction may not be possible because of regions of missing data (i.e., shadowing) while the resolution of the DEM used for the correction needs to be appropriate for the scale of features within the scene. For example, where an orthorectification is being performed on imagery where individual isolated trees or buildings are visible, then the DEM needs to have a 3D representation of these features for the imagery to be fully orthorectified. Where a suitably high-resolution DEM has not been used and the viewing geometry differed between

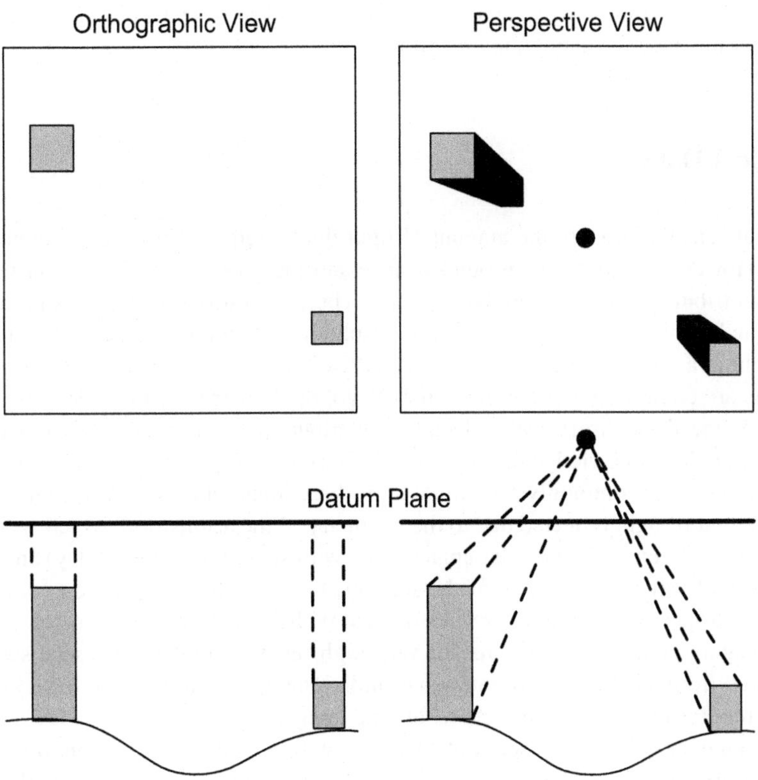

Fig. 1 Orthorectification corrects the geometry of the image with respect to the datum plane

scenes, pixel misalignments between images might be expected for these small 3D features (e.g., trees, buildings, etc.).

While recent imagery acquired from larger manned platforms has demonstrated a high degree of geometric quality, with standard and robust geometric correction routines developed, correction of data from the newer Unmanned Aerial Vehicles (UAVs or drone) platforms needs greater consideration and care. Specifically, because of the weight requirements of the UAV platforms, lighter and lower quality IMU and GPS units are fitted and therefore accuracy is lowered. Additionally, as the acquisition process involves a large number of images, with each covering small areas, an image matching process is required to create a single mosaicked image. The overlapping regions of these images can also be used to build a high-resolution DEM for the area, which can be subsequently utilised for the orthorectification of the image mosaic (Jhan et al. 2016). It is recommended that ground control points (GCPs) are acquired for ground targets unless differential GPS (dGPS) system with real-time kinematic GNSS (RTK) or post-processed kinematic (PPK) are used during the UAV acquisition. Where GCPs are used, these will subsequently need to be identified within the UAV imagery, which can be a time-consuming process. However, with the latest RTK and PPK enabled GPS systems, pixel 9 locational accuracies are commonly within ±5 cm in the x and y axis' and ±10 cm in the z axis without the need for manual intervention.

Optical Data

Optical sensors measure the amount of light that is reflected from the ground surface. However, between the ground surface and the sensor, there is an atmosphere that contributes to the measured reflectance. There are various pre-processing stages that can be applied, but removing the atmospheric and bidirectional effects is key to providing a comparable and full standardised product. However, bidirectional effects are commonly not corrected for (Nagol et al. 2015), as it can be difficult to fully define the bidirectional reflectance distribution function (BRDF). For high-resolution data, knowledge of the ground surface orientation at comparable resolutions or better is commonly not available. Bidirectional reflectance is the change in the amount of light reflected due to the geometry of the acquisition, which is attributable to differences in the solar angles (e.g., with season and time of day) and sensor geometry (i.e., view angle of the sensor). These angles are with respect to the ground surface, which themselves are defined with respect to the pixel resolution of the imagery acquired. Therefore, for very high-resolution (VHR datasets, such as acquired from a UAV), the orientation of individual leaves might need to be known to correct for bidirectional effects within the image.

When energy (in this case, light) interacts with a medium, reflection, transmission or absorbance occurs. For example, as light from the sun interacts with plant leaves, a proportion of this is reflected and transmitted and the remaining is absorbed. It is the reflected component that is measured by remote sensing instruments.

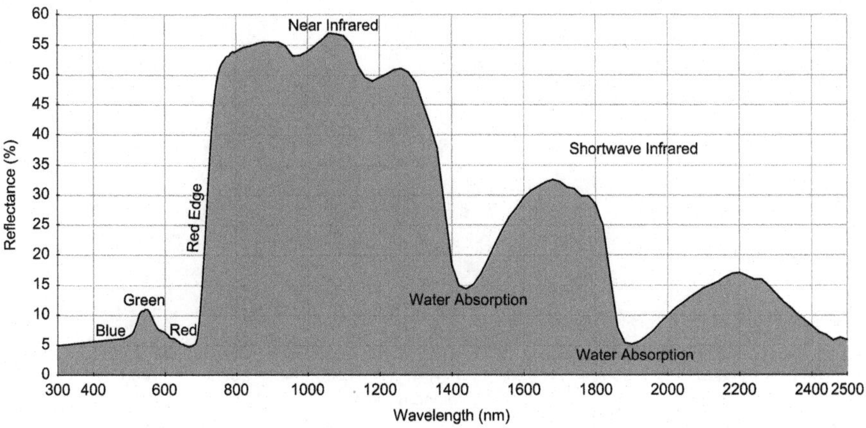

Fig. 2 Typical reflectance curve for vegetation from a field spectrometer sampling at 1 nm intervals from 300–2500 nm

The percent or proportion of energy reflected throughout the electromagnetic spectrum is commonly referred to as the reflectance curve (Fig. 2).

The spectral curve for vegetation in Fig. 2 has been measured at a very high spectral resolution (i.e., sampling at intervals every 1 nanometer; nm). Commonly, multi-spectral imagers are used for remote data acquisition and the resolution at which the reflectance of the surface is measured is therefore at a much lower spectral resolution. The resolution and sensitivity of the sensor is defined by its spectral response functions (e.g., Fig. 3a and b), with one available for each image band captured. When considering the use of an instrument for a particular application, it is the position (i.e., wavelength) of the peak of maximum sensitivity and the width of the peak that defines the measured reflectance response. The spectral response is commonly modeled as a Gaussian and therefore is quoted as the wavelength of the peak and a full-width half maximum (FWHM) of the response sensitivity. When comparing field-derived ground spectra (e.g., Fig. 2) to the signal measured by satellite or aircraft sensors, the spectral response functions need to be applied to the ground measurement (e.g., Fig. 3c).

Radiance

Optical data recorded in a particular wavelength region (λ), and obtained from the data provider, should be given in units of radiance (L_λ W m^{-2} sr^{-1} μm^{-1}). In order to compress (i.e., reduce the file size), the image is typically provided with a gain and offset to convert the pixel value, commonly referred to as the digital number (DN), to radiance where:

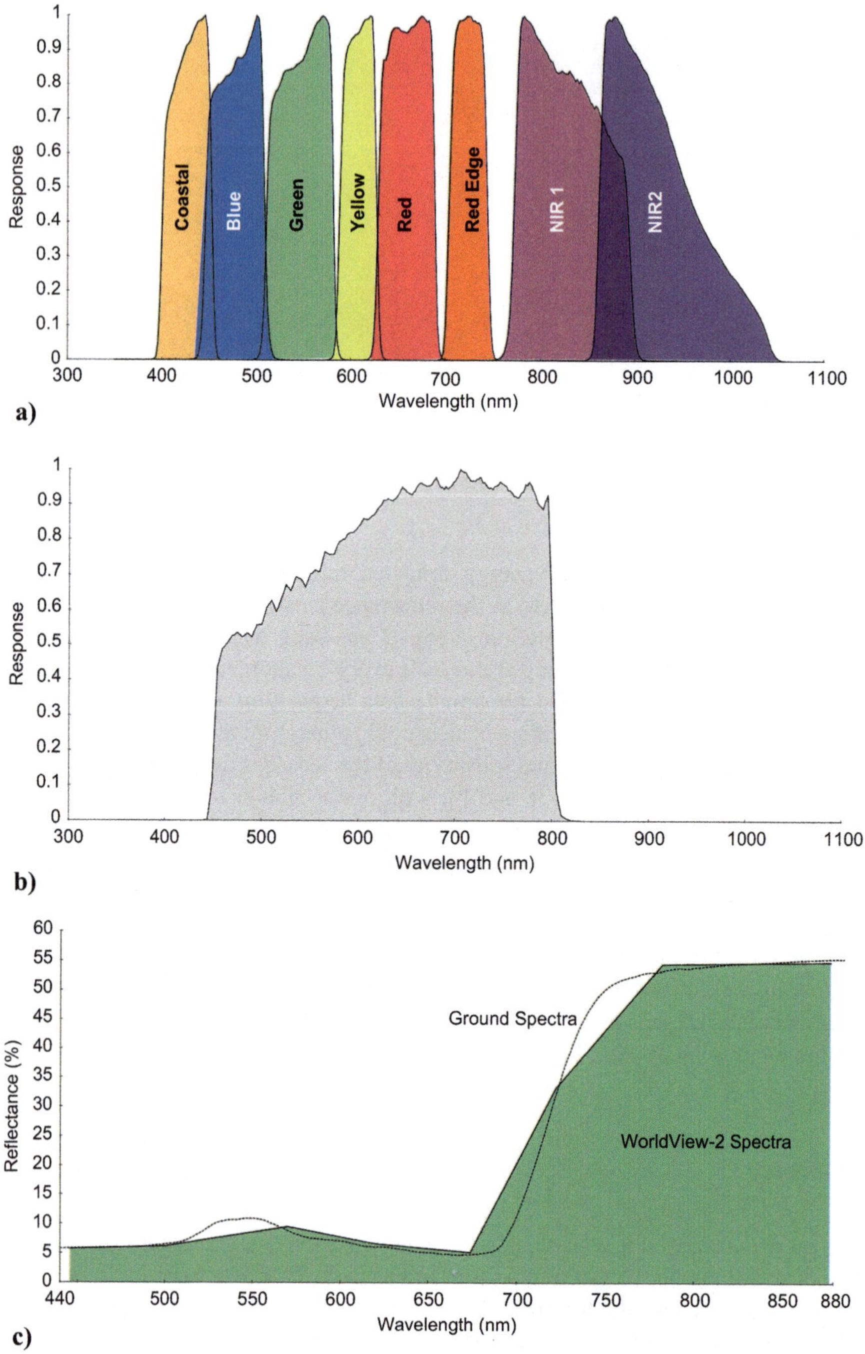

Fig. 3 (**a**) Multi-spectral image bands spectral response functions and (**b**) the panchromatic response functions for Worldview-2 data as an example. (**c**) The reflectance curve for vegetation (440–880 nm) resampled using the Worldview-2 multi-spectral functions

$$\mathbf{L}_\lambda = (\text{gain} \times \text{DN}) + \text{offset}$$

Before proceeding with further processing, these gains and offsets should be applied to your imagery, such that each pixel value represents the radiance measured by the sensor.

For UAV imagery mosaicked from many individual images, care is needed where the camera has used different exposure parameters (i.e., ISO, aperture, shutter speed). The pixels values correlating to the amount of radiance will differ and converting to radiance will not be possible once mosaicked. If correction is required for UAV imagery, then the camera parameters need to be known and ideally should be constant throughout the flight. Additionally, the camera needs to be calibrated to relate the digital number (DN) value of the camera to radiance.

At Sensor Radiance

At sensor reflectance, also referred to as top of atmosphere (TOA) reflectance, is a standard and easily calculated ratio of the incoming radiant energy (light) from the sun (ESUN) and the corresponding radiance measured by the sensor. The radiance measured at the sensor differs from the incoming signal due to the reflectance of the Earth surface and the atmosphere (or part of the atmosphere) the signal has transmitted through. Although providing a standard measure and common range of values (0–1), the reflectance measurement includes the reflectance from the atmosphere and the ground surface and therefore images taken at different times are not directly comparable. At sensor reflectance is calculated as:

$$\rho_\lambda = \frac{\pi \cdot \mathbf{L}_\lambda \cdot \mathbf{d}^2}{\mathbf{ESUN}_\lambda \cdot \mathbf{cos}\theta_\mathbf{S}}$$

where λ is the wavelength, ρ_λ is the spectral (at sensor or top of atmosphere) reflectance for wavelength λ, L_λ is the spectral radiance (W m^{-2} sr^{-1} μm^{-1}), d is the Earth-Sun distance in astronomical units, $ESUN_\lambda$ is the mean solar exoatmospheric irradiance in units of W m^{-2} μm^{-1} and θ_s is the solar zenith angle.

Surface Reflectance

Surface reflectance, also called 'bottom of atmosphere reflectance' is the ratio of incoming radiance (i.e., from the sun) with the radiance that is measured by the sensor without the atmospheric effect and should be equivalent to the signal measured if the sensor was at ground level or there was no atmosphere. To derive this

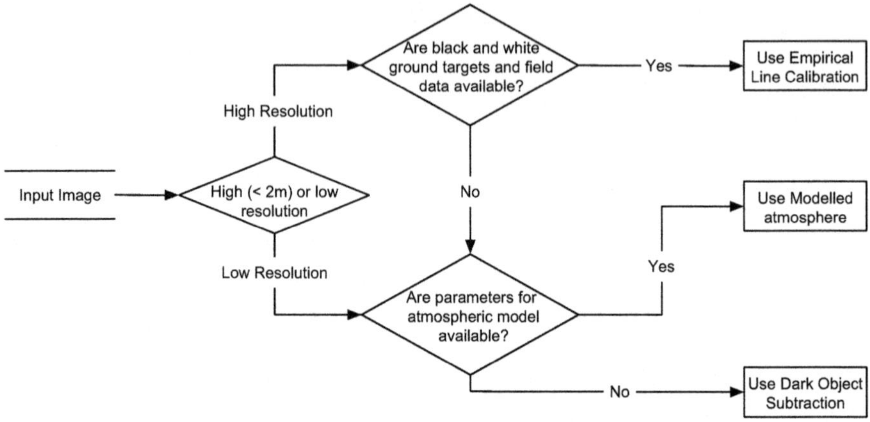

Fig. 4 Decision tree for which measure of atmospheric correction you should use

measurement, the effect of the atmosphere needs to be removed from the at sensor radiance measured at the sensor. There are a number of options (Fig. 4) for this.

The Empirical Line Calibration (Smith and Milton 2010) is commonly used to correct high-resolution airborne imagery but requires that ground data of bright and dark targets be captured at the time of the overflight. Dark Object Subtraction methods (Chavez 1988) are relatively simple and require relatively little inputs so can be easily applied to all image data but do not produce the most reliable and consistent results. It is, therefore, the method used when the others are not available. Modeled Atmospheric Correction Methods (Vermote et al. 1997; Masek et al. 2006) model reflection, absorption and scattering by the atmosphere and commonly used models include 6S (Vermote et al. 1997), LOWTRAN, MODTRAN, FLAASH, ATCOR and HYCOR. These models require many parameters to be known or estimated and can, therefore, be complex to apply. However, for lower resolution imagery or where ground spectra for targets are not available, it is the best solution. Further details on these approaches are provided in the following sections.

Empirical Line Calibration

An empirical line calibration is a simple process (Smith and Milton 2010) of collecting the ground reflectance of at least two targets that will be captured by the observing sensor, one that has a reflectance of 0% (or close to; i.e., black) and another with a reflectance of or near 100% (i.e., white). Additional targets of different shades of grey (i.e., levels of reflectance) can also be laid out to improve the reflectance estimates. The targets need to be at least three times the size of the image pixels (i.e., 1 m pixels requires at least a 3 × 3 m target) to ensure that more than one pure pixel of the target is acquired. However, larger targets producing more than one pure pixel at the pixel resolution are preferable. Another consideration is that the targets need to have a consistent reflectance across the full range of wavelengths that the sensor is measuring. Once pure pixels have been identified, a linear regression of the image pixel values to the ground reflectance measurements for each wavelength is undertaken and the resulting relationship is used to convert the image to surface reflectance. Where multiple targets at each reflectance level are available, a validation of the relationship can be carried out.

Dark Object Subtraction (DOS)

A dark object subtraction (DOS; Chavez 1988) is built on a simple assumption that the darkest pixels within the scene have little or no surface reflectance and that the radiance measured by the sensor is from the atmosphere. Therefore, while assuming the atmosphere is consistent across the scene, subtracting that atmosphere component from the whole scene can be used to convert the at-sensor reflectance values to surface reflectance. This is performed independently for each of the image bands (i.e., wavelengths). However, there is a risk that the relative relationships between the image bands can vary.

Modeled Atmospheres

Modeling the atmosphere is the most common way in which imagery is atmospherically corrected but this requires a radiative transfer (RT) model and associated parameters, many of which are supplied in the image header file from the data provider (e.g., date and time of the acquisition). However, typically you, as the end-user, would perform this analysis through a software package that aids the parameterization (e.g., automatically parses the supplied header file or associated metadata), runs the atmospheric model and applies the model outputs to the image file. There are a number of software packages and models that support this analysis (Table 2), but they each only support a defined number of sensors. These lists are being updated on a regular basis. Additionally, some products and analysis steps may not be possible for all sensors and therefore functionality may not be equal across all sensors (e.g., cirrus cloud correction uses bands only provided by Sentinel-2 and Landsat-8 instruments).

More recently, there has been some effort to standardize these processing stages and levels (Claverie et al. 2015; Feng et al. 2013; Ju et al. 2012; Roy et al. 2010) for the Landsat and Sentinel-2 imagery. The United States Geological Survey (USGS) is already supplying the Landsat archive (TM, ETM+, OSL) as an atmospherically corrected product (Masek et al. 2006) and, in time, there may be a similar service for Sentinel-2 imagery.

For the Second Simulation of the Satellite Signal in the Solar Spectrum (6S; Vermote et al. 1997) model (others models are similar), the parameters needed are given in Table 3. The sensor configuration and position parameters are well defined and known so these can be parameterized using the image header information. However, the parameters associated with the atmosphere at the time of the acquisition, specifically

Table 2 List of software packages for applying an atmospheric correction using a modeled atmosphere

Software	RT Model	Sensors	License
ATCOR-4[a] (airborne)	MODTRAN	Many – see website	Commercial
ATCOR-3[b] (satellite)	MODTRAN	Many – see website	Commercial
FLAASH[c]	MODTRAN	Many – see website	Commercial
LEDAPS[d]	6S	Landsat (TM, ETM+)	Free but closed source
SEN2COR[e]	MODTRAN	Sentinel-2	Free but closed source
ARCSI[f]	6S	Landsat (MSS, TM, ETM+, OLI), Rapideye, SPOT5, SPOT6, SPOT7, WorldView-2, WorldView-3, Pleiades, Sentinel-2	Free and Open Source

[a]www.rese-apps.com/software/atcor-4-airborn
[b]www.rese-apps.com/software/atcor-3-satellites
[c]www.harrisgeospatial.com/docs/FLAASH.html
[d]ledaps.nascom.nasa.gov
[e]step.esa.int/main/third-party-plugins-2/sen2cor/
[f]www.rsgislib.org/arcsi

the aerosol optical depth (AOD) and the total amount of water in a vertical path through the atmosphere (water vapour), are unknown and need to be provided by the user or estimated from the image for a more accurate atmospheric correction. The sensor and surface altitude parameters are defining the length of the path through the atmosphere that the signal being measured has taken (Fig. 5). The more atmosphere the signal passes through to the sensor, the larger the atmospheric effect which needs to be removed from the image (Fig. 5c).

For the dynamic components of the atmosphere, specifically the AOD and water vapour, there are various sources of information and methods that attempt to estimate those parameters from the image data itself (e.g., Masek et al. 2006). These parameters can vary over short temporal and spatial baselines while the quality of the atmospheric correction is highly sensitive (Fig. 6) to the correct estimation of these parameters. They also vary as a function of the wavelength.

The AOD is correlated with visibility (in km), and the two can be transformed from one another using the following relationship,

$$\mathrm{AOD} = \frac{3.9449}{\mathrm{vis}} + 0.08498$$

There are three main sources of AOD for parameterisation of the atmospheric model; (a) ground measurements, (b) estimates from a third party satellite or (c) estimates from the image being corrected. As the AOD varies over short temporal and spatial baselines (Wilson et al. 2014), estimates from the image being corrected will be the most reliable, both spatially and temporally, and hence this is the pre-

Table 3 Parameters for the 6S model

Parameter	Description	Known
Solar zenith	The zenith angle of the sun with respect to the earth surface for the area of acquisition.	✓
Solar azimuth	The azimuth angle of the sun with respect to the earth surface for the area of acquisition.	✓
Sensor zenith	The zenith angle of the sensor with respect to the earth surface for the area of acquisition.	✓
Sensor azimuth	The azimuth angle of the sensor with respect to the earth surface for the area of acquisition.	✓
Acquisition date and time	The exact date and time of the acquisition.	✓
Centre point of scene (lat, long)	The point on the Earth's surface for where the model is being run.	✓
Altitude of sensor	The height of the sensor above the Earth's surface.	✓
Altitude of ground surface	The height above sea level of the ground surface being measured.	✓
Atmospheric profile	The vertical distribution of the atmospheric layers at a given altitude with pressure, temperature and water vapour and ozone at that layer. However, this is commonly generalised to standard profiles for tropical, mid-latitude summer, mid-latitude winter, sub-arctic summer, sub-arctic winter. Standardised profiles can be automatically selected based on time and location.	✗ (✓)
Water vapour	The total amount of water in a vertical path through the atmosphere (in g/cm^2).	✗
Ozone	The total amount of ozone in a vertical path through the atmosphere (in cm-atm).	✗
Aerosol profile	The proportion of water-like, dust-like, oceanic-like and soot-like aerosol partials in the atmosphere. However, this is commonly generalised to standard profiles for continental, maritime, urban, desert and biomass burning. Standardised profiles can be automatically selected.	✗ (✓)
Aerosol optical depth (AOD)	The total amount of AOD in the vertical path through the atmosphere at 550 nm.	✗

ferred option. However, estimating the AOD from the image data requires some assumptions to be made to derive some estimates of the surface reflectance of the visible image bands and considerable computing resource to invert (at least partially) an atmospheric model.

Ground measurements, using a sun photometer, provide very accurate measures of AOD but these are point measurements and therefore not necessarily representative of the whole scene. Additionally, because of the sparse nature of the ground measurements, it is unlikely that data will be available for the image being processed. Some weather stations provide visibility data (e.g., in the UK) but again these are point measurements and not available everywhere.

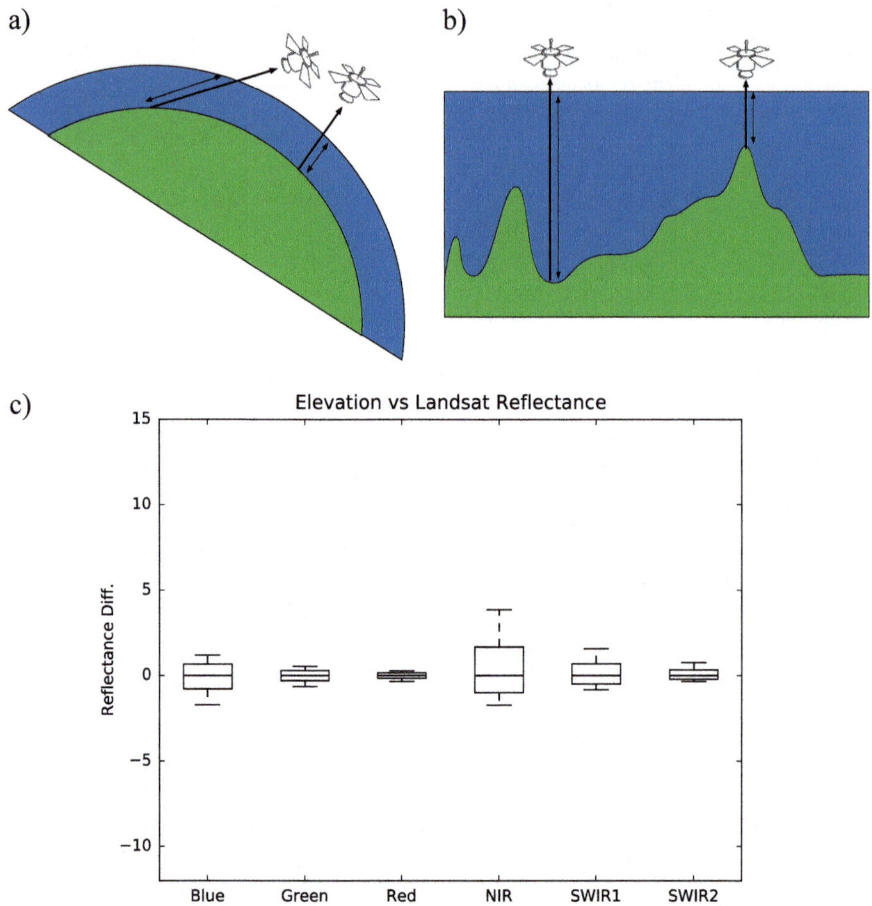

Fig. 5 Changes in the distance of the path through the atmosphere due to (a) sensor angular geometry and (b) surface altitude, which results in a variation in the outputted reflectance without appropriate correction (c) an example for the Landsat TM bands, where elevation varies from 0–5000 m

There are a number of satellite-based AOD products, primarily those derived from MODIS (Green et al. 2009). However, these products are produced at low spatial resolution (e.g., 1 km) and are generally not obtained at the same time as the image to be corrected. Furthermore, the downloading and processing of extra third party data is potentially a significant overhead for the correction of individual images. Table 4 provides an overview and reference to sources and algorithms for the retrieval of AOD.

For the correction of **atmospheric water**, there are a number of sources for water vapor within the vertical path (Table 5). The most commonly used sources are from third party satellites such as the MODIS. However, average climate data and ground-based measurements have also been used.

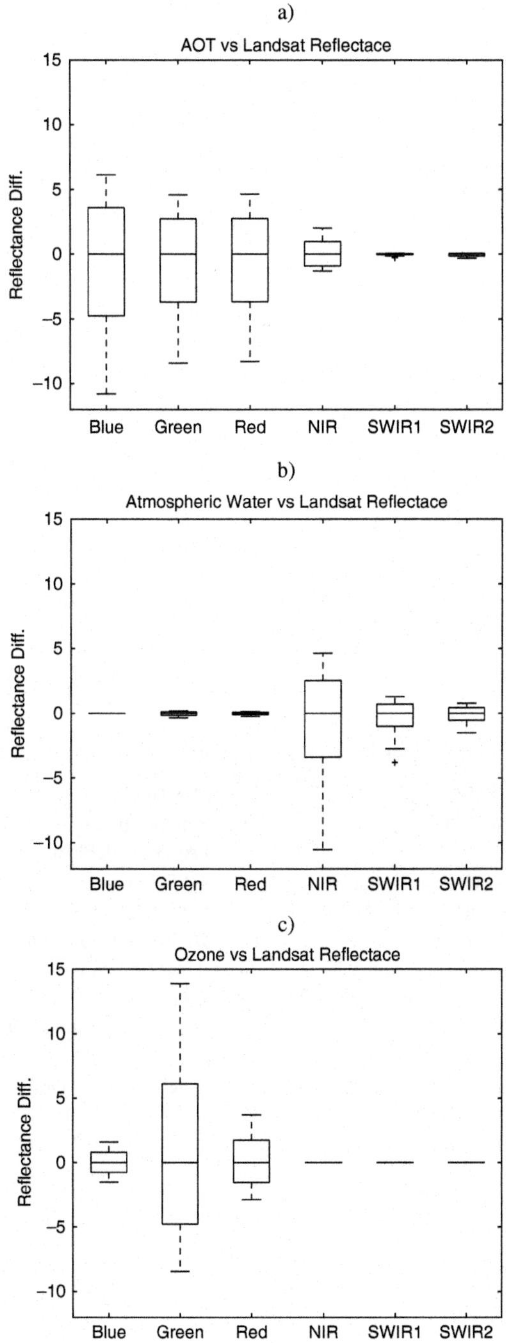

Fig. 6 Reflectance difference for Landsat TM bands with respect varying the following parameters in 6S (**a**) AOD (0.05–1.5). (**b**) vertical water column (1–15 g/cm^2) and (**c**) Ozone (0–5 cm-atm)

Table 4 Sources of measures of AOD

Source	Description
AERONET[a]	A network of sun photometers providing AOD measurements globally.
UK Meteorological Office (MIDAS)[b]	Integrated Data Archive System (MIDIS); specific to the UK, the Met Office makes ground measurements publically available.
MODIS[c]	Satellite-based measurement of AOD at 550 nm; available for free download.
LEDAPS[d]	Estimates of AOD for Landsat using dense dark vegetation (DDV) targets and relationships with the SWIR to visible wavelengths.
ARCSI[e]	Multiple algorithms, including the DDV method, but primarily uses a DOS based method to estimate surface reflectance in the blue wavelengths used for inversion.
SEN2COR[f]	Estimates AOD for Sentinel-2 using dense dark vegetation (DDV) targets and relationship from the SWIR to visible wavelengths.
Frantz et al. (n.d.)	Time series analysis to identify persistently dark targets that are used for AOD inversion.

[a]http://aeronet.gsfc.nasa.gov
[b]http://catalogue.ceda.ac.uk/uuid/220a65615218d5c9cc9e4785a3234bd0
[c]http://modis-atmos.gsfc.nasa.gov/MOD04_L2/
[d]http://ledaps.nascom.nasa.gov
[e]http://www.rsgislib.org/arcsi
[f]http://step.esa.int/main/third-party-plugins-2/sen2cor/

Table 5 Sources of measures of atmospheric water

Source	Description
MODIS[a]	Satellite-based measurement of total column water vapour. Freely available download.
Global precipitation measurement (GPM)[b]	NASA owned satellite that includes instruments for the measurement of total column water vapour.
AMSR-2[c]	JAXA owned satellite that includes instruments for the measurement of total column water vapour.
Seasonal average[d]	Where satellite estimates are not available for that date of acquisition Frantz et al. (n.d.) uses a local average.

[a]http://modis-atmos.gsfc.nasa.gov/MOD05_L2
[b]http://www.nasa.gov/mission_pages/GPM/main; Draper et al. (2015)
[c]http://suzaku.eorc.jaxa.jp/GCOM_W
[d]Frantz et al. (n.d.)

Table 6 Data and tools for establishing atmospheric ozone levels

TOMS	http://ozoneaq.gsfc.nasa.gov/data/toms/
GOME	http://www.ospo.noaa.gov/Products/atmosphere/gome/gome-A.html
OMI	http://neo.sci.gsfc.nasa.gov/view.php?datasetId=AURA_OZONE_E
NASA Ozone map tool	http://ozoneaq.gsfc.nasa.gov/tools/ozonemap/

As with water, **ozone** is commonly sourced externally from sensors such as NASA's Total Ozone Mapping Spectrometer (TOMS; 1978–2005) and ESA's Global Ozone Monitoring Experiment (GOME; 1996 to 2011). The Ozone Monitoring Instrument (OMI) has continued the time series of TOMS data since 2004 until the present. The download sites for these data are listed in Table 6, which also includes the NASA Ozone map tool that can be used to find the value of ozone from 1978 to the present based on TOMs and OMI data through a single interface.

Light Detection and Ranging (LiDAR)

Overview of Products and Software

Small footprint LiDAR data acquired via airborne platforms, primarily manned flights. However, more recently, there are small UAV octocopter based systems available (e.g., yellowscan; http://www.yellowscan.fr). LiDAR directly measures the 3D structure of a surface by way of a 3D point cloud, where other than the points returned from the same pulse, the topology of the point cloud is unknown (i.e., each pulse is independent). Multiple returns from a single fired pulse are only recorded for 'soft' targets such as vegetation (Fig. 7a) or the edges of hard targets, such as buildings (Fig. 7b). Multiple returns occur where a target causing the reflection back to the sensor is smaller than the footprint of the LiDAR. For a small footprint LiDAR, the footprint is typically around 20–30 cm. There are also so-called large-footprint LiDAR systems (e.g., ICESAT; Zwally et al. 2002), where the footprint is measured in metres, but these are not considered within this Chapter. Please refer to Shan and Toth (2009) for a discussion of large-footprint systems and their applications.

Regarding products, LiDAR produces elevation surfaces, digital terrain models (DTM) and digital surface models (DSM). The difference between the DTM and DSM provides a measure of the vertical height of features protruding from the DTM surface such as vegetation and buildings. However, to produce these products, a classification of the points associated with the ground and in some cases hard (buildings) and soft (vegetation) above the ground surface needs to be undertaken (Fig. 8). Following classification, the elevation surfaces can be interpolated to form regularly spaced raster grids. To derive other products from LiDAR, such as gap fraction (e.g.,

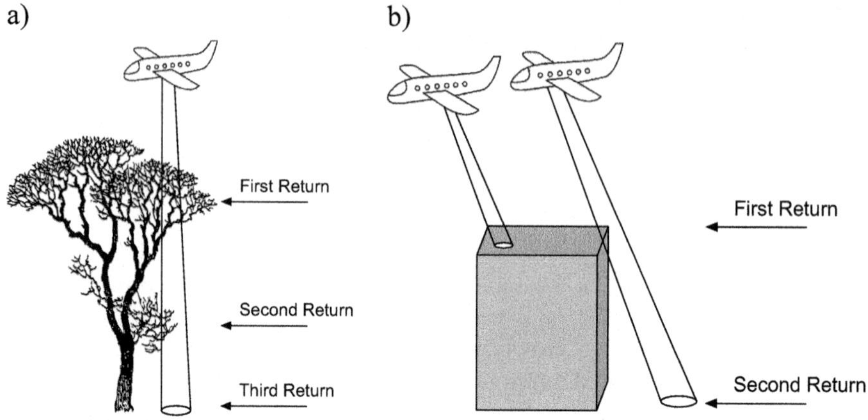

Fig. 7 (**a**) Multiple returns within a vegetation canopy and (**b**) multiple returns from the edge of a building

Armston et al. 2013; Morsdorf et al. 2006), above ground biomass (e.g., Babcock et al. 2016; Popescu 2007; Nelson et al. 1988) and other structural measures (e.g., Palace et al. 2015; Higgins et al. 2014; Zimble et al. 2003), site or region specific relationships are needed. These involve the correlation of LiDAR-derived metrics associated with the vertical structure of the vegetation with ground-based field data.

To undertake LiDAR data processing, dedicated software processing tools are required. LiDAR datasets are typically large and require a reasonable amount of computing power and storage to handle these data. Table 7 lists a number of software packages available for analyzing LiDAR data where LAStools is probably the most popular and widely used providing plugins for both ESRI ArcMap and the QGIS software packages to provide an 'easy to use' environment.

Classification of Point Clouds

To produce a DTM product from LiDAR data, the classification of ground returns and the quality of that classification is a critical processing step. There are many publications demonstrating methods for this task (e.g., Mongus and Zalik 2012; Evans and Hudak 2007; Zhang et al. 2003). However, you will most likely be limited to the algorithms implemented within the processing tools you have available.

The quality of the classification of the ground returns, and therefore the derived DTM, is limited by the number and density of the LiDAR returns that have reached the ground surface. If the LiDAR has not recorded the ground surface, then obviously the ground returns cannot be correctly identified. Where ground returns are very sparse, it is likely that they will be identified as outliers (i.e., noise) rather than true ground returns. Dense vegetation over-stories, particularly those close to the ground (i.e., 1–2 m in height), often limit the number of ground returns. However,

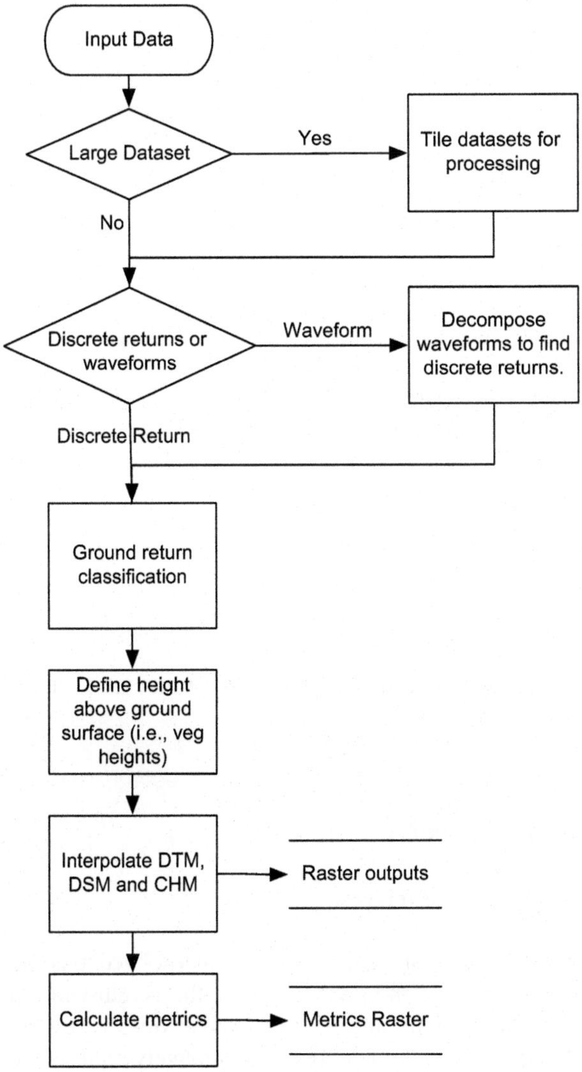

Fig. 8 Flowchart for a standard LiDAR processing chain (Adapted from Bunting et al. 2013)

if known, this can be mitigated by (a) flying higher resolution LiDAR (i.e., the number of pulses per m^2), (b) decreasing the flying height (i.e., more laser power to get weaker ground returns, but this limits the swath width), and (c) using a sensor which can differentiate returns closer to one another along the path of the pulse. Some older instruments can only differentiate returns more than 0.5–1 m from one another along the path of the pulse. Another area where ground returns can be poorly defined or classified is very steep terrain, particularly where there is also vegetation cover (Bater and Coops 2009). It is recommended that for deriving elevation models, at least 4 points per m^2 are acquired but if retrieving the vertical forest structure is of

Table 7 Software for processing LiDAR data

Software	Description	License
LAStools[a]	Becoming the most commonly used tools across the industry providing a wide range of tools. However, the free version is limited.	Commercial & Limited free version.
SPDLib[b]	Tools and file format for common LiDAR processing steps including waveform data.	Open Source
PyLiDAR[c]	A set of python modules enabling easy assesses to the LiDAR (discrete return and waveform) data as numpy arrays allowing implementation of your own algorithms.	Open Source
BCal LiDAR Tools[d]	Widely used tools, written in IDL and used through ENVI.	Open Source
Fusion[e]	US Forestry Service tools, used by many.	Free but closed source
Potree[f]	Tool for visualisation LiDAR on the web	Open Source

[a]http://lastools.org
[b]http://www.spdlib.org
[c]http://www.pylidar.org
[d]https://bcal.boisestate.edu/tools/lidar
[e]http://forsys.cfr.washington.edu/fusion/fusionlatest.html
[f]http://potree.org

interest (e.g., for establishing relationships to forest biomass), then increasing that to 8 points per m^2 is beneficial.

Standard Raster Products

Interpolation of Elevation Surfaces

The resolution at which the raster surface can be interpolated to is dependent on the density of returns that define the surface. Using the Nyquist rate, the density of returns to accurately sample the surface needs to be twice the resolution of the surface being produced to ensure all features are completely represented. However, as the ground return density varies across the scene, a compromise is usually made.

A common requirement for a DTM is that it is hydraulically correct in that it contains no holes or artificial troughs. To ensure hydraulic correctness, algorithms for filling DTM holes are applied and additional information such as break lines (e.g., river shore) can also be included in the interpolation processing.

There are many interpolation algorithms available for the generation of elevation surfaces from point cloud files including Natural Neighbour, Thin Plated Splines, Nearest Neighbour, Linear Triangulation and Inverse Distance Weighted. Bater and Coops (2009) compared a number of these algorithms and demonstrated, for a vegetation-dominated environment, that the Natural Neighbour algorithm produced high-quality results. They recommended this algorithm for general use.

Derivation of Metrics for Vertical Vegetation Structure

There are numerous statistical measures of the vertical distribution of the point cloud that have been used within the literature (e.g., Bunting et al. 2013). These include, the ratio of the number of ground returns to all returns, mean height, median height, mode height, maximum height, standard deviation of all or returns above a certain height, percentiles of height, skewness in height, Pearson mode of height, Pearson median of height and the kurtosis in height. Additionally, by filtering the returns based on their classification (e.g., ground or not-ground) or return number (e.g., first returns) etc., there are many variants of metrics which can be calculated. Your given choice of software tools will enable these metrics to be calculated. For instance, LAStools provides a command line tool to retrieve forestry metrics (lascanopy) while SPDLib provides a tool called spdmetrics. Once calculated, these metrics are commonly used within either a classification scheme to retrieve categorical classes for the scene or used within a regression analysis to field data to retrieve parameters, such as above ground biomass.

Radiometric Correction

There have been a number of attempts to radiometrically correct and/or normalise the LiDAR intensity/amplitude data (e.g., Donoghue et al. 2007). However, as of yet, there are few examples within the literature that demonstrate a clear application for this product. Therefore, for information on these processing stages, the reader is referred to Wagner (2010) and Coren and Sterzai (2007).

Standard Data Specifications

A number of organisations worldwide (e.g., the Intergovernmental Committee on Surveying and Mapping's; http://www.icsm.gov.au/elevation/) have set out standard specifications for the acquisition of LiDAR data. These specifications are commonly regarded as the minimum specification for the organisation. These specifications help ensure that data acquisitions are fit for purpose and can be used to meet the wider requirements of the organisation rather than just specific project needs. Table 8 lists a number of available specifications and, if you are acquiring LIDAR data, reference to these specifications is recommended.

Table 8 LiDAR acquisition specifications

Organisations	Location
ICSM[a]	Australia and New Zealand
British Columbia[b]	Canada
AusCover[c]	Australia
National Network of Regional Coastal Monitoring Programmes of England[d]	UK
USGS[e]	USA

[a]http://www.icsm.gov.au/elevation/
[b]http://geobc.gov.bc.ca/base-mapping/atlas/trim/specs/
[c]http://data.auscover.org.au/xwiki/bin/view/Good+Practice+Handbook/WebHome
[d]http://www.channelcoast.org/national/procurement
[e]https://lta.cr.usgs.gov/lidar_digitalelevation

Synthetic Aperture Radar (SAR)

Overview of SAR

SAR is an active instrument that sends pulses in the microwave region of the electromagnetic spectrum and records the return. The returns are processed in such a way that the movement of the instrument is used to synthesise a larger antenna than would others be physically impossible, which allows a high spatial resolution image to be produced. Forming a SAR image from the raw data is normally carried out by the data provider. The most common product from a SAR is an image of normalised radar cross section or 'backscatter' (σ^0), which is unitless. Partly because of the large range of values, σ^0 is normally expressed on a log scale in decibels (dB). However, depending on the mode and specification of the instrument, other products such as polarimetric decompositions (e.g., Pauli Decomposition; Krogager 1990) can also be generated. Where multiple acquisitions from different geometries are available (i.e., multiple satellite passes), products such as the coherence (Gaveau et al. 2003) and 3D structural information can also be derived (e.g., Ho Tong Minh et al. 2016). However, within this Chapter, just the considerations of the backscatter intensity and SAR, in general, will be discussed. For a full introduction to SAR imagery and processing, refer to Woodhouse (2005).

The intensity of σ^0 is dependent on the vertical structure (i.e., buildings and vegetation) and moisture (predominantly soil). As the size of the vertical structure increases, the magnitude of the SAR backscatter increases. For example, within a forest, pixels of higher backscatter will typical correspond with areas of larger trees. However, the background soil and vegetation moisture can also influence the signal. For example, Lucas et al. (2010) demonstrated that in dry regions of Australia, rain events can increase the SAR backscatter and therefore recommended the use of the driest scenes available when generating regional mosaics. These were identified through reference to spatial interpolations of rainfall measurements or low resolution, high-frequency AMSR-E passive microwave radiometer measures of surface

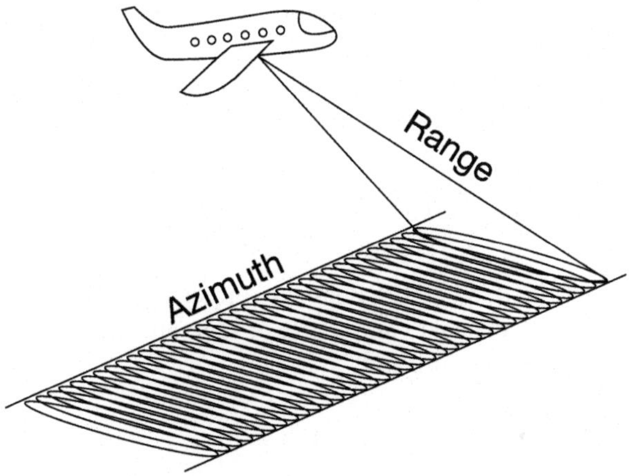

Fig. 9 The side-looking geometry of a SAR system

moisture. Reference to datasets such as these is particularly important where a time series is being constructed as scene(s) or parts of scenes might not be directly comparable in terms of vegetation change as the backscatter changes would correspond with soil and/or surface vegetation moisture amounts.

Geometric Correction

The geometry of a SAR system is quite different from that of an optical or LiDAR in that the sensor is side looking (Fig. 9) and therefore features closer to the sensor will be closer together than those further away in the raw slant-range image space. Therefore, one of the first processing stages within the geometric correction is to convert the slant-range image space into ground-range (i.e., all pixels have an equal ground cover). Following the conversion to ground range, an orthorectification is required to place the SAR imagery into the required geographic coordinate system, with consideration given to topographic relief.

Defining Sigma Nought (σ^0) and Gamma Nought (γ^0)

Typically the pixel values of the ground range images are supplied as digital numbers (DN) (to reduce the file size) and need to be converted to σ^0[dB]. To achieve this, a calibration offset (C) is applied.

$$\sigma^0[dB] = 10\log10(\mathrm{DN}) + C$$

The exact form of the equation to convert from DN to σ^0 varies for each instrument. σ^0 has an angular effect due to the variance in the incidence angles across the scene, particularly for airborne SAR where the variance in incidence angle is higher. Therefore, a correction to γ^0 can also be applied, where θ is the local incidence angle.

$$\gamma^0[dB] = \mathbf{10}\log 10\left(\frac{\mathrm{DN}}{\theta}\right) + C$$

When using σ^0 [dB] or γ^0 [dB] values for further processing, care is required when applying a process that takes an average or sums any of the pixel values, as dB is a log value. To convert from dB use:

$$\sigma^0 = \mathbf{10}\left(\frac{\sigma^0[dB]}{10}\right)$$

Following any calculations, for example calculating a mean backscatter for a set of image segments, dB values can be retrieved using:

$$\sigma^0[dB] = \mathbf{10}\log 10(\sigma^0)$$

SAR Image Filtering

SAR images contain speckle, which is noise from the image acquisition process. To reduce speckle within the scene image, filters are commonly used. Filters can be applied to a single image (e.g., Lee Filter; Lee 1981) or to time-series (Trouve et al. 2003). It is commonly recommended (Woodhouse 2005) that speckle filters are applied to SAR imagery before it is used unless the image data is being smoothed (averaged) in some way, which is the case when segmentation procedures are applied.

Conclusions

This chapter has attempted to provide an brief but wide ranging overview of the methods and processes that need to be considered when receiving remote sensing imagery prior to using the imagery for your application of interest. Once these processes are applied, the image data can be considered as analysis ready. Without satisfying the requirement of an ARD product prior to undertaking your application, is likely that your analysis will either fail or produce suboptimal results.

The deviation of ARD products can be undertaken by yourself if you have the appropriate background knowledge and software tools. However, the data provider(s) or other organisations or individuals can provide services to derive these products.

References

Armston, J., Disney, M.I., Lewis, P., Scarth, P., Phinn, S., Lucas, R.M., Bunting, P., Goodwin, N.: Direct retrieval of canopy gap probability using airborne waveform lidar. Remote Sens. Environ. **134**, 24–38 (2013)

Babcock, C., Finley, A.O., Cook, B.D., Weiskittel, A., Woodall, C.W.: Modeling forest biomass and growth: coupling long-term inventory and LiDAR data. Remote Sens. Environ. **182**, 1–12 (2016)

Bater, C., Coops, N.: Evaluating error associated with lidar-derived DEM interpolation. Comput. Geosci. **35**(2), 289–300 (2009)

Bunting, P., Armston, J., Clewley, D., Lucas, R.M.: Sorted pulse data (SPD) library—part II: a processing framework for LiDAR data from pulsed laser systems in terrestrial environments. Comput. Geosci. **56**, 207–215 (2013)

Chavez Jr., P.S.: An improved dark-object subtraction technique for atmospheric scattering correction of multispectral data. Remote Sens. Environ. **24**(3), 459–479 (1988)

Claverie, M., Vermote, E.F., Franch, B., Masek, J.G.: Evaluation of the Landsat-5 TM and Landsat-7 ETM+ surface reflectance products. Remote Sens. Environ. **169**, 390–403 (2015)

Coren, F., Sterzai, P.: Radiometric correction in laser scanning. Int. J. Remote Sens. **27**(15), 3097–3104 (2007)

Draper, D.W., Newell, D.A., Wentz, F.J., Krimchansky, S., Skofronick-Jackson, G.M.: The Global Precipitation Measurement (GPM) Microwave Imager (GMI): Instrument Overview and Early On-Orbit Performance. Sel. Top. Appl. Earth Obs. Remote Sens. IEEE J. **8**(7), 3452–3462 (2015.) http://doi.org/10.1109/JSTARS.2015.2403303

Donoghue, D.N.M., Watt, P.J., Cox, N.J., Wilson, J.: Remote sensing of species mixtures in conifer plantations using LiDAR height and intensity data. Remote Sens. Environ. **110**(4), 509–522 (2007)

Frantz, D., Röder, A., Stellmes, M., Hill, J.: An operational radiometric landsat preprocessing framework for Large-Area time series applications. IEEE Trans. Geosci. Remote Sens. 1–16 (n.d..) http://doi.org/10.1109/TGRS.2016.2530856

Evans, J.S., Hudak, A.T.: A multiscale curvature algorithm for classifying discrete return LiDAR in forested environments. IEEE Trans. Geosci. Remote Sens. **45**(4), 1029–1038 (2007)

Feng, M., Sexton, J.O., Huang, C., Masek, J.G., Vermote, E.F., Gao, F., et al.: Global surface reflectance products from Landsat: assessment using coincident MODIS observations. Remote Sens. Environ. **134**, 276–293 (2013)

Gaveau, D.L.A., Balzter, H., Plummer, S.: Forest woody biomass classification with satellite-based radar coherence over 900 000 km2 in Central Siberia. For. Ecol. Manag. **174**(1–3), 65–75 (2003)

Green, M., Kondragunta, S., Ciren, P., Xu, C.: Comparison of GOES and MODIS Aerosol Optical Depth (AOD) to Aerosol Robotic Network (AERONET) AOD and IMPROVE PM2.5 Mass at Bondville, Illinois. J. Air Waste Manage. Assoc. **59**(9), 1082–1091 (2009)

Higgins, M.A., Asner, G.P., Martin, R.E., Knapp, D.E., Anderson, C., Kennedy-Bowdoin, T., Saenz, R., Aguilar, A., Wright, S.J.: Linking imaging spectroscopy and LiDAR with floristic composition and forest structure in Panama. Remote Sens. Environ. **154**, 358–367 (2014)

Ho Tong Minh, D., Le Toan, T., Rocca, F., Tebaldini, S., Villard, L., Réjou-Méchain, M., Phillips, O.L., Feldpausch, T.R., Dubois-Fernandez, P., Scipal, K., Chave, J.: SAR tomography for the retrieval of forest biomass and height: cross-validation at two tropical forest sites in French Guiana. Remote Sens. Environ. **175**, 138–147 (2016)

Jhan, J.-P., Rau, J.-Y., Huang, C.-Y.: Band-to-band registration and ortho-rectification of multilens/multispectral imagery: a case study of MiniMCA-12 acquired by a fixed-wing UAS. ISPRS J. Photogramm. Remote Sens. **114**, 66–77 (2016)

Ju, J., Roy, D.P., Vermote, E., Masek, J., Kovalskyy, V.: Continental-scale validation of MODIS-based and LEDAPS Landsat ETM+ atmospheric correction methods. Remote Sens. Environ. **122**, 175–184 (2012)

Krogager, E.: New decomposition of the radar target scattering matrix. Electron. Lett. **26**(18), 1525–1527 (1990)

Lee, J.-S.: Speckle analysis and smoothing of synthetic aperture radar images. Comput. Graph. Image Process. **17**(1), 24–32 (1981)

Lillesand, T.M., Kiefer, R.W., Chipman, J.W.: Remote Sensing and Image Interpretation, 5th edn. Wiley, Hoboken (2004)

Lucas, R., Armston, J., Fairfax, R., Fensham, R., Accad, A., Carreiras, J., Kelley, J., Bunting, P., Clewley, D., Bray, S., Metcalfe, D., Dwyer, J., Bowen, M., Eyre, T., Laidlaw, M., Shimada, M.: An evaluation of the ALOS PALSAR L-Band backscatter – above ground biomass relationship Queensland, Australia: impacts of surface moisture condition and vegetation structure. Sel. Top. Appl. Earth Obs. Remote Sens. IEEE J. **3**(4), 576–593 (2010)

Masek, J.G., Vermote, E.F., Saleous, N.E., Wolfe, R., Hall, F.G., Huemmrich, K.F., et al.: A Landsat surface reflectance dataset for North America, 1990–2000. Geosci. Remote Sens. Lett. IEEE. **3**(1), 68–72 (2006)

Mongus, D., Zalik, B.: Parameter-free ground filtering of LiDAR data for automatic DTM generation. ISPRS J. Photogramm. Remote Sens. **67**, 1–12 (2012)

Morsdorf, F., Kötz, B., Meier, E., Itten, K.I., Allgöwer, B.: Estimation of LAI and fractional cover from small footprint airborne laser scanning data based on gap fraction. Remote Sens. Environ. **104**(1), 50–61 (2006)

Nagol, J.R., Sexton, J.O., Kim, D.-H., Anand, A., Morton, D., Vermote, E., Townshend, J.R.: Bidirectional effects in Landsat reflectance estimates: is there a problem to solve? ISPRS J. Photogramm. Remote Sens. **103**, 129–135 (2015)

Nelson, R., Krabill, W., Tonelli, J.: Estimating forest biomass and volume using airborne laser data. Remote Sens. Environ. **24**(2), 247–267 (1988)

Palace, M.W., Sullivan, F.B., Ducey, M.J., Treuhaft, R.N., Herrick, C., Shimbo, J.Z., Mota-E-Silva, J.: Estimating forest structure in a tropical forest using field measurements, a synthetic model and discrete return lidar data. Remote Sens. Environ. **161**, 1–11 (2015)

Popescu, S.C.: Estimating biomass of individual pine trees using airborne lidar. Biomass Bioenergy. **31**(9), 646–655 (2007)

Roy, D.P., Ju, J., Kline, K., Scaramuzza, P.L., Kovalskyy, V., Hansen, M., et al.: Web-enabled Landsat Data (WELD): Landsat ETM+ composited mosaics of the conterminous United States. Remote Sens. Environ. **114**(1), 35–49 (2010)

Schlapfer, D., Richter, R.: Geo-atmospheric processing of airborne imaging spectrometry data. Part 1: parametric orthorectification. Int. J. Remote Sens. **23**(13), 2609–2630 (2002)

Schwind, P., Schneider, M., Palubinskas, G., Storch, T., Mueller, R., Richter, R.: Processors for ALOS optical data: Deconvolution, DEM generation, Orthorectification, and atmospheric correction. Geosci. Remote Sens. IEEE Trans. **47**(12), 4074–4082 (2009)

Shan, J., Toth, C.K.: Topographic Laser Ranging and Scanning: Principles and Processing. CRC Press, Boca Raton (2009)

Smith, G.M., Milton, E.J.: The use of the empirical line method to calibrate remotely sensed data to reflectance. Int. J. Remote Sens. **20**, 2653–2662 (2010)

Trouve, E., Chambenoit, Y., Classeau, N., Bolon, P.: Statistical and operational performance assessment of multitemporal SAR image filtering. Geosci. Remote Sens. IEEE Trans. **41**(11), 2519–2530 (2003)

Vermote, E.F., Tanre, D., Deuze, J.L., Herman, M., Morcette, J.J.: Second simulation of the satellite signal in the solar Spectrum, 6S: an overview. Geosci. Remote Sens. IEEE Trans. **35**(3), 675–686 (1997)

Wagner, W.: Radiometric calibration of small-footprint full-waveform airborne laser scanner measurements: basic physical concepts. ISPRS J. Photogramm. Remote Sens. **65**(6), 505–513 (2010)
Wilson, R.T., Milton, E.J., Nield, J.M.: Spatial variability of the atmosphere over southern England, and its effect on scene-based atmospheric corrections. Int. J. Remote Sens. **35**(13), 5198–5218 (2014)
Woodhouse, I.H.: Introduction to Microwave Remote Sensing. CRC Press, Boca Raton (2005)
Zimble, D.A., Evans, D.L., Carlson, G.C., Parker, R.C., Grado, S.C., Gerard, P.D.: Characterizing vertical forest structure using small-footprint airborne LiDAR. Remote Sens. Environ. **87**(2–3), 171–182 (2003)
Zhang, K., Chen, S., Whitman, D., Shyu, M., Yan, J., Zhang, C.: A progressive morphological filter for removing nonground measurements from airborne LIDAR data. IEEE Trans. Geosci. Remote Sens. **41**(4), 872–882 (2003)
Zwally, H.J., Schutz, B., Abdalati, W., Abshire, J., Bentley, C., Brenner, A., Bufton, J., Dezio, J., Hancock, D., Harding, D., Herring, T., Minster, B., Quinn, K., Palm, S., Spinhirne, J., Thomas, R.: ICESat's laser measurements of polar ice, atmosphere, ocean, and land. J. Geodyn. **34**(3–4), 405–445 (2002)

Long-Term Ecological Monitoring at the Landscape Scale for Nature Conservation: The Example of Doñana Protected Area

Ricardo Díaz-Delgado

Abstract This chapter describes, as an example, the integration of a landscape scale approach based on remote sensing applications into the Long-Term Ecological Monitoring program of the Doñana protected area. I report the contribution of the landscape-scale monitoring by using remote sensing tools *sensu lato*, provided by its retrospective vision and multi-scale analysis capacity. The implemented protocols are set up as a multi-scale approach and are validated with a network of ground-truth field plots. The landscape scale monitoring program is being applied not only to habitats but also to species and to track natural and human-driven processes, mainly management actions. The approach is applied to monitor wetlands, terrestrial plant communities and geophysical processes. The chapter focuses on trends and shifts evidenced by the landscape scale approach that have been published or reported for decision making and conservation management planning. Some pros and cons are finally discussed in relation to data source availability, mission continuity and technical requirements.

Keywords Landscape scale • Remote sensing • Global change • Conservation management

Introduction

From its beginnings back in 1969, conservation in Doñana (Southwest of Spain) involved a team of researchers and managers who were aware of the importance of acquiring knowledge for the effective preservation of this protected area. In this process, the birds, Iberian (*Lynx pardinus*) and Cork Oak (*Quercus suber*) populations were critical in raising awareness of the need for systematic long-term data acquisition as a baseline for management.

R. Díaz-Delgado (✉)
Estación Biológica de Doñana (EBD), Consejo Superior de Investigaciones Científicas (CSIC), Sevilla, Spain
e-mail: rdiaz@ebd.csic.es

© Springer International Publishing AG 2017

R. Díaz-Delgado et al. (eds.), *The Roles of Remote Sensing in Nature Conservation*,
DOI 10.1007/978-3-319-64332-8_4

Many of these monitoring procedures have allowed retrospective analysis to show trends, changes and even relationships that reflect human impact. This reflects the broad consensus that long term monitoring programs are essential in addressing global change effects and in implementing decision-making (Inouye 2017; Lindenmayer et al. 2015; Magurran et al. 2010; Navarrete et al. 2010; Willig and Walker 2016).

In 2002, the Doñana Biological Station (EBD), a research centre of the Spanish Research Council (CSIC), together with the Doñana National Park Officers, proposed an integrated Long-Term Ecological Monitoring (LTEM) program. Almost a hundred monitoring protocols were defined, some to assess the status of threatened or flagship species and habitats of interest and others to track natural processes and the effects of conservation management actions (Díaz-Delgado 2010). After 15 years of implementation, the program has contributed data to provide evidence of spatial patterns and temporal trends of the target species and habitats: these have been reported in many publications. Consequently, quite a few research projects have been proposed and initiated to seek the scientific causes of the observed trends and their relationships with a range of environmental factors. Furthermore, the different management agencies have had uninterrupted access to the monitoring data and used these for decision-making in several conservation success stories at Doñana (Díaz-Delgado et al. 2016b).

The integration of the program into the Long-Term Ecological Research (LTER) network in 2008 allowed us to enhance the results and their outreach. Since then, the monitoring program and the research associated with it have benefited from the advances provided by the collaboration with the LTER networks, through participation in different research projects.

On the other hand, the program has largely benefited the available remote sensing applications by providing a full set of protocols and assisting in the interpretation of other monitoring results. All across the planet we find different conservation agencies and research projects using remote sensing tools to monitor ecosystems state and trends. Recent advances are also achieving good successes in biodiversity monitoring (Nagendra et al. 2013; Paganini et al. 2016; Pettorelli et al. 2016; Vihervaara et al. 2017; Wulder 2011).

I present an overview of some of the results obtained from the treatment, analysis and interpretation of the information collected through the monitoring program at a landscape scale. Most of these have contributed to conservation management in the Doñana protected area.

Long-Term Ecological Monitoring at Landscape Scale in the Doñana Protected Area

The landscape-scale approach integrated into the LTEM program is implemented in the Doñana Natural Space (END), south-west Spain. The approach has greatly contributed to improving our knowledge of different, ecologically relevant, natural processes (Díaz-Delgado 2010). The protocols associated with this approach are mainly applied using a combination of remote sensing images, *sensu lato*, and ground truth data from permanent field plots (Barrett 2013; Richards 2013). One advantage of this approach is the possibility to look back in time and use the first available aerial photos as a temporal reference. The current availability of long time series of images for many places in the world enables us to identify trends and shifts in different ecological processes and features (Gardiner and Díaz-Delgado 2007).

The landscape scale monitoring focuses on three different subject areas: the Doñana wetland ecosystems, including marshes and ponds, geophysical processes, and terrestrial plant communities.

Ground-Truthing

The landscape-scale approach uses plots, transects and *in situ* sampling points as ground-truth for validation purposes, which includes percentage cover for most plant communities present in Doñana.

The multi-scale approach uses involves sampling from the individual plots (microscale) through to the landscape level including mesoscale plots. When dealing with woodlands, the spatial location and associated information on individual trees (size and species) is recorded to allow scaling up using airborne (including drone) and spaceborne data.

Since 1999, we have reported trends and changes in the surface cover of dominant plant communities in Doñana. For example, there has been a general increase in the density and area occupied by Stone Pine (*Pinus pinea*) woodlands and a decrease of *Erica scoparia* healthlands replaced by more xeric species, including woodlands dominated by Juniper (*Juniperus phoenicea spp. turbinata*). These plots also provide information on habitat structure, species density, population dynamics, species abundance and diversity.

Monitoring of Doñana Wetlands

Monitoring the annual flooding and drying out processes of the Doñana natural marshes has allowed to test the 'largely assumed' hypothesis of an increase in the hydroperiod in recent decades (Díaz-Delgado et al. 2016a; Díaz-Delgado et al. 2010; Díaz-Delgado et al. 2006). Figure 1 shows the median trend estimated from the 1974–2014 time series of flooding masks retrieved from Landsat MSS, TM and ETM+ images. This approach allows us to spatially locate the temporal trends (increases or reductions) in the duration of the flooding: this is a critical variable for the biological communities of Doñana marshes. The temporal analysis reveals two significant trends: on the one hand, the hydroperiod has increased in the

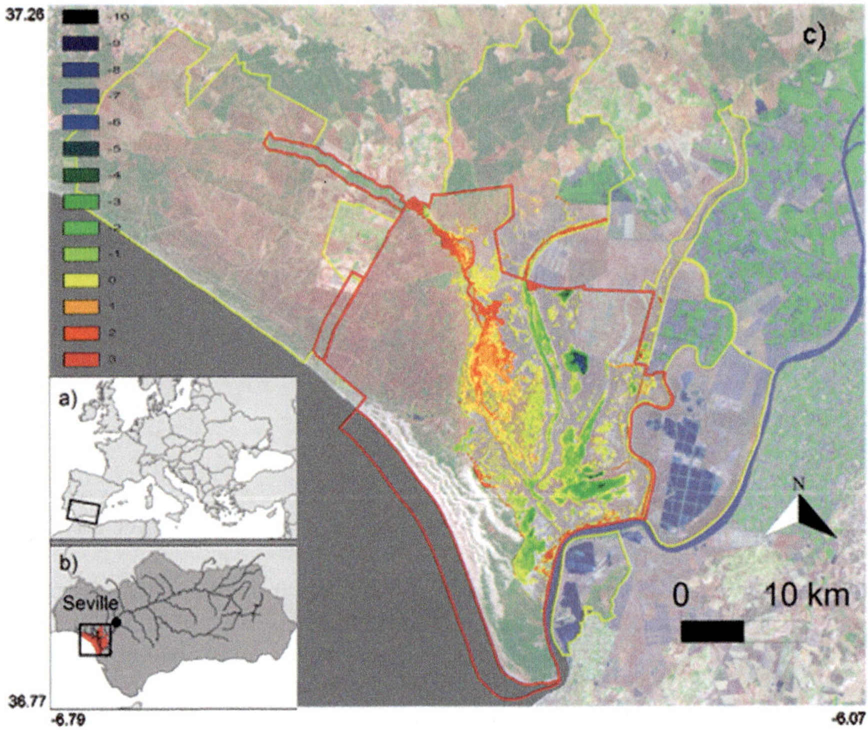

Fig. 1 The location of Doñana protected area (**a**) in Europe and (**b**) in Andalusia and (**c**) the pixel values of the hydroperiod trends (in days per year) for Doñana National (*red lines*) and Natural (*yellow line*) Parks for the 1974–2014 period. *Red and yellow colors* indicate an increase and *green/blue colors* a decrease. Background: Landsat 5 TM false color composition with bands 5-4-3

northwestern quadrant (*red and yellow colored* pixels) but has reduced in the southeast quadrant (*blue and green colors*) (Díaz-Delgado et al. 2016a).

The results from the landscape scale monitoring were used to evidence the effects of different restoration actions under the framework of the Doñana 2005 project (Chans and Díaz-Delgado 2006; Frisch et al. 2009; Santamaría et al. 2006).

Different studies have revealed the importance of the hydroperiod variable in explaining the presence, abundance and breeding success of different waterfowl (Kloskowski et al. 2009; Márquez-Ferrando et al. 2014; Ramo et al. 2013; Rendón et al. 2008; Toral et al. 2011). By using remote sensing, we were able to map the expansion of the invasive aquatic fern *Azolla filiculoides* using Landsat images and target detection techniques (Díaz-Delgado et al. 2008, 2011) enhanced by the use of hyperspectral images (Bustamante et al. 2009a). The hydroperiod was also linked to the spread of *Azolla* in the Doñana marshes (Espinar et al. 2015).

The landscape scale monitoring has benefited from the use of hyperspectral images from airborne sensors, as these have allowed more detailed mapping of the

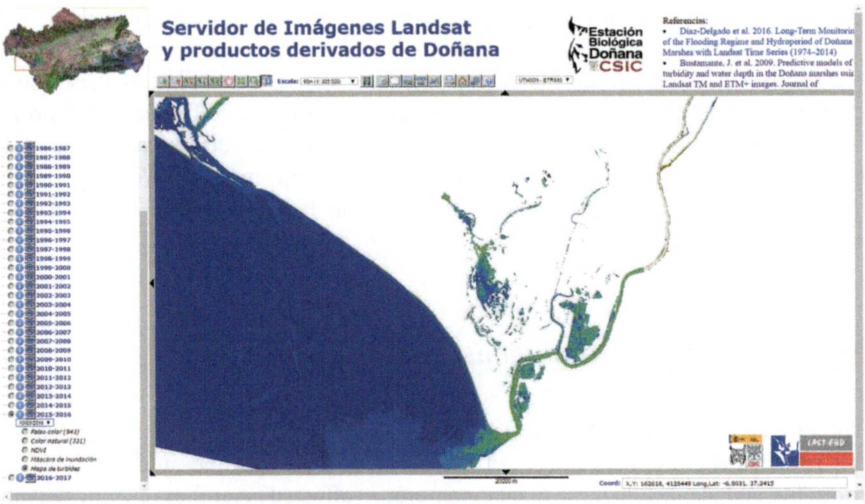

Fig. 2 The website offering the results of the Doñana wetlands monitoring program at landscape scale. The example shows the turbidity mapped for the 10th March 2016. RGB composites, flooding masks and NDVI images are also available for visualization and download

location of most of the temporal water pools in the Aeolian sands of the Doñana protected area (Gómez-Rodríguez et al. 2011). Gómez-Rodríguez et al. (2008) used this detailed cartography to produce an enhanced predictive model for the breeding habitats of the different amphibian species of Doñana. The decrease of the hydroperiod in the recent decades has been also evidenced in the Doñana water bodies (Bustamante et al. 2016) and was shown to be one of the most relevant causes of the decline in amphibian reproduction (Gómez-Rodríguez et al. 2010).

The multispectral information provided by the Landsat satellite images is systematically used to map water turbidity in both the Doñana marshes (Bustamante et al. 2009b) and the Guadalquivir estuary (Díaz-Delgado et al. 2010). The analysis of the turbidity time series has helped to quantify the spatiotemporal variability and reconstruct the turbidity regime by differentiating extreme turbid events from long-term turbid trends (Díaz-Delgado et al. 2015). Periodic flooding masks and turbidity maps from the landscape scale monitoring are accessible online through OGC Web Map Services at http://venus.ebd.csic.es/imgs/ (Fig. 2).

Monitoring of Geophysical Processes

Geophysical processes are usually quite conspicuous in remote sensing images. In the case of Doñana, the dynamics of the large sand-dune system and of the shoreline are being reconstructed and periodically mapped by applying a very simple procedure based on image segmentation (Berberoglu and Akin 2009; Pardo-Pascual et al.

2012). Shoreline regression and progradation (beach creation) are mapped along the 25 km of coastline (Díaz-Delgado 2008).

We also monitor the sedimentation processes in Doñana marshes: these data enabled us to assess the effects of the restoration actions of the Doñana 2005 project. For instance, we mapped the progressive reduction of the dejection cone caused by the *Arroyo del Partido*, one of the main tributaries to the marshes. We also located the emergence of new streams from the artificial lagoons created to retain the sediments.

Monitoring of Terrestrial Plant Communities

The application of a landscape scale monitoring approach to the Juniper woodlands has revealed the dramatic effects of the mortality events in the autumn of 2005. Mortality and damage in these plant communities affected juvenilies (canopy height lower than 1 m) more than adults (higher than 1 m). Greater damage and mortality was observed in areas with high plant density (Díaz-Delgado et al. 2014). The effects of these types of events persist during the following years despite the vigorous sprouting observed from damaged individuals (Fig. 3). In addition, a retrospective spatial analysis starting from 1956 revealed both an expansion of Juniper woodlands and a increase in plant density (García et al. 2014).

This monitoring program also benefited from using airborne hyperspectral images to map the abundances of the dominant woody species in the shrublands, by using spectral unmixing techniques (Jiménez 2011; Jiménez et al. 2005, 2011) based on spectroradiometric ground measurements (Jiménez and Díaz-Delgado 2015). The periodic detailed mapping provides evidence of the dramatic effects of drought on the *monte negro* hygrophytic plant communities and the defoliation processes in *monte blanco* xeric shrublands.

The setting-up of permanent ground-truth monitoring plots has provided a better understanding of the resilience of Doñana shrubland communities. By quantifying the functional diversity of the different plant species combined with the ground survey information on the percentage cover, structure and density of plants, we have established the great stability of shrubland communities in Doñana after extreme events (Lloret et al. 2016; Pérez-Ramos et al. 2017; de la Riva et al. 2017).

In the case of the permanent plots in the floodplain forest of Arroyo de la Rocina (tributary to Doñana marshes), we have evidenced the need for an integrated monitoring approach using field plots and remote sensing (Rodríguez-González et al. 2017). We found a consistent and decreasing trend of the area occupied by Willows (*Salix atrocinerea*) and an increase in the stem density of Narrow-leaved Ash trees (*Fraxinus angustifolia*) (Fig. 4). We found a higher percent of fallen Willow trunks in the plots. The results highlight the recent hydrological changes in one of the most important tributary rivers entering the Doñana marshes.

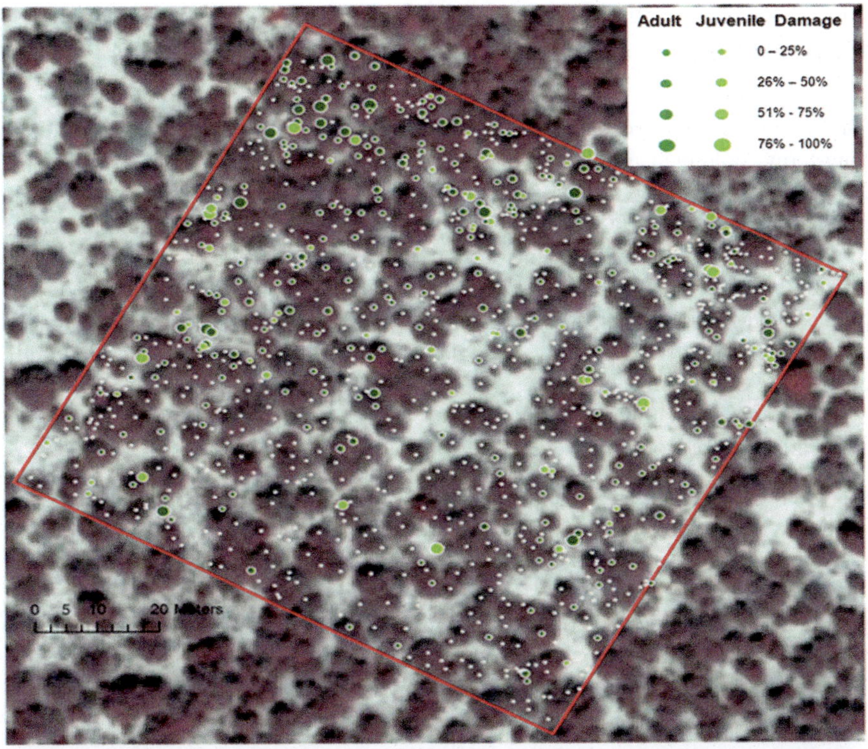

Fig. 3 An example of one of the meso-scale ground-truth plots for the monitoring of Juniper woodlands in Doñana protected area. The distribution of adult and juvenile individuals is depicted together with the percent of canopy damage measured for every individual inside the 1 ha plot

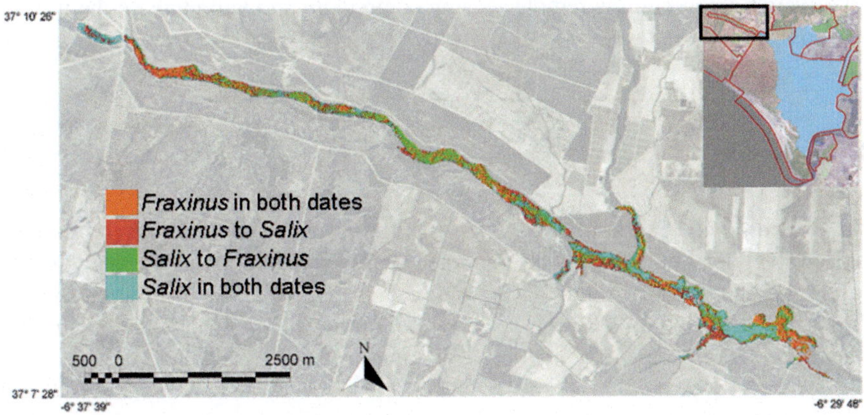

Fig. 4 Map of changes in species dominance in the Arroyo de la Rocina from the comparison of 2004 and 2015 hyperspectral image classifications. The image in the right corner shows the location of the Arroyo de la Rocina in relation to Doñana protected area (*red line*) and Doñana marshes (*blue area*)

The Pros and Cons of Integrating Remote Sensing in Nature Conservation

The use of remote sensing is strongly recommended to complement traditional long-term ecological monitoring programs based on field sampling (Gross et al. 2009; Willis 2015; Rodríguez-González et al. 2017). Monitoring at a landscape scale provides a synoptic view of natural processes and enhances the interpretation of data collected in field plots. In addition, historical time series of remote sensing images are now widely available, and this can inform our selection of reference points for restoration projects (Reif and Theel 2016) or to assess temporal trends, anomalies or shifts of relevant ecological and biodiversity indicators (Pettorelli et al. 2016; Skidmore et al. 2015).

However, the major constraints for conservation managers using remote sensing applications arise from:

- A lack of sufficient knowledge and skilled staff to integrate remote sensing techniques into decision making.
- A lack of clear guidelines on how to apply the procedures or to access the most reliable data available online.
- The lack of specific scientific and technical committees offering assessment and supervision on the optimal products, methods and technologies to be used in the long run to manage and preserve natural areas.

In addition to these specific limitations, the users of remote sensing images and products are aware of the limited lifetime of sensors and satellites: this can be an issue for the agencies responsible for maintaining continuity in their missions. Hopefully, the recent launching of Landsat 8 with the OLI instrument and the constellation of Sentinel satellites by ESA has secured the provision of Earth Observation images to the remote sensing community and will increase the use of its applications.

Conclusions

While traditional long-term ecological monitoring programs are typically underpinned solely on field sampling, a combined approach incorporating remote sensing tools is essential to providing a landscape-scale perspective. The integration should be carried out according to the specific monitoring program targets and implemented with complementary methods of ground-truth.

Acknowledgments The monitoring program at landscape scale would not be possible without the continued support and funding by *Consejería de Medio Ambiente y Ordenación del Territorio* of the Andalusian Government, *Junta de Andalucía.* The whole monitoring program was initially funded by *Organismo Autónomo de Parques Nacionales* and also supported by *Confederación*

Hidrográfica del Guadalquivir through the restoration project *Doñana 2005*. The author also wants to deeply thanks the permanent help and assistance of the staff from the Laboratory of GIS and Remote Sensing of *Estación Biológica de Doñana* (LAST-EBD) and from its Monitoring Team of Natural Processes.

References

Barrett, E.C.: Introduction to Environmental Remote Sensing. Routledge, London (2013)

Berberoglu, S., Akin, A.: Assessing different remote sensing techniques to detect land use/cover changes in the eastern Mediterranean. Int. J. Appl. Earth Obs. Geoinf. **11**, 46–53 (2009). doi:10.1016/j.jag.2008.06.002

Bustamante, J., Díaz-Delgado, R., Aragonés, D., Fernández-Zamudio, R., García Murillo, P., Cirujano, S.: Using hyperspectral sensors to map the spread of the invading water fern (Azolla filiculoides). In: Oral Presentation at the 2nd European Congress of Conservation Biology – ECCB 2009, Prague, Czech Republic (2009a)

Bustamante, J., Pacios, F., Díaz-Delgado, R., Aragonés, D.: Predictive models of turbidity and water depth in the Doñana marshes using Landsat TM and ETM+ images. J. Environ. Manag. **90**, 2219–2225 (2009b). doi:10.1016/j.jenvman.2007.08.021

Bustamante, J., Aragonés, D., Afán, I.: Effect of protection level in the hydroperiod of water bodies on Doñana's aeolian sands. Remote Sens. **8**, 867 (2016). doi:10.3390/rs8100867

Chans, J.J., Díaz-Delgado, R.: Monitoring and evaluation: the key to the Doñana 2005 Restoration Project. In: Doñana, Water and Biosphere, pp. 319–326. Confederación Hidrográfica del Guadalquivir. Ministerio de Medio Ambiente, Madrid (2006)

Díaz-Delgado, R.: Cartografía dinámica costera del Parque Nacional de Doñana. In: Actas de las Jornadas Técnicas las nuevas técnicas de información geográfica al servicio de la gestión de zonas costeras: Análisis de la evolución de playas y dunas, pp. 28–32. Universitat Politècnica de València, Valencia (2008)

Díaz-Delgado, R.: An integrated monitoring Programme for Doñana natural space: the set-up and implementation. In: Conservation Monitoring in Freshwater Habitats, pp. 325–337. Springer, Dordrecht/Heidelberg/London (2010)

Díaz-Delgado, R., Bustamante, J., Aragonés, D., Pacios, F.: Determining water body characteristics of Doñana shallow marshes through remote sensing. In: IGARSS 06. IEEE International Geoscience and Remote Sensing Symposium, pp. 3662–3664. Denver, Colorado, USA (2006)

Díaz-Delgado, R., Bustamante, J., Aragonés, D.: Caso 5. La teledetección como herramienta en la cartografía de especies invasoras: *Azolla filiculoides* en Doñana. In: Vila, M., Valladares, F., Traveset, A., Santamaría, L., Castro, P. (eds.) Invasiones Biológicas, pp. 159–163. Consejo Superior de Investigaciones Científicas, Madrid (2008)

Díaz-Delgado, R., Aragonés, D., Ameztoy, I., Bustamante, J.: Monitoring marsh dynamics through remote sensing. In: Conservation Monitoring in Freshwater Habitats, pp. 375–386. Springer, Dordrecht (2010)

Díaz-Delgado, R., Ameztoy, I., Aragonés, D., Bustamante, J.: Cartografía histórica del helecho acuático invasor *Azolla filiculoides* en la marisma de Doñana. In: Recondo Gonzáles, C., Pendás Molina, E. (eds.) Teledetección Bosques y cambio climático. Actas XIV Congreso Asociación Española de Teledetección, pp. 329–332. Asociación Española de Teledetección, Mieres del Camino (2011)

Díaz-Delgado, R., Afán, I., Silva, R.: Patrones espaciales de daño en copa en el sabinar de la Reserva Biológica de Doñana a consecuencia de un evento extremo de sequía. In: Sistemas vegetales y fauna en medios litorales: Avances en sus características, dinámica y criterios para la conservación. pp. 189–192. Rafael Cámara Artigas, Beatriz Rodríguez Pérez, Juan Luis Muriel Gómez, Sevilla, España (2014)

Díaz-Delgado, R., Aragonés, D., Afán, I., Bustamante, J.: A synoptic turbidity index for the Guadalquivir River with Landsat TM and ETM+ images: the turbiperiod. In: Fourth Recent Advances in Quantitative Remote Sensing, pp. 195–198. Publicacions de la Universitat de València, Torrent (2015)

Díaz-Delgado, R., Aragonés, D., Afán, I., Bustamante, J.: Long-term monitoring of the flooding regime and hydroperiod of Doñana marshes with Landsat time series (1974–2014). Remote Sens. **8**, 775 (2016a). doi:10.3390/rs8090775

Díaz-Delgado, R., Carro, F., Quirós Herruzo, F., Osuna, A., Baena, M.: Contribution from long-term ecological monitoring to research and management of Doñana LTSER platform. Ecosistemas. **25**, 9–18 (2016b). doi:10.7818/ECOS.2016.25-1.03

Espinar, J.L., Diaz-Delgado, R., Bravo-Utrera, M.Á., Vilà, M.: Linking Azolla filiculoides invasion to increased winter temperatures in the Doñana marshland (SW Spain). Aquat. Invasions. **10**, 17–24 (2015). doi:10.3391/ai.2015.10.1.02

Frisch, D., Arechederra, A., Green, A.J.: Recolonisation potential of zooplankton propagule banks in natural and agriculturally modified sections of a semiarid temporary stream (Doñana, Southwest Spain). Hydrobiologia. **624**, 115–123 (2009). doi:10.1007/s10750-008-9672-x

García, C., Moracho, E., Díaz-Delgado, R., Jordano, P.: Long-term expansion of juniper populations in managed landscapes: patterns in space and time. J. Ecol. **102**, 1562–1571 (2014). doi:10.1111/1365-2745.12297

Gardiner, N., Díaz-Delgado, R.: Trends in selected biomes, habitats and ecosystems: inland waters. In: Sourcebook on Remote Sensing and Biodiversity Indicators, pp. 83–102. Secretariat of the Convention on Biological Diversity, Montreal (2007)

Gómez-Rodríguez, C., Bustamante, J., Koponen, S., Díaz-Paniagua, C.: High-resolution remote-sensing data in amphibian studies: identification of breeding sites and contribution to habitat models. Herpetol. J. **18**, 103–113 (2008)

Gómez-Rodríguez, C., Bustamante, J., Díaz-Paniagua, C.: Evidence of Hydroperiod shortening in a preserved system of temporary ponds. Remote Sens. **2**, 1439–1462 (2010). doi:10.3390/rs2061439

Gómez-Rodríguez, C., Díaz-Paniagua, C., Bustamante, J.: Cartografía de lagunas temporales del Parque Nacional de Doñana. Agencia Andaluza del Agua, Consejería de Medio Ambiente, Junta de Andalucía, Sevilla (2011)

Gross, J.E., Goetz, S.J., Cihlar, J.: Application of remote sensing to parks and protected area monitoring: introduction to the special issue. Remote Sens. Environ. **113**, 1343–1345 (2009). doi:10.1016/j.rse.2008.12.013

Inouye, D.W.: Insiders' perspectives on the long-term ecological research program. Ecology. **98**, 1480–1482 (2017). doi:10.1002/ecy.1765

Jiménez, M.: Cartografía de especies de matorral de la Reserva Biológica de Doñana mediante el sistema hyperespectral aeroportado INTA-AHS. Implicaciones en el estudio y seguimiento del matorral de Doñana, (2011)

Jiménez, M., Díaz-Delgado, R.: Towards a standard plant species spectral library protocol for vegetation mapping: a case study in the Shrubland of Doñana National Park. ISPRS Int. J. Geo-Inf. **4**, 2472–2495 (2015). doi:10.3390/ijgi4042472

Jiménez, M., Díaz-Delgado, R., Soriguer, R., Fernández-Renau, A., Prado, E., Gutiérrez, O.: Aproximación de las imágenes hiperespectrales AHS a la cartografía de abundancia de matorral. Rev. Teledetec. 95–100 (2005)

Jiménez, M., Pou, A., Díaz-Delgado, R.: Cartografía de especies de matorral de la Reserva Biológica de Doñana mediante el sistema hiperespacial aeroportado INTA-AHS. Implicaciones en el estudio y segimiento del matorral de Doñana. Rev. Teledetec. 98–102 (2011)

Kloskowski, J., Green, A.J., Polak, M., Bustamante, J., Krogulec, J.: Complementary use of natural and artificial wetlands by waterbirds wintering in Doñana, south-west Spain. Aquat. Conserv. Mar. Freshw. Ecosyst. **19**, 815–826 (2009). doi:10.1002/aqc.1027

Lindenmayer, D.B., Burns, E.L., Tennant, P., Dickman, C.R., Green, P.T., Keith, D.A., Metcalfe, D.J., Russell-Smith, J., Wardle, G.M., Williams, D., Bossard, K., de Lacey, C., Hanigan, I., Bull, C.M., Gillespie, G., Hobbs, R.J., Krebs, C.J., Likens, G.E., Porter, J., Vardon, M.:

Contemplating the future: acting now on long-term monitoring to answer 2050's questions. Austral. Ecol. **40**, 213–224 (2015). doi:10.1111/aec.12207

Lloret, F., de la Riva, E.G., Pérez-Ramos, I.M., Marañón, T., Saura-Mas, S., Díaz-Delgado, R., Villar, R.: Climatic events inducing die-off in Mediterranean shrublands: are species' responses related to their functional traits? Oecologia. **180**, 1–13 (2016). doi:10.1007/s00442-016-3550-4

Magurran, A.E., Baillie, S.R., Buckland, S.T., Dick, J.M., Elston, D.A., Scott, E.M., Smith, R.I., Somerfield, P.J., Watt, A.D.: Long-term datasets in biodiversity research and monitoring: assessing change in ecological communities through time. Trends Ecol. Evol. **25**, 574–582 (2010). doi:10.1016/j.tree.2010.06.016

Márquez-Ferrando, R., Figuerola, J., Hooijmeijer, J.C.E.W., Piersma, T.: Recently created man-made habitats in Doñana provide alternative wintering space for the threatened continental European black-tailed godwit population. Biol. Conserv. **171**, 127–135 (2014). doi:10.1016/j.biocon.2014.01.022

Nagendra, H., Lucas, R., Honrado, J.P., Jongman, R.H.G., Tarantino, C., Adamo, M., Mairota, P.: Remote sensing for conservation monitoring: assessing protected areas, habitat extent, habitat condition, species diversity, and threats. Ecol. Indic. **33**, 45–59 (2013). doi:10.1016/j.ecolind.2012.09.014

Navarrete, S.A., Gelcich, S., Castilla, J.C.: Long-term monitoring of coastal ecosystems at Las Cruces, Chile: defining baselines to build ecological literacy in a world of change. Rev. Chil. Hist. Nat. **83**, 143–157 (2010)

Paganini, M., Leidner, A.K., Geller, G., Turner, W., Wegmann, M.: The role of space agencies in remotely sensed essential biodiversity variables. Remote Sens. Ecol. Conserv. **2**, 132–140 (2016). doi:10.1002/rse2.29

Pardo-Pascual, J.E., Almonacid-Caballer, J., Ruiz, L.A., Palomar-Vázquez, J.: Automatic extraction of shorelines from Landsat TM and ETM+ multi-temporal images with subpixel precision. Remote Sens. Environ. **123**, 1–11 (2012). doi:10.1016/j.rse.2012.02.024

Pérez-Ramos, I.M., Díaz-Delgado, R., de la Riva, E.G., Villar, R., Lloret, F., Marañón, T.: Climate variability and community stability in Mediterranean shrublands: the role of functional diversity and soil environment. J. Ecol. **10**(2), 281–293 (2017). doi:10.1111/1365-2745.12747

Pettorelli, N., Wegmann, M., Skidmore, A., Mücher, S., Dawson, T.P., Fernandez, M., Lucas, R., Schaepman, M.E., Wang, T., O'Connor, B., Jongman, R.H.G., Kempeneers, P., Sonnenschein, R., Leidner, A.K., Böhm, M., He, K.S., Nagendra, H., Dubois, G., Fatoyinbo, T., Hansen, M.C., Paganini, M., de Klerk, H.M., Asner, G.P., Kerr, J.T., Estes, A.B., Schmeller, D.S., Heiden, U., Rocchini, D., Pereira, H.M., Turak, E., Fernandez, N., Lausch, A., Cho, M.A., Alcaraz-Segura, D., McGeoch, M.A., Turner, W., Mueller, A., St-Louis, V., Penner, J., Vihervaara, P., Belward, A., Reyers, B., Geller, G.N.: Framing the concept of satellite remote sensing essential biodiversity variables: challenges and future directions. Remote Sens. Ecol. Conserv. **2**, 122–131 (2016). doi:10.1002/rse2.15

Ramo, C., Aguilera, E., Figuerola, J., Máñez, M., Green, A.J.: Long-term population trends of colonial wading birds breeding in Doñana (SW Spain) in relation to environmental and anthropogenic factors. Ardeola. **60**, 305–326 (2013)

Reif, M.K., Theel, H.J.: Remote sensing for restoration ecology: application for restoring degraded, damaged, transformed, or destroyed ecosystems. Integr. Environ. Assess. Manag. n/a–n/a (2016). doi:10.1002/ieam.1847

Rendón, M.A., Green, A.J., Aguilera, E., Almaraz, P.: Status, distribution and long-term changes in the waterbird community wintering in Doñana, south–west Spain. Biol. Conserv. **141**, 1371–1388 (2008). doi:10.1016/j.biocon.2008.03.006

Richards, J.A.: Remote Sensing Digital Image Analysis: An Introduction. Springer Science & Business Media, Berlin (2013)

de la Riva, E.G., Lloret, F., Pérez-Ramos, I.M., Marañón, T., Saura-Mas, S., Díaz-Delgado, R., Villar, R.: The importance of functional diversity in the stability of Mediterranean shrubland communities after the impact of extreme climatic events. J. Plant Ecol. **10**, 281–293 (2017). doi:10.1093/jpe/rtw027

Rodríguez-González, P.M., Albuquerque, A., Martínez-Almarza, M., Díaz-Delgado, R.: Long-term monitoring for conservation management: Lessons from a case study integrating remote sensing and field approaches in floodplain forests. J. Environ. Manage. **202**(2), 392–402 (2017). https://doi.org/10.1016/j. jenvman.2017.01.067

Santamaría, L., Green, A., Díaz-Delgado, R., Bravo, M.Á., Castellanos, E.: Caracoles: a new laboratory for science and wetland restoration. In: Doñana, Water and Biosphere, pp. 313–315. Confederación Hidrográfica del Guadalquivir. Ministerio de Medio Ambiente, Madrid (2006)

Skidmore, A.K., Pettorelli, N., Coops, N.C., Geller, G.N., Hansen, M., Lucas, R., Mücher, C.A., O'Connor, B., Paganini, M., Pereira, H.M., Schaepman, M.E., Turner, W., Wang, T., Wegmann, M.: Environmental science: agree on biodiversity metrics to track from space. Nat. News. **523**, 403 (2015). doi:10.1038/523403a

Toral, G.M., Aragonés, D., Bustamante, J., Figuerola, J.: Using Landsat images to map habitat availability for waterbirds in rice fields. Ibis. **153**, 684–694 (2011). doi:10.1111/j.1474-919X.2011.01147.x

Vihervaara, P., Auvinen, A.-P., Mononen, L., Törmä, M., Ahlroth, P., Anttila, S., Böttcher, K., Forsius, M., Heino, J., Heliölä, J., Koskelainen, M., Kuussaari, M., Meissner, K., Ojala, O., Tuominen, S., Viitasalo, M., Virkkala, R.: How essential biodiversity variables and remote sensing can help national biodiversity monitoring. Glob. Ecol. Conserv. **10**, 43–59 (2017). doi:10.1016/j.gecco.2017.01.007

Willig, M.R., Walker, L.R. (eds.): Long-Term Ecological Research: Changing the Nature of Scientists. Oxford University Press, Oxford (2016)

Willis, K.S.: Remote sensing change detection for ecological monitoring in United States protected areas. Biol. Conserv. **182**, 233–242 (2015). doi:10.1016/j.biocon.2014.12.006

Wulder, M.A.: Canadian Forest Service Science Highlights. Can Satellites Help Monitor Biodiversity more Effectively? Canadian Forest Service, Ottawa (2011)

Part II
Habitat Case Studies

NILS – A Nationwide Inventory Program for Monitoring the Conditions and Changes of the Swedish Landscape

Anna Allard

Abstract The National Inventory of Landscapes in Sweden (NILS) was established by the Swedish Environmental Protection Agency to provide data for policy-makers in the country. Its main role is to determine the status of (and changes in) the Swedish landscape, either as a consequence of natural and/or anthropogenic disturbances, or because of ecological processes. In 2017, data from NILS will become available on the NILS data portal for analysis by researchers and other interested parties, such as governmental bodies. NILS's data collection covers all terrestrial areas and identifies variables such as land cover, land use (including historical land use), tree, shrub, field and ground vegetation and surfaces. The program consists of two parallel inventories, a field inventory and a remote sensing inventory, with both covering the nation in five-year rotations. The field inventory employs a large group of field-workers, and all sample squares are also inventoried using aerial near-infrared stereo imagery provided by the Swedish Land Survey. The first rotations of field data will soon be available, and the data from the first rotation of the remote sensing component (2003–2007) is already available. As part of the EU, Sweden is currently updating the mapping of land cover data, using the national coverage of airborne Light Detection and Ranging (LIDAR) and the European satellites of the Sentinel series. While at the same time strengthening the statistical estimates made from the NILS data. The principal role of the NILS program is to provide the reference data for both the classification of chosen vegetation types and for the validation of the results. The data collection for this mapping will start in 2017, using remote sensing and modelling, and sampling additional inventory areas for reference data.

Keywords Inventory of landscapes • Monitoring • Remote sensing applications • National databases

A. Allard (✉)
Swedish University for Agricultural Sciences, SLU, Umeå, Sweden
e-mail: anna.allard@slu.se

© Springer International Publishing AG 2017

R. Díaz-Delgado et al. (eds.), *The Roles of Remote Sensing in Nature Conservation*,
DOI 10.1007/978-3-319-64332-8_5

Introduction

The National Inventory of Landscapes in Sweden (NILS) was established in 2003, by the Swedish Environmental Protection Agency (Swedish EPA) to provide data to support policy-makers in decisions, and for international reporting. Another goal was to create a basis for analysis in the research community: to better understand and quantify the response of the Swedish landscape to either natural or anthropogenic change at a national level. This represents a "Top-Down" approach to monitoring/reporting of nature and conservation compared to the "Bottom-Up" approach, for example, which consists of locally monitoring smaller units of a Nature reserve and reporting to a national authority. The statistical nature of the sampling means that larger areas of widely distributed habitats tend to be better represented statistically than small areas of rarer habitats. This means that some areas of the rarer and more threatened habitats need to be targeted for additional data collection, primarily to inform conservation measures. The sampling method is, however, enough to monitor changes that occur following policy decisions that affect the landscape, such as trends in people's movements, types of farming, or the culture in managing forestry. The infrastructure of the program also makes it possible for other data-collection projects to collaborate, sharing costs for staff, travel and accommodation. Examples of monitoring collaborations include; the demonstration of an integrated North-European system for monitoring terrestrial habitats (MOTH), (MOTH 2017), a follow-up of the National Survey of Meadows and Pastures (the TUVA database), and inventory programs for butterflies and bumble bees carried out in the landscape squares and commissioned by the Swedish Board of Agriculture (2017). Regional monitoring collaborations include the Remiil program, (Glimskär et al. 2016), which comprises the inventory of small biotopes in farmed fields, mires and areas of grass outside grazing grounds: this program was commissioned by eight Counties in the rural areas of central Sweden.

The NILS was motivated primarily by the need to address the national Environmental Objectives in relation to, for example, sustainable forestry and the functioning of natural/semi-natural and managed landscapes, including areas of agriculture. Different authorities are designated by the government as responsible for each objective. Examples of these include the 'Thriving Wetlands' objective (Swedish EPA responsible), which concerns the conservation and restoration of peatlands and the 'Magnificent Mountain Landscape' objective, which focuses on the natural and cultural values of landscapes that require preservation (Swedish EPA, 2017). NILS plays an active part in discussions and in the current work of developing indicators and monitoring at the authority level (Hedenås et al. 2014, 2016; Svensson et al. 2016a, 2016b, Svensson et al. 2009). Data from the program is used for different research projects, for example to inform research on the distribution and main ecosystem characteristics of wetland and peatland types in elevation zones, from the coast to the mountains (Jeglum et al. 2011). The program also contributes to the development of monitoring indicators, notably for wetlands (Berglund et al. 2016) but also to assess ecosystem services in mountainous or forested areas (Svensson et al. 2016b).

The five broad monitoring targets identified at the start of the NILS program were: (1) Landscape patterns, (2) Amount and status of sensitive or threatened habitats, (3) Land use and disturbances, (4) Structural indicators and substrates, and (5) Indicative or sensitive species. Data from the NILS program is also intended to provide Swedish input to the international (including European Union) databases. Increasingly remote sensing is providing the basis for reporting and monitoring in the EU (Zlinszky et al. 2015; Van den Borre et al. 2011).

Design of Monitoring in NILS

NILS encompasses all terrestrial habitats in Sweden, and includes land cover, land use (including historical land use), characterizing the tree, shrub, field and ground vegetation and surfaces. The inventory is designed to collect data both in the field and using remote sensing, in the same areas, as close in time as possible, enabling the capture of different resolutions of landscape. Data are collected on a total of 356 variables, 269 of which are collected during the field inventory. Definitions and instructions can be found in the manuals (NILS 2017a; Allard et al. 2010) and the program is further described in Ståhl et al. (2011).

Both the field and the remote sensing (using stereo near-infrared imagery) inventories use a rigorous set of rules and decision trees to determine the variables, and both are carried out by trained staff. In a five-year rotation, all of the 631 sample squares are sampled at both resolutions, thus capturing the "Everyday landscape" (Fig. 1). In the stratified sampling system, each sample contains a landscape square of 5 × 5 km, with an inner square of 1 × 1 km where the main inventory is carried out: this makes it possible to scale up to the national level, and estimate cover, occurrences or changes in the variables. The landscape square creates the contextual basis for the inventory.

It was important for the Swedish EPA to conform to the European standards of reporting in 2003, so the NILS program was required to describe the variables or possible groupings into the context the DPSIR Framework (Driving forces, Pressure, State, Impact, Response) used by the European Environment Agency in its reporting activities (Smeets and Weterings 1999). Table 1 shows a grouping of these variables also described by Ståhl et al. (2011). Currently, the issue of Ecological Indicators is addressed at the national level, and work is ongoing to describe the NILS data in this context, to be of use for policy and planning purposes.

To alleviate the lack of official coordinates, the program has adopted a number of Flagship squares outside of the statistical sample, subjectively chosen in discussion with governmental entities such as County Boards and Municipalities. The selection is an 'ongoing list' of add-on inventories or collaborations, and squares are added as the need occurs: the most recent being a collaboration with Swedish and Sami Authorities to establish how to use NILS data for integrated landscape planning (Hedblom et al. 2014).

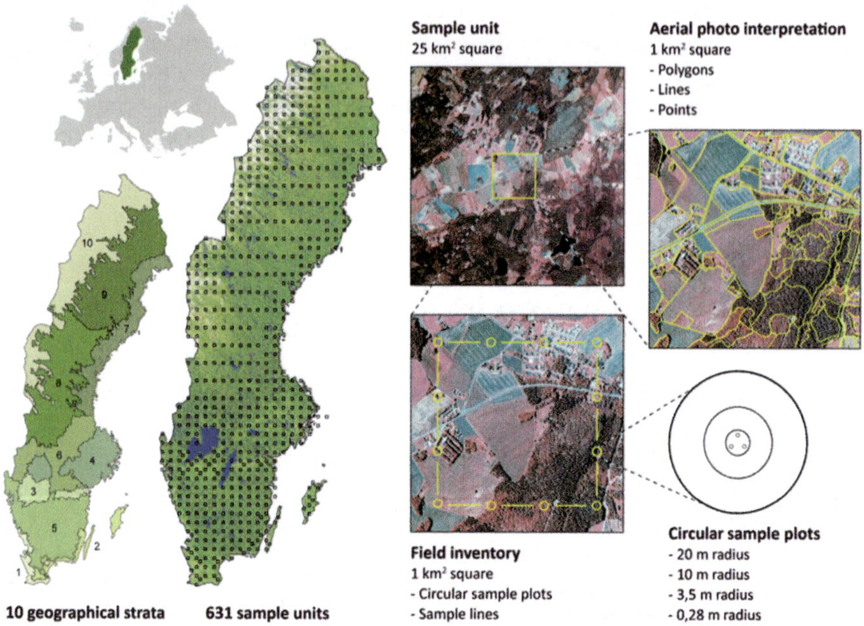

Fig. 1 The data collection design in the NILS program: 631 sample units of 1 × 1 km are monitored by aerial imagery, in polygons, lines and points. The same area is inventoried by field surveyors: in 12 circular sample plots, each comprising different radiuses, together with a line- intersect inventory along 12 lines

Table 1 An example of how NILS groups and expresses the variables in the context of the DPSIR framework

Processes **(pressure)**	*Structures* **(state)**
Ground disturbance	Vegetation structure
Hydrological changes	Dead wood and canopy structure
Grazing and mowing	Hydromorphological mire structures
Forestry	Linear and point features
Climate changes and air pollution	Soil properties
Habitats **(state)**	*Species* **(impact)**
Forest	Vegetation-forming plants
Wetlands and shores	Epiphytes
Grassland and heath	Grasslands indicators (e.g. grazing impact)
Cliffs, rocks, bare substrates	Game (droppings, etc.)
Man-made habitats (parks, etc.)	

Inventory by Remote Sensing

Aerial imagery is used in many countries to support environmental mapping and monitoring, and is discussed in conservation research (e.g. Rose et al. 2014). Examples include the UK Countryside Survey (Brandt et al. 2002; Barr et al. 2003), and the Norwegian Monitoring System, 3Q, (Dramstad et al. 2002). As with Sweden, Estonia used stereo near-infrared imagery to sample the landscape, and initial tests have been successfully carried out in Scotland for monitoring of Natura 2000-areas (Roose et al. 2007; Mattisson and Sullivan 2017). The advantage of using stereo near-infrared imagery is that different land cover (including vegetation) types can be differentiated and their health also indicated (Ihse 2007). The greatest benefit arises when these are interpreted by those with knowledge of the landscape and its ecology. Whilst individual species are often difficult to discriminate using colour infrared imagery, this can be achieved where they become dominant. However, most mapping is based on life-forms and their relative distribution in the landscape and this is generally the basis for mapping land cover, use and change as a consequence of natural process and human activities.

The military forces first established the benefits of near-infrared imagery as differences in reflectance separated camouflaged vehicles and guns from natural leaf cover. Since then, it has been used extensively in Sweden. An example is given in Fig. 2, which compares a colour infrared image with delineated land covers (Allard 2012b). The landscape is a typical setting in the mid-eastern part of Sweden, where the former rural areas have been mostly abandoned to regrowth. The foreground shows grazing land (1), with bedrock outcrops and woody shrubs on land that was never farmed or fertilized and is of comparatively high biological diversity. The area may still be grazed but the intensity is low. A second area (2) represents an area that may historically have been used to produce hay but where the unevenness of the ground hindered ploughing and hence only grazing has taken place. Recent abandonment of the area is indicated by the regenerating deciduous trees and some elderly birches and oaks. The third area (3) represents a farmed field that has been abandoned but where the addition of fertilizers has led to differences in the composition of herbaceous (grasses, forbs) and woody shrubs. The final area (4) is a mix of former grazing land but also mature coniferous forests that has been managed for timber (house and tool construction). Each of these units can be differentiated within the colour aerial photography.

Throughout Sweden, changes in the landscape are varied but can be quantified from the time-series of aerial images available: the nation was fully covered by stereo images in the 1950s–1960s and again in the 1970s–1980s, which presents good opportunities to 'hindcast' the estimations backwards in time. To date, all of the sample squares of the 1970s–1980s in farming areas have been fully inventoried: pastures and artificial land types (e.g. built-up areas) and agricultural areas have been compared (Christensen et al. 2015), and there is an on-going inventory that goes further back in time into the 1950–1960s. The results of this inventory will be compared to European efforts (Köhler et al. 2006). As examples of hindcasting,

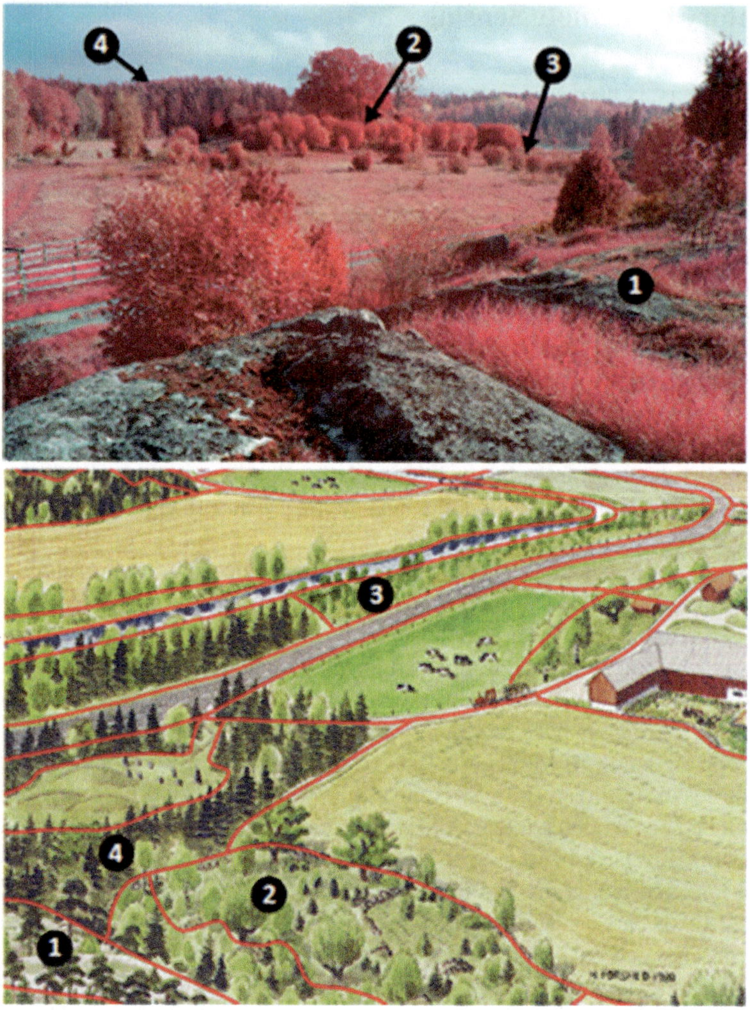

Fig. 2 Example of how to delineate land use, land cover and historical depth in a Swedish rural landscape, see text for explanations. The upper photo is taken by Clas Hättestrand, Stockholm University and the drawing is an excerpt from the NILS Interpretation manual

Fig. 3a illustrates an environment much discussed in Sweden, the peri-urban areas including urban sprawl. This environment is undergoing rapid change, with farmed and grazed fields abandoned as a consequence of new land use; where people work in towns or are employed by forestry instead of farming. Both squares are situated in the mid/southern part of Sweden, and convey areas that were farmed in the 1980s and are now road construction (No. 1), golf courses (No. 2) surrounded by non-managed grass/shrublands and non-used wooded grazing lands reverting to forest.

Fig. 3 An example of two NILS squares, showing peri-urban landscape change over roughly 20 years. The delineated polygons have been classified into an overall descriptor variable, see the legend for types

The approach of using variables instead of predefined classes enables the program to conform by conversion into other classifications, both nationally and internationally, including the European Environment Agency EUNIS habitat type classification (Davies et al. 2004), the Food and Agriculture Organisation (FAO) Land Cover Classification system, LCCS (Di Gregorio and Janssen 2005). The harmonization effort of the European Biodiversity Observation Network (EBONE), represents the first work with conversion of the NILS remote sensing data (Ortega et al. 2012; Allard 2012a). A presentation of the national NILS remote sensing data, in the land use classification system of the governmental authority Statistics Sweden, is given in Fig. 4. The data is from the first rotation, 2003–2007 and compares well to the official data from 2005, as is shown on the NILS on-line data portal (NILS 2017b). Sweden is dominated by forest, wetland and mountainous areas, although almost all forests are actively managed, with rotations of clear-cuts and plantations of new saplings, and should not be considered as "natural land". This classification uses a concept of forests as areas covered densely by trees, a closed canopy. Since then, Statistics Sweden has adopted the classification system of the FAO, where forest is most often classified as a 10% cover of trees (and a potential of reaching 5 m in height), and no obvious other land use. Using the FAO classification Sweden has a cover of "forest" reaching 79% (Statistics Sweden 2013).

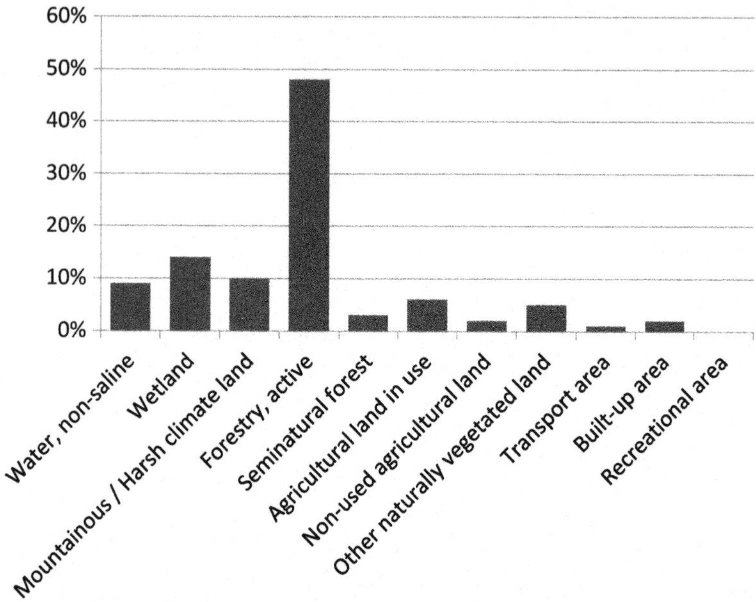

Fig. 4 Land use in Sweden, using NILS-data converted to conform to the definitions used by 'Statistics Sweden'. Note that forest/forestry in this classification system does not conform to the European FAO-class "forest", where forest would cover 79% of the land surface. Using the variables in the NILS, the collected data can be reclassified into many different classification systems

For the data to be consistent across Sweden, the ancillary digital data covers should also be consistent. That is seldom possible over a large country, so controlled deviations are allowed in some cases: for example, LIDAR data and National Reference maps are registered at lower resolution over the Mountains parts. During the run of the program, research has been undertaken on several occasions to provide reliable data at low cost over the landscape square, mostly as hybrid methods, using different digital layers of national data together with complementary interpretation of the stereo imagery. Updates to the classification procedures, including calibration and validation, are provided in Lindgren et al. (2015). A new digital terrain model has also been integrated to produce wetness indices that cover most water courses, including those within forests (Lestander et al. 2015). Many of the smaller water courses in mountainous areas are also included in the mapping.

The NILS program is also looking forward and maximising the use of data from the new European Sentinel-2 optical sensors, which became available from 2015. As providers of data to others, SLU is currently collaborating with Metria in Stockholm, to design a new scheme for collecting reference data (for training classification algorithms as well as validation of the product) for the new National Land Cover Data of Sweden: a continuation of a project called CadasterENV, financed by the European Space Agency (ESA). This Land Cover Data will be made publicly available and, when combined with other statistical data, has the potential to be used for landscape analysis and for planning of nature conservation actions. The different types of forest are well understood, as are built-up areas, so the development part of the collaboration entails updates on methodology for large-scale classifications of open land into varied field layers, starting with the mountain range and moving on to the rest of the country (Metria 2017; Metria and SLU 2017). The work will be done during 2017–2019 and includes collating existing data from the inventories, and creating new data for any vegetation type that is not captured well enough in the samples (for example seminatural grasslands or some of the more uncommon types of vegetation in the mountainous areas). The method will encompass a modified version of modelling of desired vegetation types, using Sentinel-2 data, surface layers and wetness indices from; LIDAR data. The sampling design will then be modified to better capture the vegetation types, both inside and between the existing NILS squares (Svensson et al. 2017; Reese et al. 2011). The available infrastructure of annual field workers in NILS enables relatively easy collection of ground truth-data, and the integration of stereo interpretation ensures that any new plots are suitable for remote sensing purposes (e.g. being sufficiently large and homogenous). The long-term aim is to build a database of reference data, which is accessible to both researchers and authorities for classification purposes.

Challenges in Using Remote Sensing for Monitoring

In any national inventory program, a number of challenges arise. Sweden is an elongated country with a diversity of ecosystems and biogeographical regions, therefore those interpreting the remote sensing data need a wide breadth of knowledge. This is often acquired by living in a region or by developing an interest in particular ecosystems and/or how these are used or can be best managed for differing gains. Recruiting people with sufficient knowledge can sometimes prove difficult.

Whilst remote sensing data provide a valuable source of information for monitoring landscapes, the technologies are constantly evolving: this can compromise the consistency of observations generally required by monitoring systems. In some cases, changes may be the result of a difference in observation modes and time periods. Furthermore, some datasets have only been acquired on relatively few occasions (e.g., the national LIDAR survey) and hence the lack of repeat coverage may limit change detection. Alternative methods for retrieving biophysical attributes (e.g., image matching based on aerial photography) may be used but errors in retrieval are often introduced (Granholm et al. 2015, 2017). Monitoring programs, therefore, have to be consistent in terms of the data used and knowledge available, as well as flexible in response to changing technologies and ideas.

References

Allard A.: Variables in Environmental Monitoring Enables a Multitude of Classifications – A Way to Incorporate Inventory Data from National Programs into Harmonization Efforts. Abstract, In Proceedings of Planet Under Pressure 2012, London, England (2012a)

Allard, A. (ed.): Instruktion för flygbildsinventeringen vid Nationell Inventering av Landskapet i Sverige, NILS 2007, (in Swedish, an English version of the first year, 2003, is available at: http://www.slu.se/sv/centrumbildningar-och-projekt/nils/publikationer) (2012b)

Allard, A., Glimskär, A., Svensson, J., Christensen, P.: Monitoring landscape and vegetation in the Swedish NILS-program. In: Bryn, A., Dramstad, W. and Fjellstad, W, (red.) Mapping and Monitoring of Nordic Landscapes, Conference Proceedings, Viten fra Skog og landskab, 2010:1: 5–8 (2010)

Barr, C.J., Bunce, R.G.H., Clarke, R.T., Firbank, L.G., Gillespie, M.K., Howard, D.C., Petit, S., Smart, S.M., Stuart, R.C., Watkins, J.W.: Methodology of Countryside Survey 2000 Module 1: Survey of Broad Habitats and Landscape Features, Final Report. Centre for Ecology and Hydrology, England (2003)

Berglund, H., Sundberg, S., Adler, S.: Art- & naturtypsstatistik – frågeställningar, indikatorer och presentationer för våtmarker. Swedish Species Information Centre, Artdatabanken (2016.) (In Swedish)

Brandt, J.J.E., Bunce, R.G.H., Howard, D.C., Petit, S.: General principles of monitoring land cover change based on two case studies in Britain and Denmark. Landsc. Urban Plan. **62**, 37–51 (2002)

Christensen, P., Eriksson, Å., Sandring, S.: Jordbrukslandskapet – Tillstånds- och förändringsanalyser baserade på data från NILS. SLU, Department of Forest Resource Management, Work report No. 445. (2015) (In Swedish)

Davies, C.E., Moss, D., Hill M.O.: EUNIS habitat classification revised 2004, European Environmental Agency, European Topic Centre for Nature Protection and Biodiversity. Retrieved from http://eunis.eea.europa.eu/upload/EUNIS_2004_report.pdf (2004)

Di Gregorio, A., Janssen, L.J.M.: Land Cover Classification System (LCCS), version 2, Classification concepts and user manual, FAO, *Environmental and Natural Resources Series*, No. 8. Rome (2005)

Dramstad, W.E., Fjellstad, W.J., Strand, G.H., Mathiesen, H.F., Engan, G., Stokland, J.N.: Development and implementation of the Norwegian monitoring programme for agricultural landscapes. J. Environ. Manag. **64**, 49–63 (2002)

Glimskär, A., Arlt, D., Grandin, U., Kindström, M., Kindström, S., Wikberg, S., Gunnarsson, U., Hedenbo, P., Rygne, H., Göthlin, E.: Results for Small biotopes, Grasslands and Mires in Regional Environmental Monitoring, (in Swedish). The County Board of Örebro, Report 2016:35. Can be accessed at: http://www.remiil.se/Sv/publikationer/Pages/default.aspx (2016)

Granholm, A.-H., Olsson, H., Nilsson, M., Allard, A., Holmgren, J.: The potential of digital surface models based on aerial images for automated vegetation mapping. Int. J. Remote Sens. **36**, 1855–1870 (2015). doi:10.1080/01431161.2015.1029094

Granholm, A.-H., Lindgren, N., Olsson, K., Nyström, M., Allard, A., Olsson, H.: Estimating vertical canopy cover using dense image-based point cloud data in four vegetation types in southern Sweden. Int. J. Remote Sens. **38**, 1820–1838 (2017). doi:10.1080/01431161.2015.1029094

Hedblom, M., Hedenås, H., Allard, A., Svensson, J., Jougda, L. Methods and possibilities to apply NILS data for applied integrated landscape planning: Reindeer Husbandry Plans, UAV low altitude aerial photos and dialogue. Baltic Landscape Project, Report No 27, Work Package 4 (2014)

Hedenås, H., Christensen, P., Svensson, J.: Evaluation of NILS data in the Scandinavian Mountain Range. SLU, Department of Forest Resource Management Work report No. 427 (2014) (In Swedish)

Hedenås, H., Christensen, P., Svensson, J.: Changes in vegetation cover and composition in the Swedish mountain region. Environ. Monit. Assess. **188**, 452 (2016)

Ihse, M.: Colour infrared aerial photography as a tool for vegetation mapping and change detection in environmental studies of Nordic ecosystems: a review. Nor. J. Geogr. **61**(4), 170–191 (2007). doi:10.1080/00291950701709317

Jeglum, J., Sandring, S., Christensen, P., Glimskär, A., Allard, A., Nilsson, L., Svensson, J. Main Ecosystem Characteristics and Distribution of Wetlands in Boreal and Alpine Landscapes in Northern Sweden Under Climate Change, Ecosystems Biodiversity, Oscar Grillo and Gianfranco Venora (Ed), ISBN: 978–953–307-417-7 (2011). InTech, Available from: http://www.intechopen.com/articles/show/title/main-ecosystem-characteristics-and-distribution-of-wetlands-in-boreal-and-alpine-landscapes-in-north

Köhler, R., Olschofsky, K., Gerard, F. (eds.): Land Cover Change in Europe from the 1950'ies to 2000 –Aerial photo interpretation and derived statistics from 59 samples distributed across Europe. Institute for World Forestry, University of Hamburg. ISBN: 80-89088-46-5 (2006)

Lestander, R., Löfgren, S., Henrikson, L., Ågren, A.M.: Relationship between structural features and water chemistry in boreal headwater streams—evaluation based on results from two water management survey tools suggested for Swedish forestry. Environ. Monit. Assess. **187**, 190 (2015). doi:10.1007/s10661-015-4385-x

Lindgren, N., Christensen, P., Nilsson, B., Åkerholm, M., Allard, A., Reese, H., Olsson, H.: Using optical satellite data and airborne Lidar data for a Nationwide sampling survey. Remote Sens. **7**, 4253–4267 (2015). doi:10.3390/rs70404253

Mattisson, A., Sullivan, G.: Trial mapping of upland Annex I and EUNIS habitats using stereo colour near-infrared aerial imagery. Scottish Natural Heritage Commissioned Report No. 767 (2017)

Metria: The CadasterENV project. http://www.cadasterenv.se/. Accessed 20 Apr 2017 (2017)

Metria & SLU: Slutrapport: Metodutveckling inom kartering av markslag i Fjällen. Report to Swedish Environmental Protection Agency (2017) [In Swedish]

MOTH: Demonstration of an Integrated North-European System for Monitoring Terrestrial Habitats; www.slu.se/centrumbildningar-och-projekt/moth. Accessed 30 Mar 2017 (2017)

NILS: National Inventory of Landscapes in Sweden. http://www.slu.se/nils. Accessed 30 Mar 2017 (2017a)

NILS: Online Data portal: http://akka.srh.slu.se/datavardskap/ (2017b)

Ortega, M., Metzger, M.J., Bunce, R.G.H., Wrbka, T., Allard, A., Jongman, R.H.G., Elena-Rosselló, R.: The potential for integration of environmental data from regional stratifications into a European monitoring framework. J. Environ. Plan. Manag. **5**, 39–57 (2012)

Reese, H., Allard, A., Nilsson, M., Olsson, H.: Varying training data set size for supervised classification of alpine vegetation, manuscript in Reese, H., Classification of Sweden's Forest and Alpine Vegetation Using Optical Satellite and Inventory Data, Doctoral Thesis No 2011:86, Faculty of Forest Sciences, SLU, Umeå (2011)

Roose, A., Sepp, K., Saluveer, E., Kaasik, A., Oja, T.: Neighbourhood-defined approaches for integrating and designing landscape monitoring in Estonia. Landsc. Urban Plan. **79**, 177–189 (2007)

Rose, R.A., Byler, D.J., Eastman, R., Fleishman, E., Geller, G., Goetz, S., Guild, L., Hamilton, H., Hansen, M., Headley, R., Hewson, J., Horning, N., Kaplin, B.A., Laporte, N., Leidner, A., Leimgruber, P., Morisette, J., Musnski, J., Pintea, L., Prados, A., Radeloff, V.C., Rowen, M., Saatchi, S., Schill, S., Tabor, K., Turner, W., Vodacek, A., Vogelmann, J., Wegmann, M., Wilkie, D., Ten, W.C.: Ways remote sensing can contribute to conservation. Conserv. Biol. **29**, 350–359 (2014)

Smeets, E., Weterings, R. Environmental indicators: Typology and overview. European Environmental Agency, Copenhagen, Technical Report No 25 (1999)

Ståhl, G., Allard, A., Esseen, P.-A., Glimskär, A., Ringvall, A., Svensson, J., Sundquist, S., Christensen, P., Gallegos Torell, Å., Högström, M., Lagerqvist, K., Marklund, L., Nilsson, B., Inghe, O.: National Inventory of landscapes in Sweden (NILS) – scope, design, and experiences from establishing a multiscale biodiversity monitoring system. Environ. Monit. Assess. **173**, 579–595 (2011)

Statistics Sweden: Land Use in Sweden, 6th edn. Official Statistics of Sweden, Statistics Swede, regions and Environment Department, Stockholm (2013.) (In Swedish)

Svensson, J., Allard, A., Christensen, P., Eriksson, Å., Esseen, P.-A., Glimskär, A., Sandring, S.: Landscape Biodiversity Monitoring in the Swedish NILS Program. XIII World Forestry Congress, Buenos Aires (2009)

Svensson, J., Christensen, P., Hedenås, H.: NILS för miljömålet Levande skogar. SLU, Department of Forest Resource Management Work report No. 450 (2016a) (In Swedish)

Svensson, J., Mikusinski, G., Esselin, A., Adler, S., Blicharska, M., Hedblom, M., Hedenås, H., Sandström, P., Sandström, S., Wardle, D.: National Monitoring for Assessing Ecosystem Services in Fennoscandian Alpine and Boreal Landscapes, NILS ESS. Swedish Environmental Protection Agency, Report 6754 (2016b) (In Swedish, with summary in English). http://naturvardsverket.se/Om-Naturvardsverket/Publikationer/ISBN/6700/978-91-620-6754-0/

Swedish Board of Agriculture: The TUVA Database. http://www.jordbruksverket.se/. Accessed 30 Mar 2017 (2017)

Swedish Environmental Protection Agency: Environmental Objectives Portal. http://www.swedishepa.se/Environmental-objectives-and-cooperation/Swedens-environmental-objectives/ Accessed 30 Mar 2017 (2017)

Van den Borre, J., Paelinckx, D., Mücher, C.A., Kooistra, L., Haest, B., De Blust, G., Schmidt, A.M.: Integrating remote sensing in Natura 2000 habitat monitoring: prospects on the way forward. J. Nat. Conserv. **19**, 116–125 (2011). doi:10.1016/j.jnc.2010.07.003

Zlinszky, A., Heilmeier, H., Balzter, H., Czúcz, B., Pfeifer, N.: Remote sensing and GIS for habitat quality monitoring: new approaches and future research. Remote Sens. **7**, 7987–7994 (2015)

Mapping Coastal Habitats in Wales

Gwawr Jones, Peter Bunting, and Clive Hurford

Abstract Many areas across Europe are mapped and monitored using a large range of different data types, sources and classification schemes leading to gaps in the knowledge required to fulfill the European Council's Habitats Directive (1992). The Earth Observation Data for Habitat Monitoring (EODHaM) system, developed during the EU FP7 BioSOS project, introduces a systematic, hierarchical approach that is applicable to all sites and available as a standard, providing classifications of high value for conservation and biodiversity purposes (Lucas et al. Int J Appl Earth Observ Geoinf 37:17–28, 2015). The system is built on the Land Cover Classification System (LCCS) developed by the FAO for use in the field. The aim of this project is to generate accurate maps of the location, extent and condition of coastal Annex I habitats at Kenfig Burrows Special Area of Conservation (SAC), using VHR Worldview-2 data.

Indices, such as Normalized Difference Vegetation Index (NDVI) allow straightforward visual threshold determination in the rule base, classifying LCCS Level 3 with accuracies of 90% and above. Beyond Level 3, *in situ* data is key for training and validating EO data to determine if (a) lifeforms/habitats are separable with the available EO data, and (b) suitable thresholds can be determined for classification. Numerous indices can be calculated, and using the GPS point training data, a separability analysis based on Analysis of Variance (ANOVA) allows those with the highest separation scores to be chosen as layers for classification. By plotting the

G. Jones (✉)
Joint Nature Conservation Committee (JNCC), Peterborough, UK

Aberystwyth University, Aberystwyth, UK
e-mail: gwawr.jones@jncc.gov.uk

P. Bunting
Earth Observation and Ecosystem Dynamics Group, Department of Geography and Earth Sciences, Aberystwyth University, Aberystwyth SY23 3DB, UK
e-mail: pfb@aber.ac.uk

C. Hurford
Natural Resources Wales, Cardiff, Cardiff, UK
e-mail: churford5@gmail.com

© Springer International Publishing AG 2017

R. Díaz-Delgado et al. (eds.), *The Roles of Remote Sensing in Nature Conservation*,
DOI 10.1007/978-3-319-64332-8_6

training data sets into boxplots, suitable thresholds are determined. The appropriateness of LCCS here varies with specific sites; for example, slack habitat in sand dune ecosystems can be accurately mapped from contextual information derived from slope (calculated using VHR LiDAR data) and can therefore be translated to habitat from LCCS Level 3. Classifications are therefore translated from land cover to habitat after LCCS Level 3 instead of following the hierarchy to Level 4 and beyond.

Once the broad habitat baseline is mapped, thresholds become restricting as they set clear straight lines in the feature space when classifying, therefore machine learning techniques such as random forest and/or support vector machines are more suitable for determining whether dominant species within broad habitat classes can be separated and classified accurately. By classifying dominant species, condition of habitats can be inferred. With accuracies of classifying some habitats higher than others when implementing EO data into a monitoring system, field surveying can never be ruled out to attain the knowledge required for the habitats directive. However, surveying can be applied specifically to those habitats that EO data cannot sufficiently classify.

Keywords Annex I • Habitat mapping • Land Cover Classification System • EODHaM system • Machine-learning

Introduction

The heterogeneous nature of habitats, particularly beyond the broad habitat level, presents complications when classifying from remotely sensed imagery, as these habitats are often difficult to separate spectrally due to low inter-class separation and high intra-class variability. Many classification methods are available within the remote sensing community, which address the complexity of habitat mapping, but conservation bodies use few of these, as the products produced by Earth Observation (EO) are often either not detailed or accurate enough for purpose. Therefore, the chosen methods need to be as automated as possible and readily interpreted with simple user-defined parameters that can be adjusted easily.

This study aimed to test the ability of remote sensing to map the extents of protected habitats, and mainly those listed as Annex I (of European importance) by the EC's Habitat Directive (of Annex I habitats of European importance listed in the EC Habitat's Directive). The study site comprises terrestrial coastal habitats. The study also tested the ability to infer habitat condition from EO data, mainly using the presence of a dominant species as a proxy for condition.

Background

Habitat Mapping with Remote Sensing

Remote sensing has enormous potential as a source of information on landscape patterns, habitats and dominant species (Bock et al. 2005). Many advantages of remote sensing data for improving the efficiency for habitat mapping and monitoring have already been discussed, and similar to the advantages mentioned in Alexandridis et al. (2009) can be summarised as: High Resolution (HR) coverage of large areas at low cost; observation at several non-visible wavelengths of the spectrum, and; more consistent processing across a study area. Mapping of broad habitat types as generic land cover classes is a common practice using remote sensing and is done on a very coarse scale (Wulder et al. 2004). At a global scale, land cover mapping has been accomplished by utilising the Moderate Resolution Imaging Spectroradiometer (MODIS) satellite at 500 m resolution (Friedl et al. 2010), while country and regional level land cover classifications have been accomplished using medium resolution sensors. The two main types of satellite imagery at this resolution, which are more commonly used in ecological remote sensing, are the Landsat images and the SPOT system, neither of which, with resolutions of 30 m and 10 m respectively, are capable of providing the quality needed for the whole range of habitats mapped. This is particularly true if also trying to detect components associated with habitat condition – always a requirement for conservation management and reporting.

The first satellite-derived pixel-based land cover map of the UK was generated using Landsat TM in 1990 as part of the United Kingdom Land Cover Map (UK LCM) (Fuller et al. 1994). Another map was generated in 2000 using an object-based approach (Fuller et al. 2002) but these maps have been used reluctantly by the ecological community due to, mainly, low resolutions and inaccuracies, but also due to the lack of understanding amongst users of the limitations of remote sensing. The updating of the Phase 1 survey in Wales in 2010 (Lucas et al. 2011) also used EO data in the form of SPOT-5, ASTER and IRS time-series as a repeat field survey was deemed unlikely. However, many of the previous problems still existed even with higher resolutions and better accuracies. On the other hand, this study created the first national habitat map (as opposed to land cover) generated through the implementation of EO data, and can potentially be adapted to allow continual monitoring of the extent and condition of habitats (Lucas et al. 2011).

The prospect of monitoring vegetation phenology from EO platforms is also a key area of interest when discussing the use of EO data and habitat monitoring. With the emergence of long time data records from sensors it is now possible to observe variations in phenological parameters, such as length of the growing season. Visible changes in vegetation phenology may be important indicators of climatic change, as phenology responds to the effect of several physiological and biogeochemical factors of the ecosystem (Menzel 2002). For example, time-series NDVI data have been used to indicate changes in LAI globally, with these reflecting human-induced and natural events and processes, including those related to climactic fluctuation (Liu et al. 2010). Phenology is also important for estimating biological productivity, understanding land-atmosphere interactions and the management of vegetative resources (Lieth 1971; Taylor 1974; Sarmiento and Monasterio 1983). Therefore, it is a key factor in mapping to the species level using EO data and also essential for monitoring change. Additionally, timing the acquisition of remotely sensed datasets to coincide with critical phenological stages of flowering or leaf senescence is very important when mapping invasive species (He et al. 2011).

In recent years, Very High Resolution datasets have increased in popularity as there is the potential to resolve more habitat categories, but the lack of shortwave-infrared band in these datasets has significantly hampered their potential for monitoring complex environments. However, sensors such as Worldview-2 and Worldview-3, with additional coastal, yellow, red edge and near infrared bands are anticipated to provide benefits over other VHR sensors such as IKONOS, Quickbird and GeoEye. Techniques and software for processing these data, in addition to SAR and LiDAR data, are also likely to become more available in future years, and the increase in open source material will benefit many managers of protected areas in countries where funding is more limited (Nagendra et al. 2013). However, it is recognised that more effort should be put into developing a coherent and operational method which produces most, or all, relevant parameters to contribute to assessment of conservation status (Corbane et al. 2015).

It is also important to have well designed programmes of field data collection that maximise the use of data for remote sensing interpretation and conservation assessments. *In situ* field sampling networks therefore, need to be designed in combination with remote sensing using, for instance, stratified sampling designs to carefully assess species distributions across different habitat types and enhance interpretative power (Nagendra 2001). Unless the spatial grain, extent and timing of remote sensing data, *in situ* data and models are well matched to these features of the habitats, the robustness of conclusions on management effectiveness, and the interpretive power of the analytical techniques used, will be limited. Remote sensing interpretation needs to be grounded in field data, as this is critical for effective adaptive management and monitoring (Nagendra et al. 2013).

Image Analysis Techniques

Image analysis within the remote sensing community traditionally refers to image classification, which is described as the systematic grouping of classes or themes extracted from remotely sensed data: it is a preferred technique because the methods are well known and widely used within the community. The output is generally simple to understand, and the accuracy of the results can be assessed quantitatively and qualitatively (McDermid et al. 2005).

Traditional classification methods can generally be referred to as either unsupervised or supervised. As stated from the name, unsupervised classification does not require prior knowledge of the area, whereas supervised classification does require *a priori* knowledge for training the classifier. Classes can then be assigned based on predicted variables measured from the training dataset (Cerna and Chytry 2005). Examples of unsupervised algorithms include K-means and ISODATA, and are often used for thematic mapping on a large scale as they do not require spatially detailed ground data initially, and produce useful information by clustering spectrally similar pixels (Tso and Olsen 2005; Giri et al. 2011). The algorithms are also often widely available in image processing and statistical software packages (Langley et al. 2001). However, the benefits of unsupervised techniques are often outweighed by the difficulty of post classification labelling, which does require ground information (McDermid et al. 2005).

The most widely used supervised classification technique is the Maximum Likelihood Classifier (MLC). It has been used to successfully map areas with high classification accuracies (MacAlister and Mahaxay 2009; Laba et al. 2008) but works on the assumption that input data follow a Gaussian distribution. MLC may perform poorly in the presence of non-parametric distributions (Peddle 1995) as it is heavily reliant on the distribution of the training data. Another method includes the calculation of spectral indices to characterise specific attributes of plants (DeFries et al. 1995). The most widely used spectral index is the NDVI and is based on the fact that vegetation is highly reflective in the near infrared and has a high absorption rate in the visible red wavelengths. The contrast between these wavelengths can be used as a biophysical parameter that correlates with photosynthetic activity of vegetation, which is an indication of 'greenness' (Wang and Tenhunen 2004). Therefore, NDVI is a good indicator to reflect dynamic changes of different vegetation groups (Geerken et al. 2005). This area of spectral indices has expanded rapidly and numerous indices are available for different biophysical parameters (e.g, soil moisture, plant senescence).

Another approach to mapping in comparison to pixel based classification is object-based image analysis (OBIA). Environmental objects are parts of the real world for which information can or should become available (e.g, a tree). As soon as the real world part becomes a formal object in a spatial dimension then it can be subject to some kind of classification (Bock et al. 2005). However, object based analysis is very much dependent on the spatial resolution of available imagery. The follow-up map for the UK LCM (Fuller et al. 2002) was one of the principal drivers in the development of OBIA in ecological remote sensing.

The advantage of the object-based approach is that it offers new possibilities for image analysis as image objects can be characterised not only by spectral values but by texture, shape, context, relationship and thematic information supplied by ancillary data (Bock et al. 2005). Additional layers of knowledge can be valuable to a classification system especially when certain habitat types do not have very distinct spectral features. Many segmentation algorithms are not adapted to detect the variety of geographical entities comprising a complex scene and while they perform well in delineating some landscape objects, this is rarely true for all objects of interest (Marceau et al. 1994). The issue of scale is problematic for segmentation algorithms as segments in an image will never represent meaningful objects at all scales which is required for many applications, although multi-scale segmentation approaches may be able to combat this (Blaschke et al. 2001).

Traditional methods are able to learn automatically from clustering or training data, while knowledge based classifiers require user-defined thresholds to determine class relationship. While training data is not necessarily required, expert knowledge of the area of interest is needed, which is why knowledge based classifiers require information from ecologists for vegetation characterisation. The use of extensive field knowledge and auxiliary data and the use of empirical rules to extract thematic features has proved successful, and can even improve classification accuracy (Gad and Kusky 2006; Shrestha and Zinck 2001). The most common knowledge-based classifiers are therefore, rule-based, and are commonly accompanied by image segmentation and OBIA approaches. Lucas et al. (2011) applied a rule-based approach to update the Phase 1 habitat of Wales, by utilising satellite imagery and ancillary data. Approaches like this enable full user control but gathering specific knowledge and obtaining ancillary data is often seen as an enormous task and can be very costly and too time consuming (Xie et al. 2008). However, in countries such as the UK, much of this data is already acquired by conservation bodies, and better communication between the remote sensing and ecological communities is encouraged (Lucas et al. 2007).

In recent years, when faced with large dimensional and complex data spaces, machine-learning algorithms have emerged as more accurate and efficient alternatives to the traditional classification techniques used within the remote sensing community (Roudriguez-Galiano et al. 2012). The algorithms used typically involve statistical pattern recognition, the theory of which was mostly developed in the 1960s and 1970s. Some major developments include the Bayes decision theory problem (Chow 1957), nearest neighbour decision rules (Cover and Hart 1967) and supervised and unsupervised learning (Fukunaga 1990). During the latter part of the 1980s, artificial neural networks and support vector machines were developed as statistical classifiers and these had a significant impact on the remote sensing community (Bishop 2006). The early assumption that each pixel in a multispectral image has a histogram with an approximate Gaussian distribution was made by Fu (1982) and became most popular in classifying multispectral data with use of the maximum likelihood algorithm for example. Even with the development of new sensors and the expanded applications of remote sensing, the Gaussian assumption remains a good approximation (Chen and Ho 2008), which explains their popularity. However, most parametric classifiers are heavily reliant on this assumption and data may not always represent normal distributions.

A variety of nonparametric machine algorithms exist including k-Nearest Neighbour (Gabroswki et al. 2003), Bagging (Breiman 1996), Adaboost (Freund and Schapire 1996), Decision Trees (Breiman 1984), and Support Vector Machines (Mountrakis et al. 2011).. An ensemble of classifiers based on decision trees are also proving popular such as Random Forest (Breiman 2001). As Chan and Paelinckx (2008) stated, no single algorithm has demonstrated clear superiority for all problems therefore it is recommended that multiple techniques are investigated for mapping applications.

Study Area

Coastal and near shore environments contain a wide range of habitats that are often characterised by high biodiversity, including rocky cliffs, shore platforms, sandy beaches, dunes, salt marshes and mudflats. Wales has a total coastline length of approximately 1600 km (Brazier et al. 2007). Geologically, north and west Wales are mostly composed of relatively hard rocks such as slates, mudstones and sandstones, while south and south-east Wales is dominated by softer rocks such as shales, coal measures and limestone (Webb et al. 2010). The Welsh coastline supports a wealth of habitats and c.70% is protected by environmental or conservation designations (Williams and Davies 2001). This study focuses on a sand dune ecosystem in south east Wales, called Kenfig Burrows, The Kenfig Special Area of Conservation (SAC) is a site of European importance (Fig. 1), a National Nature Reserve (NNR) and and a UK Site of Special Scientific Interest (SSSI). The Annex I habitats that are contained are listed in Table 1.

The dune system at Kenfig is stabilizing and suffering an ongoing loss of successionally-young habitats, including successionally-young humid dune slack vegetation which supports populations of the rare and declining fen orchid, *Liparis loeselii*. The entire UK population of *L.loeselli* var. ovata (which is listed in Annex II of the EC Habitats Directive) occurs at Kenfig.' This species is also protected by the 1992 Bern Convention and Wildlife and Countryside Act, 1992.

Slack mowing, managed by Bridgend Borough Council, combined with stock grazing, occurs on a regular cycle at Kenfig. The blowout enlargements and rejuvenation areas at Kenfig are trial management interventions that have the potential to improve the mobility of sand at Kenfig, though any sand movement to date has been local in nature. The fen orchid restoration scrapes have been more successful in recreating successionally-young humid dune slack vegetation at Kenfig, and c.15% of these had been recolonised by fen orchids at the time of writing. These scrapes are unlikely to increase sand mobility of the site, but they do provide more opportunities for the spread of the stress tolerating species associated with the successionally-young slack habitats. A need for increased grazing levels stimulated the building of a 3.7 km fence in the north of the site, which excludes areas of the dune system. However, this has had a negligible impact on the availability of bare sand (Pye and Blott 2011).

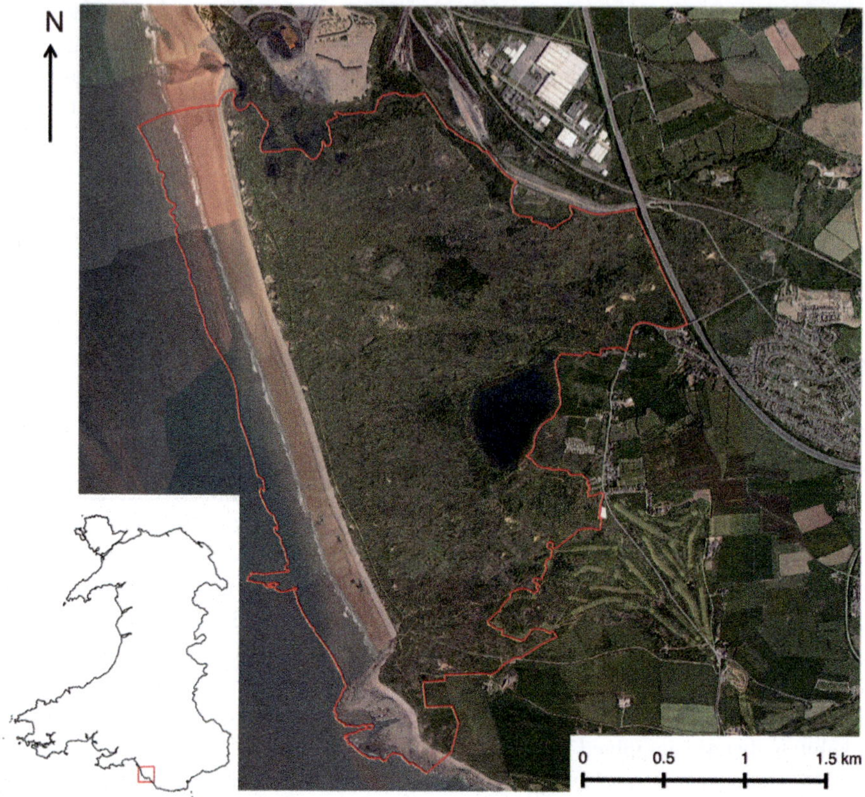

Fig. 1 Location of Kenfig Burrows site and aerial photography of the site acquired in 2006 before any major management measures were performed. The extent (*in red*) is the SAC boundary. Note the near absence of bare sand

Table 1 List of features that are present at Kenfig Burrows SAC

Designation	Feature
Annex I habitats that are primary reason for the selection of this site	2130 – Fixed dunes with herbaceous vegetation ('grey dunes');
	2170 – Dunes with *Salix repens ssp. argentea (Salicion arenariae);*
	2190 – Humid dune slacks;
	3140 – Hard oligo-mesotrophic waters with benthic vegetation *Chara spp.*
Annex I habitats present as a qualifying feature, but not primary reason for the selection of this site	1330 – Atlantic salt meadows (*Glauco-Puccinellietalia maritimae*)
Annex II species that are primary reason for the selection of this site	1395 – Petalwort *Petalophyllum ralfsii*;
	1903 – Fen orchid *Liparis loeselii*

Methods

Data Collection and Pre-processing of Earth Observation (EO) Data

A wide variety of remotely sensed data have provided observations of the earth's environments, with the availability of satellite images increasing exponentially in recent years. It is, however, very important that datasets are selected based on the capability of the data to perform the task at hand. This section describes the satellite and airborne data used in this study.

Very High Resolution Satellite Imagery

One of the many advantages of using satellite imagery for mapping and monitoring is the possibility of acquiring data regularly with identical sensor specifications. This increases the potential use, particularly within vegetation monitoring systems, as the repeatability of surveys is often deemed problematic and renders many methods of data acquisition unsuitable. With a range of Very High Resolution (VHR) satellites (<2 m pixel resolution) now available, the level of detail seen from space provides greater opportunities to map and monitor habitats at finer scales.

Images gathered by Worldview-2 are often used to provide detailed classifications of landscapes. This satellite was launched in 2009 and observes in 8 spectral bands at a spatial resolution of 2 m: a 0.46 m panchromatic band is also available. In addition to the red, green, blue and Near Infra-Red (NIR) bands, there are four additional bands that have been created to support specific applications. The coastal (400–450 nm), yellow (585–625 nm), red edge (705–745 nm) and NIR2 bands (860–1040 nm) all support vegetation identification and analysis. The red edge band is particularly important for the analysis of vegetation condition, as changes in chlorophyll production in this region can indicate plant health. Furthermore, Worldview-2 can revisit any site location within one day and is capable of imaging 975,000 km^2 of the land surface on a daily basis (Digital Globe 2009).

As a commercial satellite, Worldview-2 allows users to task image acquisition by setting the area of interest (minimum of 10 $km^{2)}$ and the time window for image capture. The minimum recommendation for time window duration is 6–8 weeks, which increases the likelihood of a cloud free acquisition. This amount of control allows users to capture seasonal variability within the landscape, which is important for vegetation monitoring and can be used to map habitats that are only spectrally unique at certain times of the year. For European environments, Lucas et al. (2015) suggested to use (as a minimum) imagery acquired during the pre- and peak-flush periods, where the vegetation is relatively stable for extended periods (e.g., no or full leaf cover). However, additional discrimination can be provided in the transitions from the pre- to the peak-flush and the peak to the post-flush period. The latter period can be particularly useful for discriminating different plant species. It should be noted that in some environments (e.g., the southern Mediterranean), the pre- and peak-flush periods may be associated with summer and winter whereas in northern Europe, this is reversed.

For this project, a Worldview-2 image was captured at peak (July 2012) and post flush (September 2012). Unfortunately, satellite imagery is not provided as ready-to-use data and a set of processing steps needs to be undertaken before any further analysis such as classification. Several factors can affect satellite measurements, from variations in atmospheric conditions to viewing geometry, and these effects need to be corrected to ensure comparability of different images and sensors within and between years. The main corrections applied to imagery are atmospheric, radiometric and orthorectification and have been reviewed in Chap. 3.

LiDAR

Light Detection and Ranging (LiDAR) data is often used (for) to create VHR Digital Terrain Models (DTMs) and Digital Surface Models (DSMs). In the UK, the Environment Agency Geomatics Group routinely collects airborne LiDAR data from coastlines and catchments in England and Wales to generate DTMs and derived products at 1–2 m spatial resolution. As of September 2015, processed Digital Terrain and Surface Models and point cloud 3D datasets are freely available to anyone under a governmental open-data scheme in the UK. (Available from Data.gov.uk). For this project, products such as slope and aspect were also calculated from this dataset in addition to the DTM.

Data from Unmanned Aerial Vehicles (UAVs) and Aerial Photography

Recent advances in UAV technologies and software have facilitated routine and exponential use of these data for environmental applications. However, for habitat monitoring, data needs to be flown at consistent heights under optimal conditions for comparability. For this reason, users need to consider their options carefully and plan flights in a manner that aids the processing of data after acquisitions (e.g., by placing light and dark tarpaulins on sites that are sufficiently large to capture atmospheric conditions for correction, which might not be suitable for small protected sites). The UAV data used for this project were acquired at low cost over multiple days and were mosaicked and georeferenced subsequently to create a visual single snapshot product at 40 cm pixel resolution.

Aerial photography is still extensively used by nature conservation bodies in the UK. The images are typically acquired every 5 years, in the red, green and blue channels with a pixel resolution of <1 m, depending on flight altitude. Data acquired in the NIR channels are also becoming more popular but processing usually involves mosaicking of different flight paths and colour rendering and matching to remove variability within the imagery as, during national acquisitions, data are captured on different days. These images, therefore, are not calibrated and cannot be used to retrieve biophysical attributes. This type of product can only be used visually and the boundaries of sites and habitats are usually digitised manually by monitoring staff. Aerial photography can also be used as a training and validation dataset for products generated from satellite sensor data, such as land cover and habitat maps.

Field Data

The decision support tools used for vegetation management require accurate information on the spatial array of different plant communities. The use of Global Positioning Systems (GPS) allows habitat data to be collected with location accuracy of below 1 m, these data can include habitat type, presence of species and photographs. Ideally, to accommodate errors in satellite image processing at 2 m resolution and GPS accuracy, habitats need to be homogenous across a 6 m^2 area to ensure that data associated with the GPS points are truly representative of that specific location. When using coarser pixel resolution data, this area needs to be increased. However, with rare or sparse habitats, this may not be possible and the heterogeneous and nature of vegetation (e.g., those occurring in fine-grained mosaics) needs to be at the forefront of any considerations for collecting training points and validation datasets.

Staff who are undertaking site monitoring routinely georeference ground data collections using a GPS. However, these point datasets tend to include only habitats of interest and very rarely include other classes that are in abundance on sites and these are often ignored (i.e., are not monitored). For a dataset to be suitable for training and validating satellite products, both types of habitat need to be represented. For this study, those points were collected from the UAV data and aerial photography and combined with field-measured data. To ensure an even spatial distribution of points into training and accuracy classes, a sampling grid was generated at different resolutions (5 m, 10 m, 15 m, 20 m) and split accordingly. (Fig. 2).

Index Calculation

For characterizing vegetation health and condition, a number of spectral indices are commonly used. These include (a) the Normalised Difference Vegetation Index (NDVI), which represents a ratio of productivity from vegetation, (b) the Normalised Difference Wetness Index (NDWI), which calculates the proportion of water in leaves and c) the Plant Senescence Reflectance Index (PSRI), which calculates the proportion of senescent or non-photosynthetic vegetation. The indices that are available are sensor dependent and the Index Database (2016) is a particularly useful resource for making a selection. For Worldview-2, there are 134 indices that can be calculated. For this study, two Worldview-2 images were obtained in the peak and post-flush period, allowing these indices to be calculated and compared over time.

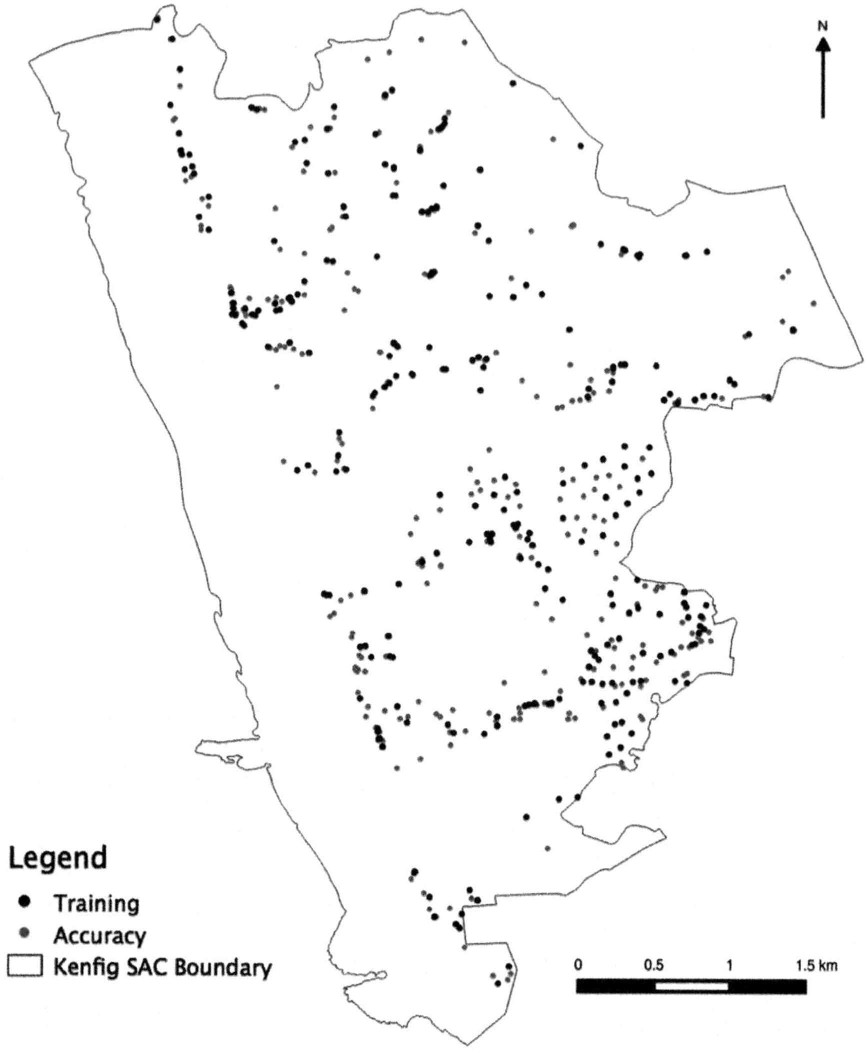

Fig. 2 Allocation of ground measurements according to their role in training and validating habitat classifications

EODHaM System and ANOVA Separation

Most areas in Europe are mapped and monitored using a range of different data sources and methods and with varying degrees of success: this has led to large gaps in the knowledge required to fulfill the requirements of the EC Habitats Directive (Borre et al. 2011). The only project that has attempted to create a systematic approach for mapping habitats from EO data to date is the FP7-funded biodiversity multi-source monitoring system: from space to species (BIOSOS) project, which focused on a multidisciplinary approach to bridge the gap between remote sensing scientists and ecologists, allowing funding to be used effectively for both monitoring and management practices. The project also developed the Earth Observation for Habitat Monitoring (EODHaM) system (Lucas et al. 2015), which provided a standardised framework for consistent land cover and habitat mapping and for monitoring Natura 2000 sites. A key component of the system is the inclusion of decision rules within a hierarchical classification structure generated by experienced ecologists and remote sensing scientists, and the system's use of the Land Cover Classification System (LCCS). The LCCS was developed by the Food and Agricultural Association (FAO) and represents a common classification scheme that can be used anywhere on global to local scales (Fig. 3).

The EODHaM system was utilised for the first stage of mapping (Fig. 4) as it allows the use of multiple data sources and types and is run within open source software that is freely available.

For the first stage of classification (LCCS level 3), a rule-based classification based primarily on thresholds of spectral data and indices is used. First, vegetated areas were separated from non-vegetated areas, which created a classification with an overall accuracy of 98%. All discrepancies encountered were explained by the presence of short sparse vegetation in primarily sandy areas known as dune annuals communities, shifting dune and semi-fixed dune habitat. However, as the spectral diversity within the vegetated areas is high, a more robust method was developed to select the data layers that would provide the best separation for distinguishing the more complicated classes such as lifeform.

An Analysis of Variance (ANOVA) was performed on all possible data layers, from the 134 indices calculated per Worldview-2 scene to the DTM and derived measures such as slope. The training dataset allowed analyses to test if a significant difference (at $\rho < 0.05$) existed between data layers for each lifeform category. The ANOVA coefficient or F-ratio is an extension of Fisher's discriminant and provides a measure of separability between multiple classes (Scheffe 1959). The magnitude and significance level associated with each ANOVA F value were used to determine the most effective layers in separating categories where the larger the F value, the more likely it is that the null hypothesis of no differences between group means is false, indicating greater separation. Although, one of the assumptions of this method is that data are normally distributed, numerous studies have demonstrated that the data's distribution has very little effect on the F-ratio (Tiku 1971; Box and Watson 1961). Once the best layer or layers were selected, boxplots were then used to determine the optimal threshold values for classification.

An example of a successful category classified using this approach is the woody (trees) lifeform class. The class is separable by using two data layers and was classified with an overall accuracy of 84%. However, some classes were deemed inseparable based on ANOVA as the distribution of classes in n-dimensional space becomes too complicated Therefore, alternative methods of classification were considered, and described in the next section (Figs. 5 and 6).

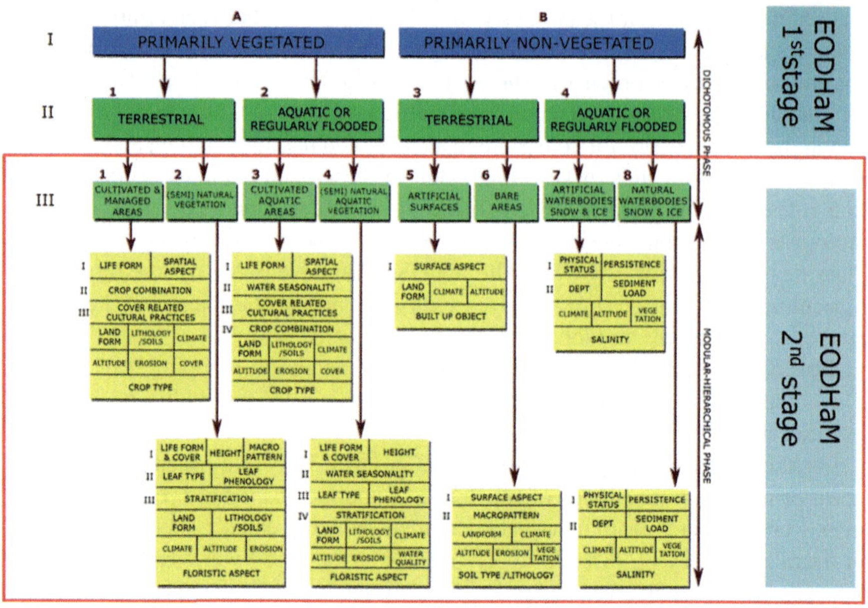

Fig. 3 The FAO LCCS encompassing levels 1–3 and beyond

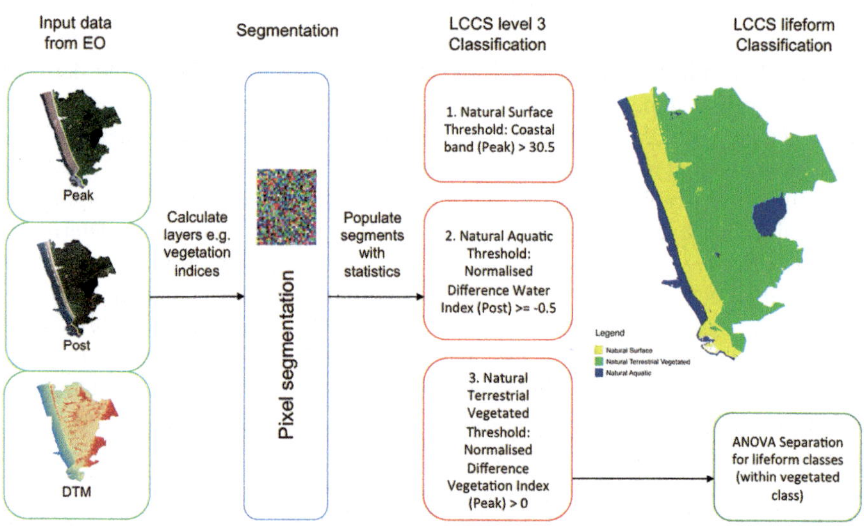

Fig. 4 The EODHaM system and thresholds for level 3 mapping using the FAO Land Cover Classification System (LCCS). With this hierarchical system, each object (which can include pixels) is classified only once into one of eight level 3 broad categories. Natural terrestrial vegetated is classified last and the threshold for this is low to capture all the variability within the class

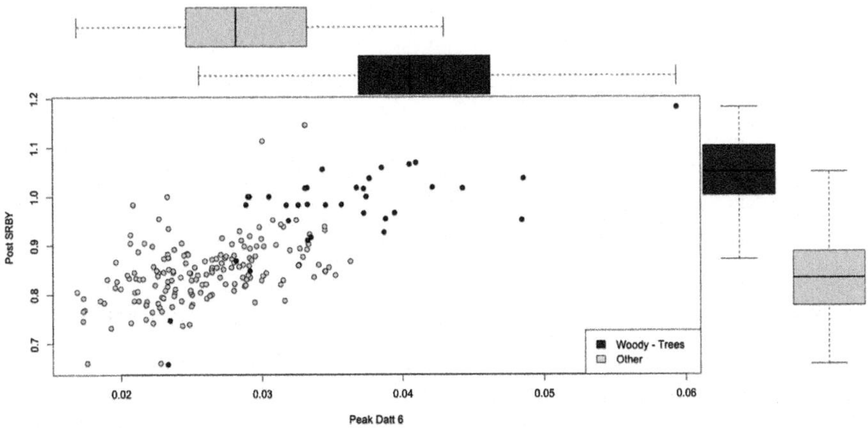

Fig. 5 Example of training data projected into an n-dimensional space between the two most separable indices where the classes are deemed separable. Indices are: Datt 6 (Datt 1998); *SRBY* Simple ratio of blue and yellow bands. Chosen thresholds are Peak Datt 6 > 0.03 and Peak SRBY > 0.9

Detailed Habitat Mapping

For the next stage of mapping, machine learning algorithms were used. Fig. 7 shows the method used to calculate multiple layers for each habitat class. This is necessary to test which algorithm would be most suitable, which input features are the most significant and which layer shows the highest correspondence with actual spatial distributions of classes in the field.

Algorithm Selection

Table 2 provides brief descriptions of algorithms tested before a selection of the best performing were chosen for further classification. An open source python library called scikit-learn was utilised to perform all classifications and for more information on calculating these algorithms see (Scikit-learn website 2016).

The algorithms that were finally selected were all nonparametric, as there is a known bias in the training dataset caused by the presence of sparse or rare habitats and a lack of representative point data in those classes. Reasons to not adopt algorithms for further analysis include overestimation of some classes leading to others not being classified, which is to be expected with assumptions of Gaussian distributions that are not present in the data. The chosen algorithms were mainly ensemble classifiers as they produce higher accuracies and are particularly more robust than, for example, the decision tree algorithm. Also, a group of classifiers has been found to perform more accurately than any single classifier (Ghimire et al. 2006). The support vector machine (SVM) also proved robust, although the selection of parameters

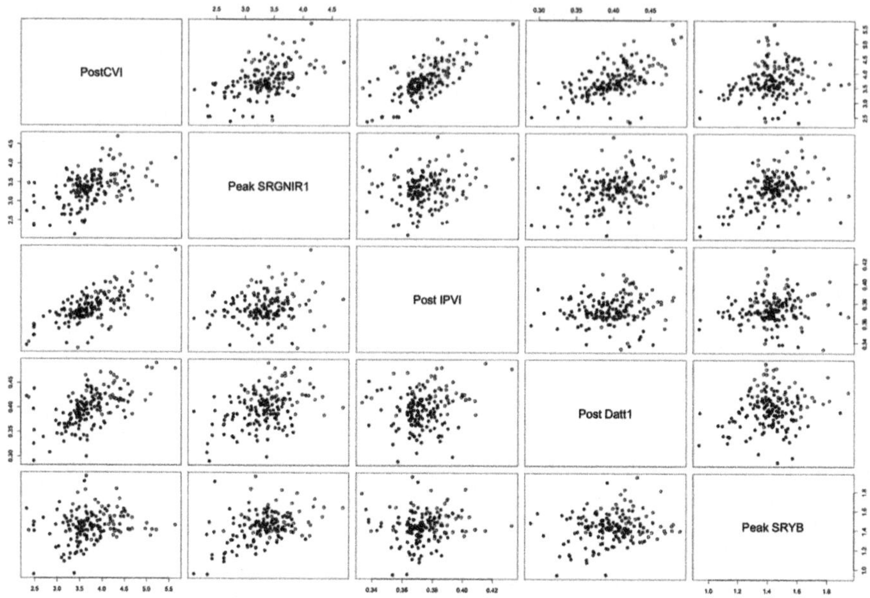

Fig. 6 Example of an n-dimensional space between the most separable indices where classes (bracken (*orange*), other vegetation (*green*)) are deemed inseparable. Indices are: *CVI* Chlorophyll Vegetation Index (Hunt et al. 2011); *SRGNIR1* Simple ratio of green and Near-Infrared 1 band; *IPVI* Infrared Percentage Vegetation Index (Crippen 1990); Datt1 (Datt 1999); *SRYB* Simple ratio of yellow and blue bands

was particularly difficult. As SVMs are traditionally binary classifiers, kernels are necessary to combat the multivariate problem, and the chosen kernels for this study were linear and used a radial basis function. The multivariate problem also includes a choice between 'one against the rest' and 'one against one' methods, where the latter was chosen based on its preference within the literature (Pal and Mather 2005; Mountrakis et al. 2011).

Input Feature Selection

Feature selection to provide inputs for predicting classification algorithms is also a key consideration that could affect the efficacy and accuracy of mapping outputs. Another reason to be selective about input features is the curse of high dimensionality, or the Hughes phenomenon, where classification performance will reach a peak without proportional increase in the training sample size, beyond which performance degrades (Landgrebe 2003). Therefore, high dimensions in the data need to be reduced to ensure the predictive power of algorithms. Techniques such as Principal Components Analysis (PCA) or Independent Component Analysis (ICA) are often employed for this purpose where only the first few components are used as feature vector input to the classifiers. However, in this process, some useful

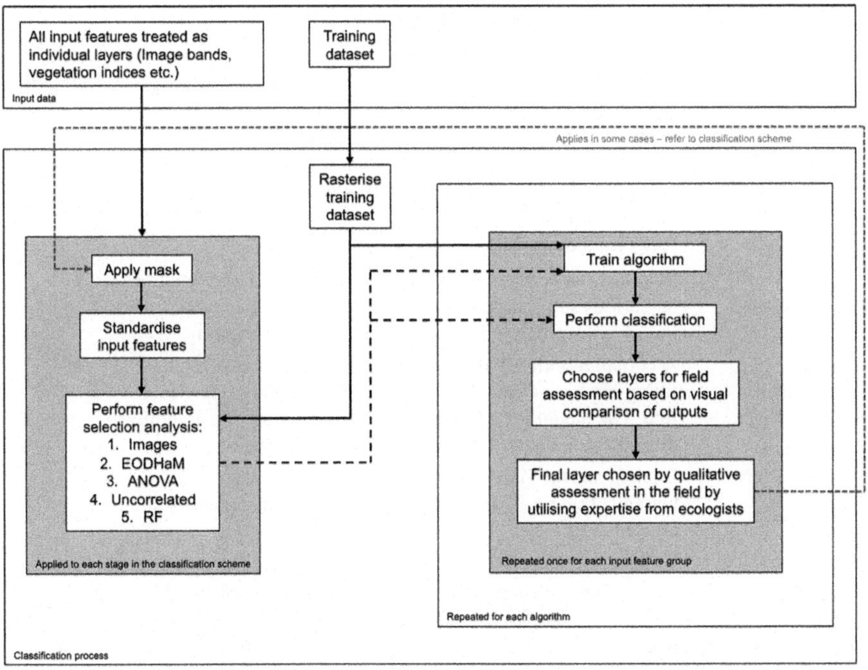

Fig. 7 Schematic of the classifying process integrating machine learning algorithms

information may be lost (Chen and Ho 2008) and these approaches were therefore not considered further. Instead, five different scenarios where used:

1. Image bands and DTM layers such as slope and aspect only (known as Images in Fig. 6);
2. Indices chosen on their recommendation from the EODHaM system and literature (known as EODHaM in Fig. 6);
3. 20 indices with the highest F scores from the separation analysis (ANOVA) per class (known as ANOVA in Fig. 6);
4. Indices that have a low correlation value for each class (i.e., every observation that has a correlation of 0.75 or above is discarded from further analysis (known as Uncorrelated in Fig. 6);
5. 20 indices with the highest calculated importance for each class, as delineated from the Random Forest (RF) algorithm (known as RF in Fig. 6).

Classification Structure and Validation

To preserve the hierarchical nature of the EODHaM system, a similar structure was formulated where classes were determined by outputs from algorithms instead of manually inputted thresholds (Fig. 8). This structure would still allow other datasets

Table 2 Algorithms that were investigated and subsequently chosen for further classification analysis

Algorithm	Description	Chosen
AdaBoost	Meta-estimator that begins by fitting a classifier on the original dataset and then fits additional copies of the classifier on the same dataset but where the weights of incorrectly classified instances are adjusted such that subsequent classifiers focus more on difficult cases.	✓
Decision tree	A decision tree classifier.	
Extremely random forest	Meta-estimator that fits a number of randomised decision trees on various sub-samples of the dataset and uses averaging to improve the predictive accuracy and control over-fitting.	✓
Linear discriminant analysis	A classifier with a linear decision boundary, generated by fitting class conditional densities to the data and using Bayes' rule. The model fits a Gaussian density to each class, assuming that all classes share the same covariance matrix.	
Gaussian Naïve Bayes	Implements the Gaussian Naïve Bayes algorithm for classification where the likelihood of the features is assumed to be Gaussian.	
Nearest neighbour	Classifier implementing the k-nearest neighbours vote.	
Quadratic discriminant analysis	A classifier with a quadratic decision boundary, generated by fitting class conditional densities to the data and using Bayes' rule. The model fits a Gaussian density to each class.	
Random forest	Meta-estimator that fits a number of decision tree classifiers on various sub-samples of the dataset and use averaging to improve the predictive accuracy and control over-fitting. The sub-sample size is always the same as the original input sample size but the samples are drawn with replacement if bootstrap is used.	✓
Support vector machine	Support vector classification where the multiclass support is handled according to one against one scheme.	✓

such as vector maps to be inputted as masks. As the basic land cover classes generated using the EODHaM system were sufficiently accurate, these outputs where used as baseline masks. Another key feature of maintaining the hierarchical nature of the EODHaM system is the ability to classify the same class on different land covers. For example, some of the discrepancies in the basic land cover classification were attributed to the presence of short sparse vegetation in primarily sandy areas known as dune annuals communities, shifting dune and semi-fixed dune habitat. By running that class on the bare ground mask, all those areas were targeted and mapped regardless of the known error in masks created from the EODHaM system. The class was then combined to create the final Annex I habitat map (Fig. 13), which in this instance was the Shifting Dunes along the shoreline with *Ammophila arenaria* ("white dunes"). In addition, the dominant species within the Annex I habitats were also mapped, which provides a proxy for condition and can help inform management decisions for maintaining the site. An example of where knowing the dominant species in a slack habitat indicates poorer condition is the presence of the grass *Calamagrostis epigejos*, as it can become a near-monoculture and prevent other species from thriving.

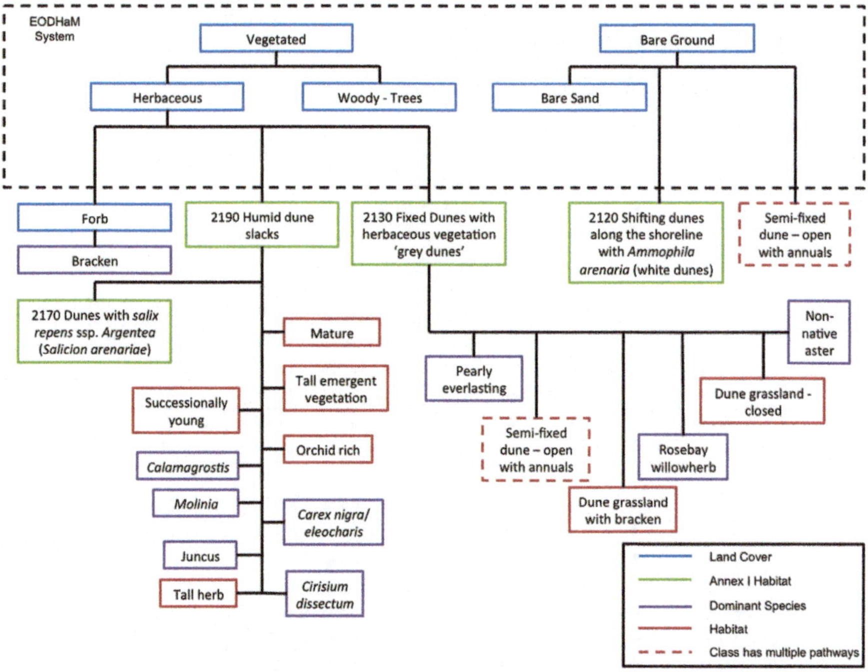

Fig. 8 Classification hierarchical structure for Kenfig. Each box that splits into nodes is used as a mask each time the algorithms are run

A number of layers were selected (see example in Table 3) based on a visual interpretation for further field analysis as quantitative methods of validating mapping products from EO are not sufficient at validating spatial aspects of outputs (Foody 2002). These outputs were entered into a Global Positioning System (GPS) and were further scrutinised in the field to provide a qualitative validation and examine which layers represented the most suitable spatial distribution. The final layer for each class was chosen in the field. An analysis of overall accuracies determined from error matrices was also carried out.

Results

The results show that no single layer was a perfect fit for all classes within the masked output. Table 4 shows the final layer chosen at each stage of the classification hierarchy and a different algorithm with a different input feature were chosen each time, while Fig. 9 shows an illustration of these layers.

Figure 10 shows the overall point accuracy for layers generated from the herbaceous mask. There are some that clearly performed better than others according to this method of determining accuracy. The layer deemed to best represent classes in the field was the Extremely Random Forest algorithm with the BioSOS input features. There

Table 3 An example of the method for choosing which layers were most suitable for further field assessment before choosing the final layer for mapping

Input Features	Classifier	Distribution of habitats[a]			Comment
		FDG	Slack	Bracken	
ANOVA	AdaBoost	↑	↓	↑	Slack features are missing; bracken is over-represented
	RF	↑	↓	□	Slack features are missing
	ERF	↓	↑	□	Mixing of FDG and Slack features
	SVM rbf	↓	↑	↑	Mixing of FDG and Slack features; bracken is over-represented
	SVM linear	□	□	□	Chosen for field assessment
EODHaM	AdaBoost	↑	↓	□	Slack features are missing
	RF	□	□	□	Chosen for field assessment
	ERF	□	□	□	Chosen for field assessment
	SVM rbf	↓	↑	□	Slack features are over-represented
	SVM linear	↓	□	↑	Mixing of FDG and bracken
Images	AdaBoost	↑	↓	↓	FDG is over-represented and bracken nearly absent
	RF	↓	↑	□	Slight mixing of FDG and Slack features
	ERF	□	□	□	Chosen for field assessment
	SVM rbf	↓	↑	□	Mixing of FDG and Slack features
	SVM linear	↓	↑	↑	Mixing of FDG and Slack features; bracken is over-represented
Uncorrelated	AdaBoost	↑	↓	□	Slack features are missing
	RF	↓	□	↑	Bracken over-represented
	ERF	↓	↑	↑	Slack features and bracken are dominant
	SVM rbf	↓	↑	↑	Slack features and bracken are dominant
	SVM linear	↓	↑	↑	Mixing of FDG and Slack features; bracken is over-represented
Random Forest	AdaBoost	↑	↓	□	Slack features are missing
	RF	↑	↓	□	Slack features are missing
	ERF	□	□	□	Chosen for field assessment
	SVM rbf	↓	↑	↑	Slack features and bracken are dominant
	SVM linear	↓	↑	↑	Mixing of FDG and Slack features; bracken is over-represented

Interpretation Keys: ↑ Over-representation; ↓ Under-representation; □ Correct representation; O Absent; ✖ Difficult to comment

[a] Habitat Keys: FDG = Fixed Dune Grassland

were nine layers that performed better than the chosen final layer. However, these methods of accuracy are based on point data. To test the accuracy of the qualitative methods of validation using the expertise of site managers and ecologist, GPS areas were acquired of certain slack features on site (Fig. 11). The overall areas of these features were calculated and the chosen layer had a higher percentage of slack classified than any other layer (Fig. 12). This proved that testing mapped outputs in the field, in addition to relying on point data to represent spatial accuracy, is key and important for the uptake of products derived from EO in monitoring solutions (Fig. 13).

Discussion and Conclusions

This study has evaluated and tested a method designed to generate high-resolution maps for monitoring a network of protected designated sites. It has then integrated approaches that can be more automated to enhance the method and ensure better

Table 4 Chosen layers for final map production

Input features	AdaBoost					Random forest					Extremely Random Forest					SVM – linear kernel					SVM – rbf kernel				
	A	E	I	R	U	A	E	I	R	U	A	E	I	R	U	A	E	I	R	U	A	E	I	R	U
Herb mask												✓													
Slacks mask													✓												
Sand mask																				✓					
FDG mask																		✓							

A ANOVA, *E* EODHaM, *I* Images, *R* RF, *U* Uncorrelated

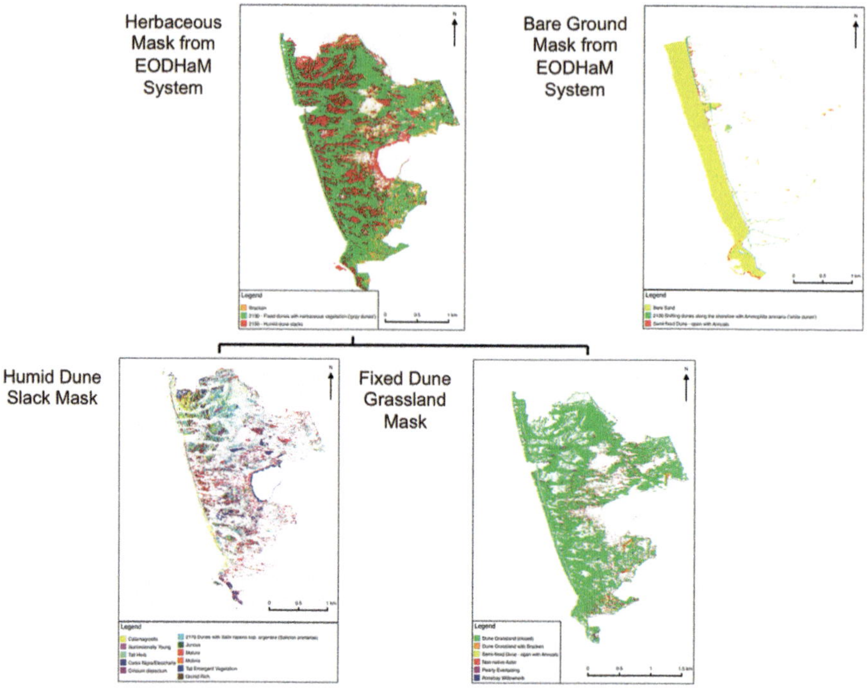

Fig. 9 Layers used for final map creation

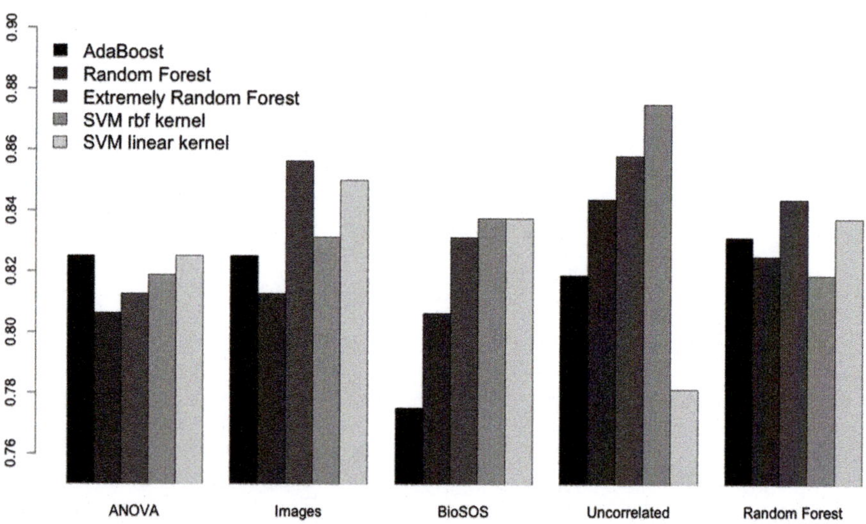

Fig. 10 The overall point accuracies for all classification layers within the herbaceous mask. The combined average is 80% accuracy

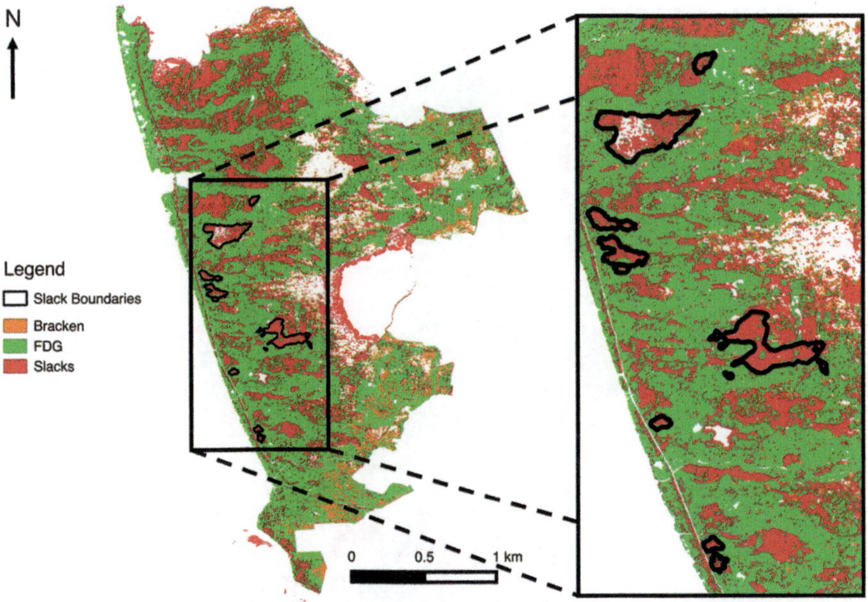

Fig. 11 Location of the slack boundaries collected using a hand-held GPS device in the field

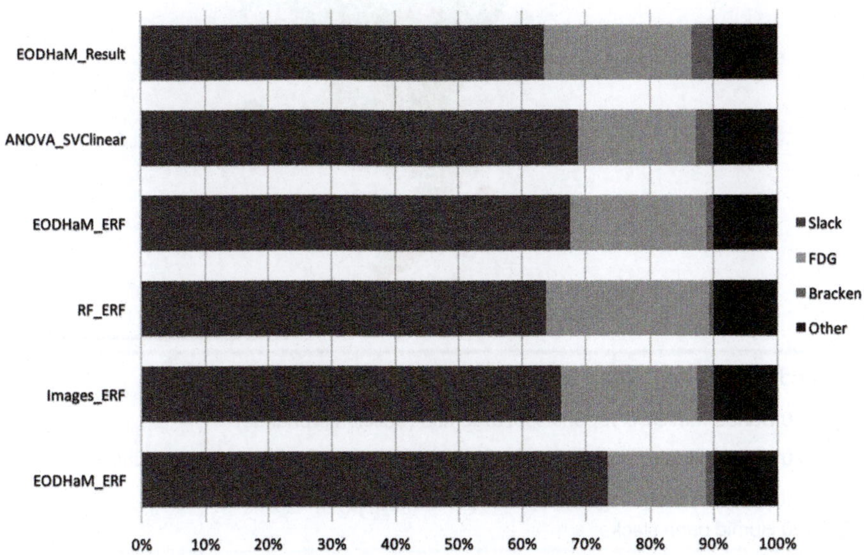

Fig. 12 Percentage of pixels classified correctly within the slack boundaries seen in Fig. 10 for all layers chosen for further field assessment. The chosen layer is EODHaM_ERF

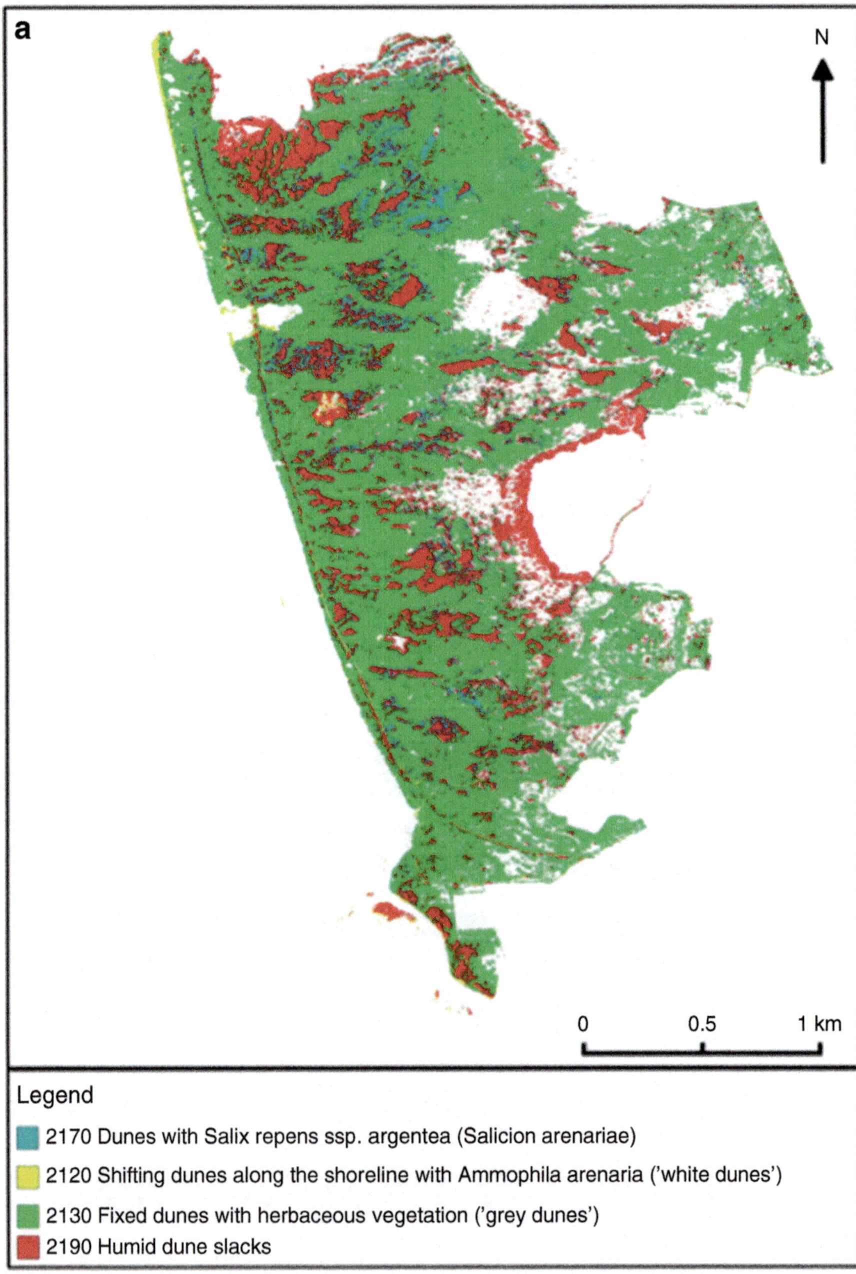

Fig. 13 Final maps created for Kenfig SAC where (**a**): Annex I habitat map; (**b**): Dominant species within the Humid dune slack class; (**c**): Dominant species within the Fixed dune grassland class

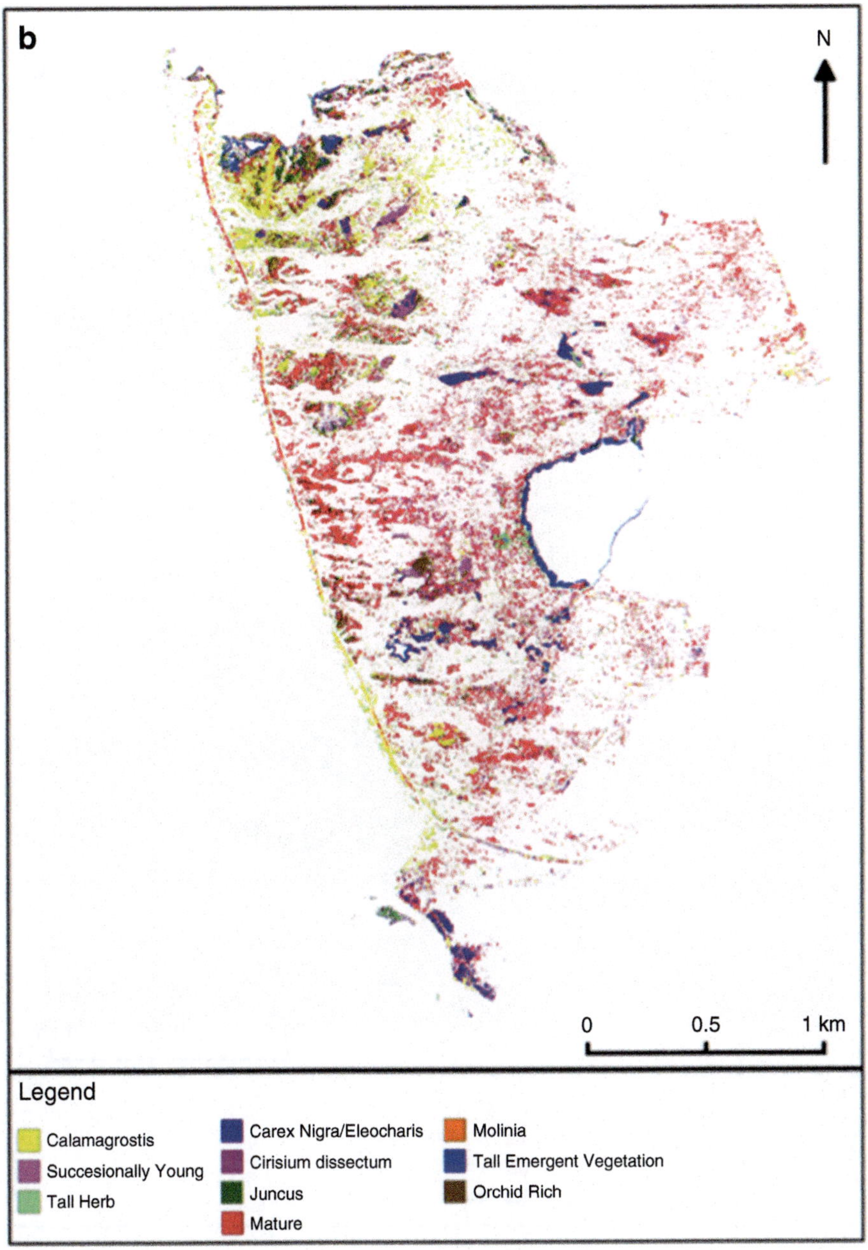

Fig. 13 (continued)

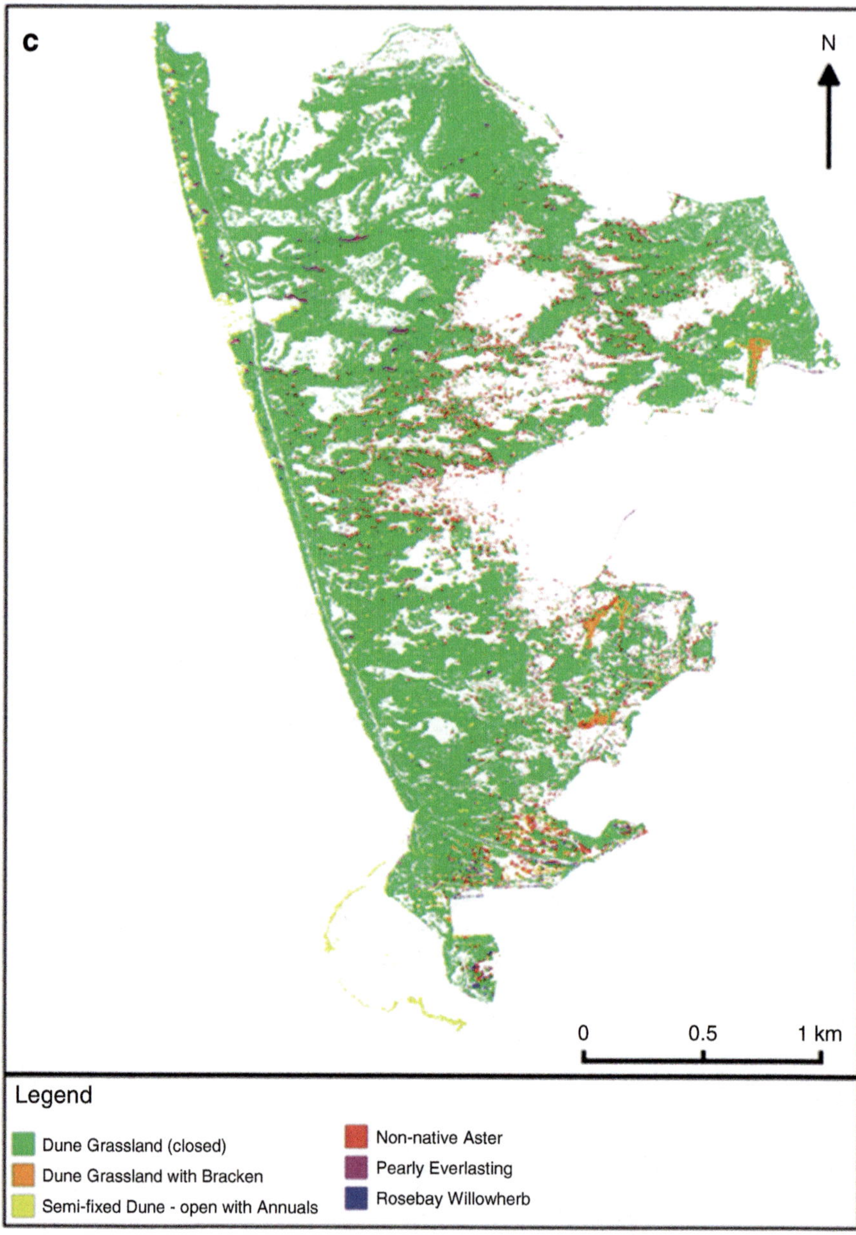

Fig. 13 (continued)

quality and efficacy. Very few studies have considered a integrating two very different classification methods into a hierarchical system, preferring instead to individually test methods and select that which provides better overall performance. The benefit of the EODHaM system is its ability to integrate any dataset or layer and the hierarchical structure aids separation, particularly if classes of spectral similarity are present on the areas of interest.

This study aimed to map the extent of habitats that are protected by law, and to provide both a baseline map to monitor the habitats and evidence to inform policy. Now that high-resolution boundaries have been established for this site, other EO datasets, such as those generated by the Sentinel satellites (from the European Space Agency's Copernicus Programme), can be used to identify areas of change. As the data from the Sentinel programme is distributed at no cost, the expense of running this method as an operational system has been reduced considerably. The monitoring aspect is envisaged to be a risk system where it will be up to site managers to interpret whether the change seen is a normal seasonal change or a risk worth investigating further with a field visit.

This system is semi-automated and relies on the ecological contextual information to generate a hierarchy that works for each site: this means that it can easily be enrolled UK wide (depending on availability of EO and field data) and readily implemented in other countries. Using the VHR satellite data proved effective for mapping the Annex I habitats at this site and even went a step further by separating dominant species. However, although the resolution of the EO data was deemed efficient for this site, we cannot assume that it will be effective for all Annex I habitats. For example, this method would not perform well for habitats concealed under canopies or fine-grained habitats with a patch size smaller than the pixel size of the data and the corresponding GPS error. For more suggestions on the habitats that can or cannot be mapped with EO data, see the Crick Framework (Medcalf et al. 2014 and Chap. 7).

References

Alexandridis, T.K., Lazaridou, E., Tsirika, A., Zalidis, G.C.: Using earth observation to update a Natura 2000 habitat map for a wetland in Greece. J. Environ. Manag. **90**(7), 2243–2251 (2009)

Bishop, C.M.: Pattern Recognition and Machine Learning. Springer, New York (2006)

Blaschke, T., Strobl, J., et al.: Whats wrong with pixels? Some recent developments interfacing remote sensing and gis. GeoBIT/GIS. **6**(1), 12–17 (2001)

Bock, M., Xofis, P., Mitchley, J., Rossner, G., Wissen, M.: Object-oriented methods for habitat mapping at multiple scales–case studies from northern Germany and Wye Downs, UK. J. Nat. Conserv. **13**(2), 75–89 (2005)

Borre, J.V., Paelinckx, D., Mücher, C.A., Kooistra, L.: Integrating remote sensing in Natura 2000 habitat monitoring: prospects on the way forward. J. Nat. Conserv. **19**(2), 116 (2011)

Box, G., Watson, G.S.: Robustness to non-normality of regression tests. Biometrika. **49**, 93–106 (1961)

Brazier, P., Birch, K., Brunstrom, A., Bunker, A., Jones, M., Lough, N., Salmon, L., Wyn, G.: When the Tide Goes Out: the Biodiversity and Conservation of the Shores of Wales–Results from a 10 Year Intertidal Survey of Wales. Countryside Council for Wales, Bangor (2007)

Breiman, L.: Classification and Regression Trees. Chapman & Hall/CRC, London (1984)
Breiman, L.: Bagging predictors. Mach. Learn. **24**, 123–140 (1996)
Breiman, L.: Random forests. Mach. Learn. **45**, 5–32 (2001)
Cerna, L., Chytry, M.: Supervised classification of plant communities with artificial neural networks. J. Veg. Sci. **16**(4), 407–414 (2005)
Chan, J.C., Paelinckx, D.: Evaluation of random forest and adaboost tree-based ensemble classification and spectral band selection for ecotope mapping using airborne hyperspectral imagery. Remote Sens. Environ. **112**, 2999–3011 (2008)
Chen, H., Ho, P.: Statistical pattern recognition in remote sensing. Pattern Recogn. **41**(9), 2731–2741 (2008)
Chow, C.K.: An optimum character recognition system using decision functions. IEEE Trans. Electronic Comput. **EC-6**, 247–254 (1957)
Corbane, C., Lang, S., Pipkins, K., Alleaume, S., Deshayes, M., Millan, V.E.G., Strasser, T., Borre, J.V., Toon, S., Michael, F.: Remote sensing for mapping natural habitats and their conservation status–new opportunities and challenges. Int. J. Appl. Earth Obs. Geoinf. **37**, 7–16 (2015)
Cover, T.M., Hart, P.E.: Nearest neighbor pattern classification. IEEE Trans. Inf. Theory. **13**(1), 21–27 (1967)
Crippen, R.E.: Calculating the vegetation index faster. Remote Sens. Environ. **34**(1), 71–73 (1990)
Data.gov.uk, 2016.: URL: https://www.environment.data.gov.uk/ds/survey/#/survey (2016)
Datt, B.: Remote sensing of chlorophyll a, chlorophyll b, chlorophyll a+b and total carotenoid content in Eucalyptus leaves. Remote Sens. Environ. **66**(2), 111–121 (1998)
Datt, B.: Remote sensing of water content in Eucalyptus leaves. Aust. J. Bot. **47**(6), 909–923 (1999)
DeFries, R., Hansen, M., Townshend, J.: Global discrimination of land cover types from metrics derived from avhrr pathfinder data. Remote Sens. Environ. **54**(3), 209–222 (1995)
Digital Globe: Digital Globe Website, URL: https://www.digitalglobe.com/about/our-constellation (2009)
Foody, G.M.: Status of land cover classification accuracy assessment. Remote Sens. Environ. **80**(1), 185–201 (2002)
Freund, Y., Schapire, R.E.: Experiments with a new boosting algorithm, Machine Learning. Proceedings of the Thirteenth International Conference, pp. 148–156 (1996)
Friedl, M.A., Sulla-Menashe, D., Tan, B., Schneider, A., Ramankutty, N., Sibley, A., Huang, X.: MODIS collection 5 global land cover: algorithm refinements and characterization of new datasets. Remote Sens. Environ. **114**(1), 168–182 (2010)
Fu, K.S.: Application of pattern recognition to remote sensing. In: Fu, K.S. (ed.) Applications of Pattern Recognition. CRC Press, Boca Raton (1982)
Fukunaga, K.: Introduction to Statistical Pattern Recognition, 2nd edn. Academic, Orlando (1990)
Fuller, R., Groom, G., Jones, A.: Land cover map of Great Britain. An automated classification of landsat thematic mapper data. Photogramm. Eng. Remote. Sens. **60**(5), 553–562 (1994)
Fuller, R.M., Smith, G.M., Sanderson, J.M., Hill, R.A., Thomson, A.G.: The UK land cover map 2000: construction of a parcel-based vector map from satellite images. Cartogr. J. **39**(1), 15–25 (2002)
Gabroswki, S., Jozwik, A., Chen, C.H.: Nearest neighbor decision rule for pixel classification in remote sensing. In: Chen, C.H. (ed.) Frontiers of Remote Sensing Information Processing, pp. 315–327. World Scientific Publishing, Singapore (2003)
Gad, S., Kusky, T.: Lithological mapping in the eastern desert of Egypt, the Barramiya area, using Landsat Thematic Mapper (TM). J. Afr. Earth Sci. **44**(2), 196–202 (2006)
Geerken, R., Zaitchik, B., Evans, J.: Classifying rangeland vegetation type and coverage from NDVI time series using Fourier filtered cycle similarity. Int. J. Remote Sens. **26**(24), 5535–5554 (2005)
Ghimire, P.O., Benediktsson, J.A., Sveinsson, J.R.: Contextual land-cover classification: incorporating spatial dependence in land-cover classification models using random forests and the Getis statistic. Remote Sens. Lett. **1**, 45–54 (2006)

Giri, C., Ochieng, E., Tieszen, L.L., Zhu, Z., Singh, A., Loveland, T., Masek, J., Duke, N.: Status and distribution of mangrove forests of the world using earth observation satellite data. Glob. Ecol. Biogeogr. **20**(1), 154–159 (2011)

He, K.S., Rocchini, D., Neteler, M., Nagendra, H.: Benefits of hyperspectral remote sensing for tracking plant invasions. Divers. Distrib. **17**(3), 381–392 (2011)

Hunt, E.R., Daughtry, C.S.T., Eitel, J.U.H., Long, D.S.: Remote sensing leaf chlorophyll content using a visible band index. Agron. J. **103**(4), 1090–1099 (2011)

Index Database: Index Database Website, URL: http://www.indexdatabase.de/ (2016)

Laba, M., Downs, R., Smith, S., Welsh, S., Neider, C., White, S., Richmond, M., Philpot, W., Baveye, P.: Mapping invasive wetland plants in the Hudson river national estuarine research reserve using Quickbird satellite imagery. Remote Sens. Environ. **112**(1), 286–300 (2008)

Landgrebe, D.A.: Signal Theory Methods in Multispectral Remote Sensing. Wiley, New York (2003)

Langley, S.K., Cheshire, H.M., Humes, K.S.: A comparison of single date and multi-temporal satellite image classifications in a semi-arid grassland. J. Arid Environ. **49**(2), 401–411 (2001)

Lieth, H. The phenological viewpoint in productivity studies. In: Productivity of Forest Ecosystems. Proceedings of the Brussels Symposium by UNESCO, pp. 71–83 (1971)

Liu, S., Liu, R., Liu, Y.: Spatial and temporal variation of global lai during 1981–2006. J. Geogr. Sci. **20**(3), 323–332 (2010)

Lucas, R., Rowlands, A., Brown, A., Keyworth, S., Bunting, P.: Rule-based classification of multi-temporal satellite imagery for habitat and agricultural land cover mapping. ISPRS J. Photogramm. Remote Sens. **62**(3), 165–185 (2007)

Lucas, R., Medcalf, K., Brown, A., Bunting, P., Breyer, J., Clewley, D., Keyworth, S., Blackmore, P.: Updating the Phase 1 habitat map of Wales, UK, using satellite sensor data. ISPRS J. Photogramm. Remote Sens. **66**(1), 81–102 (2011)

Lucas, R., Blonda, P., Bunting, P., Jones, G., Inglada, J., Arias, M., Kosmidou, V., Petrou, Z.I., Manakos, I., Adamo, M., Charnock, R., Tarantino, C., Mücher, C.A., Jongman, R.H.G., Kramer, H., Arvor, D., Honrado, J.P., Mairota, P.: The Earth Observation Data for Habitat Monitoring (EODHaM) system. Int. J. Appl. Earth Obs. Geoinf. **37**, 17–28 (2015)

MacAlister, C., Mahaxay, M.: Mapping wetlands in the lower Mekong Basin for wetland resource and conservation management using Landsat ETM+ images and field survey data. J. Environ. Manag. **90**(7), 2130–2137 (2009)

Marceau, D.J., Howarth, P.J., Gratton, D.J.: Remote sensing and the measurement of geographical entities in a forested environment. 1. The scale and spatial aggregation problem. Remote Sens. Environ. **49**(2), 93–104 (1994)

McDermid, G.J., Franklin, S.E., LeDrew, E.F.: Remote sensing for large-area habitat mapping. Prog. Phys. Geogr. **29**(4), 449–474 (2005)

Medcalf, K.A., Parker, J.A., Turton, N., Bell, G.: Making Earth Observation Work for UK Biodiversity Conservation Phase 1, JNCC Report, 495. Joint Nature Conservation Committee, Peterborough (2014)

Menzel, A.: Phenology: its importance to the global change community. Clim. Chang. **54**(4), 379–385 (2002)

Mountrakis, G., Im, J., Ogole, C.: Support vector machines in remote sensing: a review. ISPRS J. Photogramm. Remote Sens. **66**, 247–259 (2011)

Nagendra, H.: Using remote sensing to assess biodiversity. Int. J. Remote Sens. **22**(12), 2377–2400 (2001)

Nagendra, H., Lucas, R., Honrado, J.P., Jongman, R.H., Tarantino, C., Adamo, M., Mairota, P.: Remote sensing for conservation monitoring: assessing protected areas, habitat extent, habitat condition, species diversity, and threats. Ecol. Indic. **33**, 45–59 (2013)

Pal, M., Mather, P.M.: Support vector machines for classification in remote sensing. Int. J. Remote Sens. **26**(5), 1007–1011 (2005)

Peddle, D.R.: Knowledge formulation for supervised evidential classification. Photogramm. Eng. Remote. Sens. **61**(4), 409–417 (1995)

Pye, K., Blott, S.J.: Kenfig Sand Dunes – Potential for Dune Reactivation (2011).
Roudriguez-Galiano, V.F., Ghimire, B., Rogan, J., Chica-Olmo, M., Rigol-Sanchez, J.P.: An assessment of the effectiveness of a random forest classifier for land-cover classification. ISPRS J. Photogramm. Remote Sens. **67**, 93–104 (2012)
Sarmiento, G., Monasterio, M.: Life forms and phenology. Ecosyst. World. **13**, 79–108 (1983)
Scheffe, H.: The Analysis of Variance (1959)
Scikit-learn website.: URL: http://www.scikit-learn.org/stable/ (2016)
Shrestha, D.P., Zinck, J.A.: Land use classification in mountainous areas: integration of image processing, digital elevation data and field knowledge (application to Nepal). Int. J. Appl. Earth Obs. Geoinf. **3**(1), 78–85 (2001)
Taylor Jr., F.G.: Phenodynamics of production in a mesic deciduous forest. In: Phenology and Seasonality Modeling, pp. 237–254. Springer, Berlin (1974)
Tiku, M.L.: Power function of the F-test under non-nominal situations. J. Am. Stat. Assoc. **66**, 913–916 (1971)
Tso, B., Olsen, R.C.: Combining spectral and spatial information into hidden markov models for unsupervised image classification. Int. J. Remote Sens. **26**(10), 2113–2133 (2005)
Wang, Q., Tenhunen, J.D.: Vegetation mapping with multitemporal NDVI in north eastern China transect (NECT). Int. J. Appl. Earth Obs. Geoinf. **6**(1), 17–31 (2004)
Webb, H., Pye, K., Huckle, J., Blott, S.: Beach Topographic Variability in Relation to Significant Biological. Countryside Council for Wales, Bangor (2010)
Williams, A., Davies, P.: Coastal dunes of wales; vulnerability and protection. J. Coast. Conserv. **7**(2), 145–154 (2001)
Wulder, M.A., Hall, R.J., Coops, N.C., Franklin, S.E.: High spatial resolution remotely sensed data for ecosystem characterization. Bioscience. **54**(6), 511–521 (2004)
Xie, Y., Sha, Z., Yu, M.: Remote sensing imagery in vegetation mapping: a review. J. Plant Ecol. **1**(1), 9–23 (2008)

Integrated Monitoring for Biodiversity Using Remote Sensing: From Local to Regional – A Case Study from Norfolk

Katie Medcalf, Jacqueline Parker, Gemma Bell, Paul Robinson, Samuel Neal, Martin Horlock, and Johanna Breyer

Abstract The integration of ecological knowledge into remote sensing classification schemes can support operational biodiversity conservation work. Focusing on the county of Norfolk in the UK, this chapter describes a Defra-JNCC funded project (Making Earth Observation Work for UK Biodiversity) that aimed to develop cost effective tools and techniques for mapping semi-natural habitats, thereby supporting the work of the Norfolk Biodiversity Information Service. To meet the breadth of needs of habitat practitioners in the county, maps were produced across scales ranging from a county-level map to those providing more detail, as required by site managers. The finer scale maps also provided information on habitat condition. This approach to mapping complements other habitat surveillance methods and can form part of an ecological mapping and monitoring toolbox of techniques.

Each of the maps was generated using Object Based Image Analysis (OBIA) techniques that incorporated ecological knowledge into a rule-based classification. As input, we used images from earth observation platforms of varying spatial resolution, spectral range and seasonal frequency. Ecological principles and local knowledge guided the classifications. We used available ground truth data to perform targeted validation and to guide revisions to the maps. The continual enhancement and updating of the maps results in a 'living map' that benefits from the learning process.

The maps have subsequently been used by local partners on a diverse range of projects, from pollution and sediment modelling to supporting analysis of bat distribution. The regional maps are being further developed and enhanced by ecologists on the ground to record the presence and ecological status of habitats in the county. This chapter also explores how this continual enhancement and updating

K. Medcalf (✉) • J. Parker • G. Bell • J. Breyer
Environment Systems Ltd, Aberystwyth, UK
e-mail: katie.medcalf@envsys.co.uk

P. Robinson
Joint Nature Conservation Committee (JNCC), Peterborough, UK

S. Neal • M. Horlock
Norfolk Biodiversity Information Service, Norwich, UK

© Springer International Publishing AG 2017

R. Díaz-Delgado et al. (eds.), *The Roles of Remote Sensing in Nature Conservation*,
DOI 10.1007/978-3-319-64332-8_7

develops as a 'living' map and allows learning from all of the projects to feed back into the map.

Keywords Habitat mapping • Ecological rule base • Remote sensing of vegetation • Norfolk • Multi-scale classification • Living map • Making Earth Observation Work • Object based image analysis (OBIA)

A 'Multi-scale' Approach with Ecological Input

Why Adopt a Multi-scale Approach to Mapping?

Information on the extent, location and condition of semi-natural habitats is essential to meet UK reporting obligations and commitments. These include legislative obligations in the EU Habitats Directive as well as domestic commitments to monitor the status and condition of priority habitats or provide audits of ecosystem function and service provision, as set out in the Convention for Biodiversity (2010); Strategic Plan for Biodiversity (2011–2020[1]). Gaps remain in our knowledge about the location and condition of semi-natural habitats of high nature conservation value (Medcalf et al. 2011) and, in particular, where they are located outside of the statutory site series.

A suite of Defra[2]-JNCC[3] funded projects, Making Earth Observation Work for UK Biodiversity (Medcalf et al. 2011, 2013, 2015; Gerard et al. 2015; Rowland and Morton 2013) and follow-on mapping work[4] (Bell et al. 2015) set out to provide a platform and improve the capacity and capability, particularly at the local level, to use Earth Observation (EO) and geoinformation techniques to inform the conservation and management of natural resources.

The Norfolk Biodiversity Information Service (NBIS) collates, manages and supplies biodiversity data to a wide range of users. Maps were produced for the NBIS using a range of remotely sensed imagery, with the outputs mapped to 'real world' Ordinance Survey (OS) features that are both recognizable and practical.[5] A multi-scale approach addressed the needs for biodiversity data at the regional, sub-regional and site scales.

The regional scale maps provided information about the location and extent of semi-natural habitats in Norfolk and were designed to:

---

[1] See: https://www.gov.uk/government/uploads/system/uploads/attachment_data/file/69446/pb13583-biodiversity-strategy-2020-111111.pdf

[2] Department for Environment, Food and Rural Affairs.

[3] Joint Nature Conservation Committee

[4] A paper by Bell et al. (2015) provides the results of a follow-on project to map the whole of Norfolk.

[5] https://www.ordnancesurvey.co.uk/business-and-government/products/mastermap-products.html

- Support a range of strategic policy applications (e.g., for land use planning, ecosystem restoration, targeting agri-environment schemes and planning integrated landscapes);
- Provide a source for collating regional information for reporting purposes;
- Help direct fieldwork where the specific habitat type is uncertain;
- Provide a baseline against which to monitor change;
- Provide an information source to feed into other analyses (e.g. green infrastructure and connectivity modelling).
- Provide a baseline for ecosystem service mapping

The sub-regional and more localised scale maps focused on areas known to be important for biodiversity and included the Norfolk Broads and the North Norfolk Coast. These maps were developed to:

- Inform site-based management for specific plant communities and the species that depend upon them;
- Increase understanding of how the different components of the habitats identified in remote sensing imagery relate to habitat condition;
- Provide a cost-effective alternative to traditional survey techniques – dynamic habitats, fragmented habitats and those that occur in complex mosaics and in inaccessible locations are difficult and costly to map by fieldwork.

The site level maps provided more detailed information on habitats and features in designated sites and were intended to:

- Be used by site managers
- Help managers to understand the condition of habitats and other natural features on the site
- Build an understanding of the extent of alien species invasion

Mapping Approach

The project evaluated the transferability of mapping techniques that use the Object Based Image Analysis (OBIA) method developed for Wales (Lucas et al. 2011), on the premise that if the same techniques could be used in a very different biogeographical area of the United Kingdom, then it may be possible to produce nationally consistent map data. The project also had a specific focus on ensuring knowledge exchange, enabling habitat practitioners to better understand EO imagery, products and techniques.

Principles Guiding Mapping at all Scales

Spectral and Textural

Ecologists can use phenotypic characteristics of individual species, including the combination of species in different parts of the canopy, including the understorey, to help identify plant communities and habitats. With optical remote sensing, this is not the case, and the features identifiable are only those seen from above, therefore all (or with a sparse canopy, most) of the return from the signal comprises the canopy species. With a dense canopy no component of the understorey will be visible in the signal. Therefore, habitats can only be identified clearly from EO if they are distinguished in botanical terms by their main canopy dominant species. This knowledge of the dominant cover species of each habitat is considered in terms of how it is manifest in the imagery in terms of:

- Vegetation productivity (estimated using the Normalised Difference Vegetation Index (NDVI) (Tucker 1979);
- Vegetation wetness/dryness (which requires the Short Wave Infra-Red band (SWIR) (Gao 1996);
- The amount of living and dead vegetation and proportion of non-vegetated areas (Tucker 1979);
- The structure of the vegetation, including woodiness (Kerr and Ostrovsky 2003);
- Variation in the spectral characteristics of different communities at different stages in the growing season (Cole et al. 2014).

These factors determine the type, scale and amount of imagery needed to map the habitats.

The Crick Framework

The Crick Framework, was developed as part of the project (Medcalf et al. 2011, 2013); it brings together ecological and Earth Observation knowledge and serves to:

- Categorise habitats in terms of their ability to be mapped remotely; and,
- Provide detailed descriptions of the capacity of EO to support the identification of habitats.

A wide range of interacting factors has been considered along with ecological knowledge, to develop a generic classification system that proposes categories (tiers) of habitat groups (Fig. 1). Habitats are described in terms of spectral characteristics and the spatial detail needed to map them. For instance, small scale or narrow habitats can only be mapped with spatially detailed image data (of a finer scale than the object itself, so it is clearly identifiable in the imagery), and are therefore placed in Tier 3b.

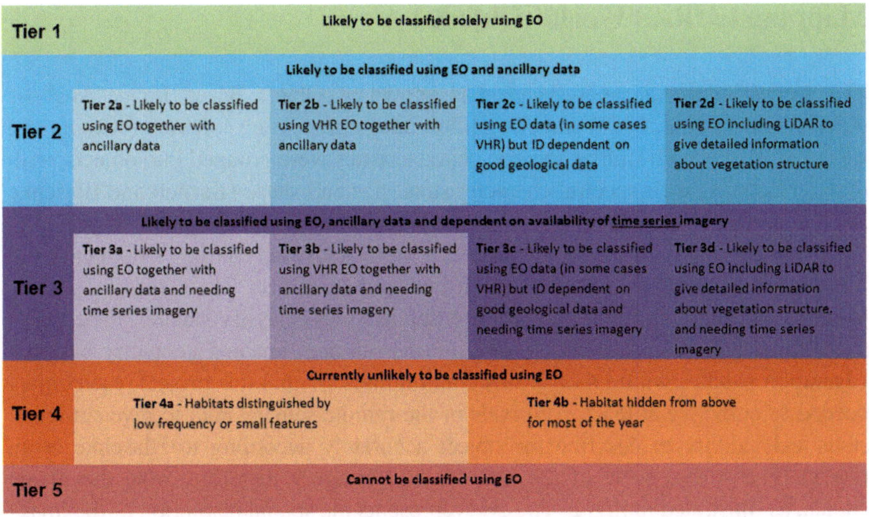

Fig. 1 The Crick Framework: The Tiers categorise habitats based on the EO and ancillary data required to map them. *VHR* Very High Resolution (<10 m per pixel)

Similarly, certain habitats are only associated with particular geological substrate conditions and are therefore assigned to Tier 2c or 3c, which require more ancillary information. A User Manual, aimed at habitat specialists provides a detailed description of all aspects of the Crick Framework (Environment Systems 2012). This is supported with examples, explanations and illustrative scenarios of how the framework can be used to support the evaluation of opportunities for mapping different types of habitats using a range of types of EO and ancillary data.

To apply the Crick framework to Norfolk, a detailed understanding of the habitats in an area, their species assemblages and scale of variation was required. The NBIS provided a list of habitats known to be present in the region and field visits were conducted to gather 'training data' so that the spectral properties of the main habitat types in the imagery could be described. The field work concentrated on those features of the vegetation communities that were *a priori* considered as critical to distinguish communities from one another in the imagery and included:

- Differences in sward structure;
- Differences in sward heterogeneity;
- The productivity differences and annual growth cycles;
- The amount of bare ground and the colour of the soil (particularly those with a red coloration);
- The amount of woody material and vegetation structure.

These factors determined the type, scale and amount of imagery needed to map the habitats.

Mapping to ‘Real World’ Objects

The process of image analysis used is known as Object Based Image Analysis (OBIA) (Lucas et al. 2007) and eCognition image processing software (eCognition) was used. Pixels with similar spectral characteristics were grouped into objects, with the user defining their spectral characteristics, size and shape (Burnett and Blaschke 2003; Karl and Maurer 2010). For Norfolk, a multi-resolution -stage ‘segmentation – classification – re-segmentation’ approach was used where large ‘woodland’ and ‘field-sized’ objects were firstly delineated and classified. Then, the image was re-segmented within these boundaries to delineate and classify smaller, irregularly-shaped polygons which were commensurate in dimensions and geometry with the vegetation patches within the various communities. Within this project, each object needed to comprise at least five pixels of the remote sensing data to give sufficient statistical validity to describe the object accurately according to Blaschke et al. (2008). The spatial scale of the features and imagery therefore drive the image choice, as illustrated in Fig. 2. The grid represents the spatial resolution of the imagery, with each grid square being an image pixel. The controlling factor in the identification of wet grassland is the combination of pixel size (i.e., image resolution) and the relative size of the woodland, grassland and scrub. In Fig. 2a, all the wet grassland can be clearly discriminated from the surrounding vegetation as it is represented by many pixels in high resolution imagery. In Fig. 2b, it would be possible, but more difficult, to identify as the pixel at the boundary of the wet grassland receives spectral contributions from the surrounding dry woodland. In Fig. 2c, the grassland would not be distinguishable as it is of a smaller dimension than the pixel.

The description of real world objects within the habitat map was assisted by using ancillary information. For example, the OS MasterMap[6] topography layer with field boundaries was used within the segmentation to set rules to restrict habitats to occur only within particular polygons, based on ecological knowledge (e.g. small areas of scrub woodland at the edge of, or within, an agricultural field). Using this

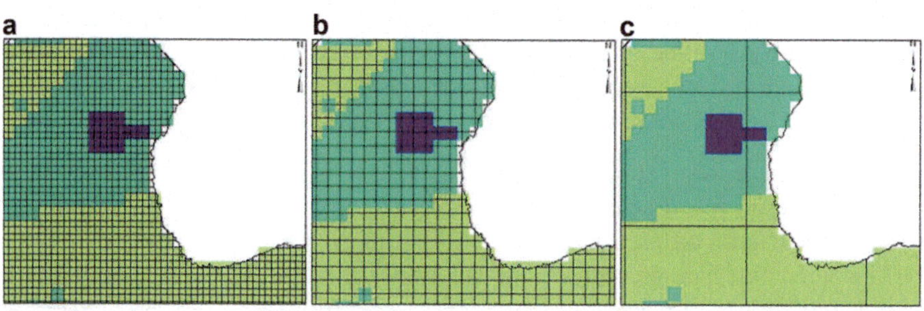

Fig. 2 Diagrammatic representation of the effect of pixel size, (a) 30 m, (b) 100 m and (c) 1000 m, on habitat feature recognition with areas of wet grassland (purple) surrounded by dry woodland (blue) and scrub (yellow)

[6] https://www.ordnancesurvey.co.uk/business-and-government/products/topography-layer.html

type of data also ensures that the outputs are of practical use at the local level and enabled the maps produced at differing scales in Norfolk to be aligned and comparable. This allows key features (e.g., roads, rivers and field boundaries) within the landscape to be consistently mapped at all scales.

Landscape Context: Confining the Search for Habitats

Once real world objects are defined, these can be classified using the rule-based approach which is informed by ecological knowledge. Often there is insufficient spectral information within the EO data to allow classification of habitats given the diverse range of land cover and habitats types present in the extensive areas covered by a satellite image. To overcome this, the component elements in the landscape are separated using the ecological properties of the main habitat types. This 'landscape context' of the habitat (Lucas 2011) is significant when mapping at the regional scale because some habitats only occur in specific locations in the landscape. For example, Floodplain and Coastal Grazing Marsh is only found within the floodplain or next to brackish waters which are seasonally inundated. The location of these habitats therefore has a direct geographic range. By first considering and examining this 'macro-scale' the ecologist is able to define the 'ecologically coherent' distinct landscapes where particular habitats are known to be located. The EO specialist can then exploit this information together with knowledge of the spectral properties of the habitats to produce the classification. This is shown in Fig. 3 (see Bell et al. 2015; Medcalf et al. 2011; Medcalf et al. 2013).

Placing Boundaries in Ecotones

Placing boundaries between habitats on maps is challenging whether mapping by fieldwork or by remote sensing as the habitats are not always clearly delineated. In some environments, a transition area occurs where habitats meet and integrate, which is referred to as an ecotone. In the coastal areas of north Norfolk, for example, saltmarsh grades into grazing marsh which in turn grades into natural grasslands as the distance from the sea and height of the land increases. Within these habitats, each of these 'ecotones' is represented in the imagery by a change in productivity and wetness in the spring. Distinguishing one habitat from another therefore depends on understanding the relationship between the ecotones and making a decision about where a 'hard mapped' boundary would best be placed along the continuum (Natural England 2011a, b, c; 2012).

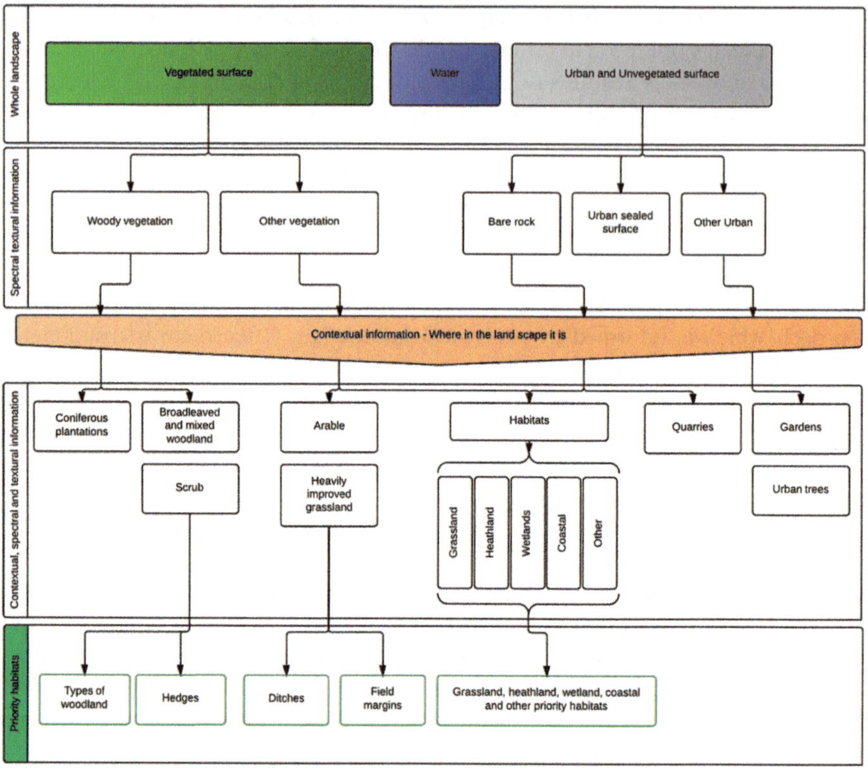

Fig. 3 Diagrammatic representation of the process of 'landscape splitting' for Norfolk

Selecting Imagery for Classification

The number of images needed and the timing of these is driven by the particular biogeography of the environment. In the example above, a spring image is necessary to separate the coastal grassland types. Other habitats needed a combination of 'leaf-on' and 'leaf-off' imagery. Norfolk is a diverse and dynamic environment, where cycles of arable cropping dominate the landscape and factors such as topography can limit the times of year when imagery should be acquired.

One of the key findings of the project was recognition of the effects on the mapping approach arising from differences in the spatial scale at which semi-natural habitats occur in different parts of the UK. In Wales, where the techniques were first developed (Lucas et al. 2011), semi-natural habitats can occur in large upland blocks: mosaics of semi-natural habitats form spatial patterns across whole valley systems or hillsides. By contrast, in Norfolk, the pressures on the land for arable use, and the character of the aquifer-fed habitats have resulted in greater fragmentation of semi-natural habitats with these then covering small areas and occurring in far more intricate mosaics. For this reason, reliable identification of semi-natural habitats requires the inclusion of higher resolution (more detailed) imagery especially when considering sub-regional and site scales (Bell et al. 2015). The complete table of imagery used for the Norfolk mapping at regional and sub-regional scale is shown in Fig. 4, with this ranging from Landsat and SPOT to RapidEye and GeoEye.

Local to Regional Scale Outputs in Norfolk

To map at the regional scale (Figs. 5 and 6), we first identified areas of arable land, to avoid confusion with semi-natural habitats that potentially share spectral characteristics at certain points in a crop growth cycle (Franke et al. 2012). In areas where there are many different cropping cycles, several images at different times of year are needed to identify land that has been ploughed or re-seeded. For semi-natural habitats, an image acquired both during a 'leaf-on' period and when the vegetation is 'leaf-off' provides key information for the OBIA rule development as the differences in spectral properties over time often helps to distinguish one community from another. For example, bracken has very distinct spectral characteristics in

Sensor	Spatial Resolution (m)	Scene size	Spectral resolution	January	February	March	April	May	June	July	August	September	October	November	December
Landsat	30	180x170km	B, G, R, NIR, SWIR				2011	2011	2011	2011		2011	2011		
DMC	30	600x600km	G, R, NIR, SWIR (pre-2008), TIR		2011										
IRS	20	150x150km	G, R, NIR, SWIR					2011							
ASTER	30	60x60km								2011	2005	2003			
SPOT	10	60x60km	G, R, NIR, SWIR			2011 2012									
RapidEye	6.5	Swath width ~80km not sold by scene	B, G, R, Red Edge, NIR						2011						
GeoEye-1	1.65	5x10km, 5x10km, 5x15km, 5x5km	B, G, R, NIR								2011		2011		

Fig. 4 Satellite imagery acquired for the Norfolk pilot study

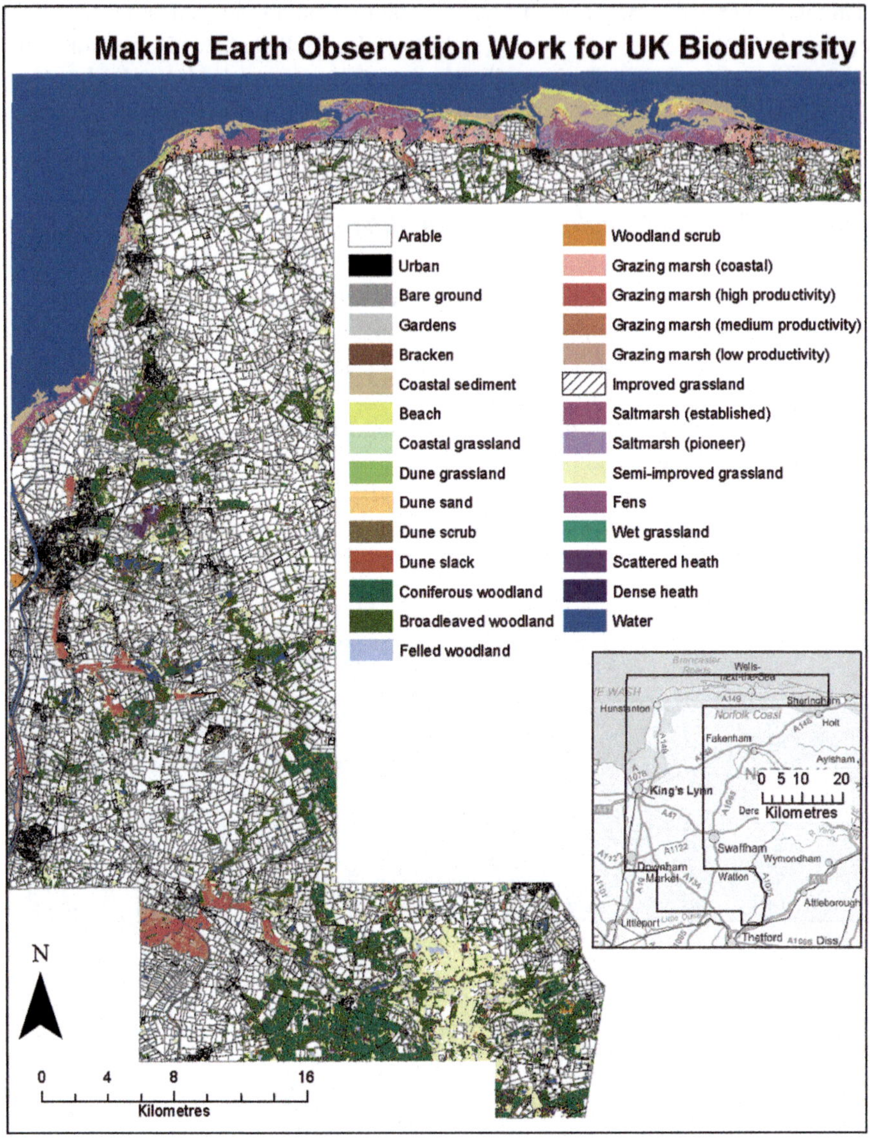

Fig. 5 Regional scale mapped output for an area of Norfolk

winter, with the dead leaves having a noticeable 'orange/brown' colour and there is no photosynthetic vegetation present in the canopy. By contrast, in summer, bracken has very high productivity and a closed canopy. Using the ratio between these two states allows bracken to be identified with relative ease. The timing of the 'leaf-on' and 'leaf-off' imagery however needs to be considered in the biogeographical context. In Wales a 'leaf-off' image before March contains too much shadow from the hills to be useful. However, in Norfolk, March is often well into the growing season and a February 'leaf-off' image was found to be more useable.

Within the Norfolk study, classification followed a multi-stage stage process: a draft map was created, which was checked by members of NBIS using randomly assigned ground survey points (Medcalf et al. 2011). The feedback from NBIS was used to refine the classification rulebase, especially for features that had not been well identified in the initial classification. This cycle of refining the maps and field checking was carried out over several iterations. The final outputs of the mapping are reported in Medcalf et al. (2013) and show that, following field survey by the NBIS, the accuracy was found to dependent on the input imagery available. For the eastern study area, the overall accuracy was 89% but this was lower for the western study area (78%). The greater classification accuracy in the eastern study area arose because of the greater temporal spread of imagery available, their higher pixel resolution and greater spectral range.

The errors in the classification are not randomly distributed when the rule based mapping is used but were spatially concentrated in:

- in areas obscured or shaded by clouds;
- at the boundaries of images; or
- areas with a less than ideal time series of imagery.

At the regional scale in Norfolk, a range of high priority habitats (BAP and Annex I) were identified using OBIA and rule-based classification (Fig. 6). Splitting the landscape into its component parts also allowed the classification of habitats, including the floristically and structurally complex grazing marshes and Breckland heathlands. Based on the Crick Framework, the priority habitats that were not fully identified were generally of Type 4a. However, for these habitats, the segmentation and classification approach was useful for generating 'areas of search' as the broader or 'parent' habitats are identifiable and can be delineated. At the landscape scale, wet heathland types (required for mapping Annex I habitats) could often, but not always, be mapped with the use of contextual data. The landscape scale work, using SPOT and IRS imagery, allowed the broad saltmarsh communities (water, sediment, vegetated saltmarsh) to be separated. However, to distinguish particular components of vegetation that are relevant to specific priority or Annex 1 habitats within these more broadly defined classes, high resolution imagery such as GeoEye (1.65 m) and a high quality Digital Terrain Model (1 m or better) or LiDAR were required.

Sub Regional Mapping: Considerations and Outputs

At a sub-regional scale, finer resolution satellite data are required to map features such as dykes, small pockets of scrub and wet grasslands. In addition to the satellites that provided spectral information, LiDAR data was used to give a structural component to the landscape. This allowed the separation of features such as reed beds

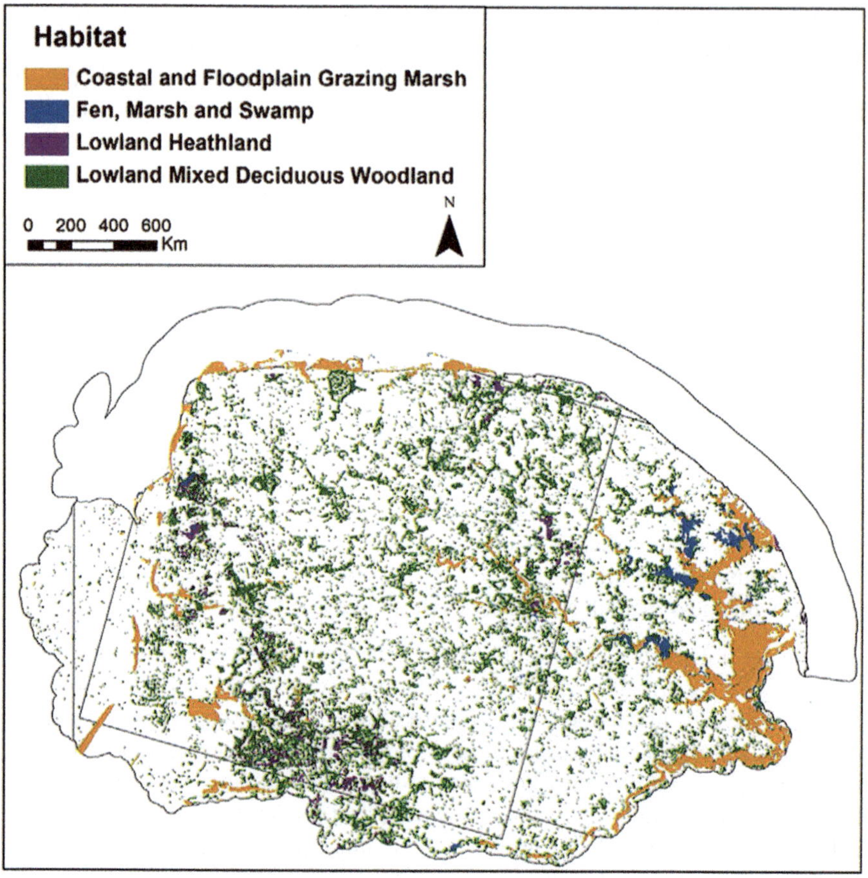

Fig. 6 The distribution of selected habitats of biodiversity importance in Norfolk at a regional scale

from other homogeneous productive grassland types, based upon their height and adjacency to water.

The Norfolk coastal zone, within which sub-regional mapping took place, was defined by distance from the sea, height above sea level and the presence of sand dune features, such as dune slacks. This was achieved using spectral rules together with ancillary data on elevation. In addition, LiDAR data were used to provide a structural component for features from the saltmarsh communities. Figure 7 illustrates the additional detail of habitat mapping that was achievable with the introduction of the higher resolution imagery and LiDAR data.

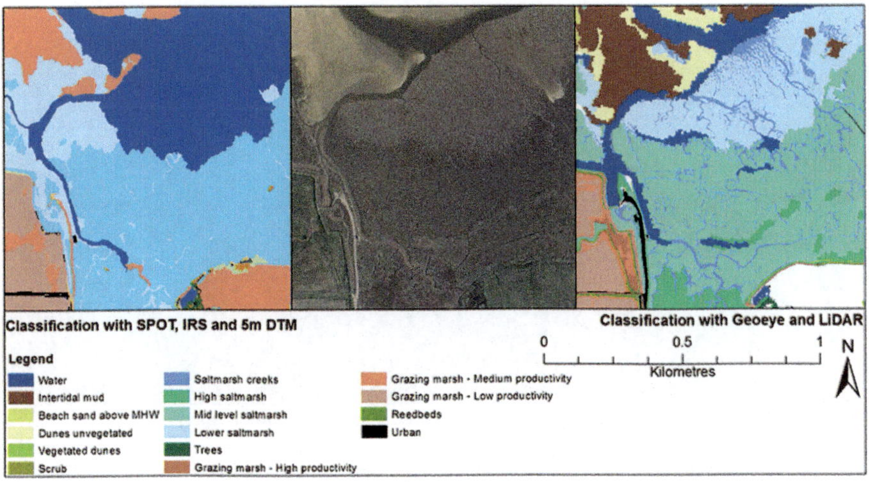

Fig. 7 Classification of saltmarsh showing the difference in spatial scale and classification detail achievable using regional scale mapping and data (left hand image) and sub-regional scale ultra-high resolution imagery (Geoeye and LiDAR – right hand image). The water level varies with the imagery captured at different states of the tide

Site Level Mapping: Considerations and Outputs

Choosing imagery for mapping at the site level, again, is driven by the detail needed and the scale and intricacies of the habitat mosaics. At an individual site level, data from Unmanned Aerial Vehicles (UAVs) red, green and blue (RGB) and near infra-red (NIR) imagery and a Digital Surface Model (DSM, vertical resolution of 10 cm) derived from stereo images can be used. The maps produced are extremely detailed and are considered by the NBIS to be particularly useful for areas that are difficult to access safely (e.g., salt marshes) or that cannot be visited at certain times (e.g., due to nesting birds on heaths). Using this approach, habitats can be mapped using EO to distinguish vegetation at a similar scale to an NVC survey.

A limitation of using UAV data is that only one overflight might be possible. Again, the choice of timing of the image acquisition is determined by the knowledge of the ecological and spectral characteristics of the features of most interest. In the Norfolk Broads, the invasive Himalayan Balsam is a significant problem and its distinctive pink blooms show up clearly, and can be mapped from the August acquisitions of UAV imagery (Fig. 8).

Using UAV data at a wetland bog complex at Dersingham, individual ash trees within the heathland were identifiable, which could potentially indicate that the heathland might be experiencing shrub invasion as a consequence of reduced grazing activity. In order to maintain the ecological value of the site, selected trees may have to be removed. The regional level mapping shows the main expanse of gorse scrub, but not the detail of individual young trees; therefore for scrub monitoring, knowledge of the individual site and the species posing the problems should be sought

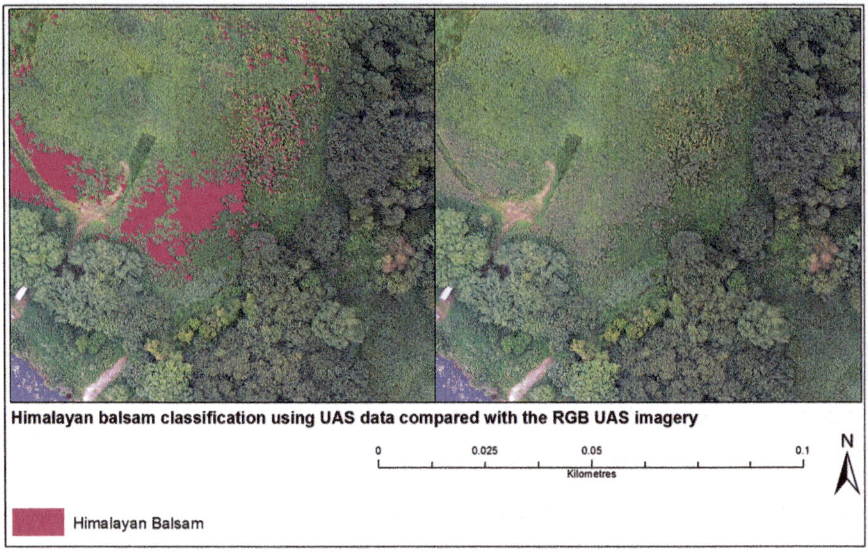

Fig. 8 Classification of Himalayan Balsam from UAV imagery in the Norfolk Broads

before deciding on the imagery to use. On the Breckland heathland sites, grazing is important in maintaining the condition of the sward, with patches of bare ground and very short swards maintaining the biodiversity of these habitats. Whilst both regional and site level maps showed large patches of bare ground, the UAV data revealed the smaller patches which are most important for maintaining the communities.

At a site level the maps are suitable for understanding how a specific site or species interacts in relation to its surroundings (e.g., on wetland bog complexes and a Breckland heathland site). The features that could be identified are of value for:

- Planning scrub management
- Quantifying grazing pressure
- Understanding differences in wetness within sites
- Mapping bare ground and invasive species.

Examining the relationships between habitats at a site and within the surrounding areas can indicate where there are risk factors to the habitat condition or conversely, opportunities for habitat expansion (Breyer et al. 2016). This is especially informative for protected sites management (Fig. 9). In several areas of the Norfolk Broads, protected sites (such as SSSIs) were mapped using UAVs. Measures of vegetation productivity (NDVI) from lower resolution satellite data and the regional scale mapping were produced rapidly and at low cost, as the regional map was complete. Spectral measures, such as the NDVI, assisted site managers in identifying how management of the land surrounding a protected site impacts upon its ecological condition (e.g., through nutrient deposition from heavy use of fertiliser around the site itself).

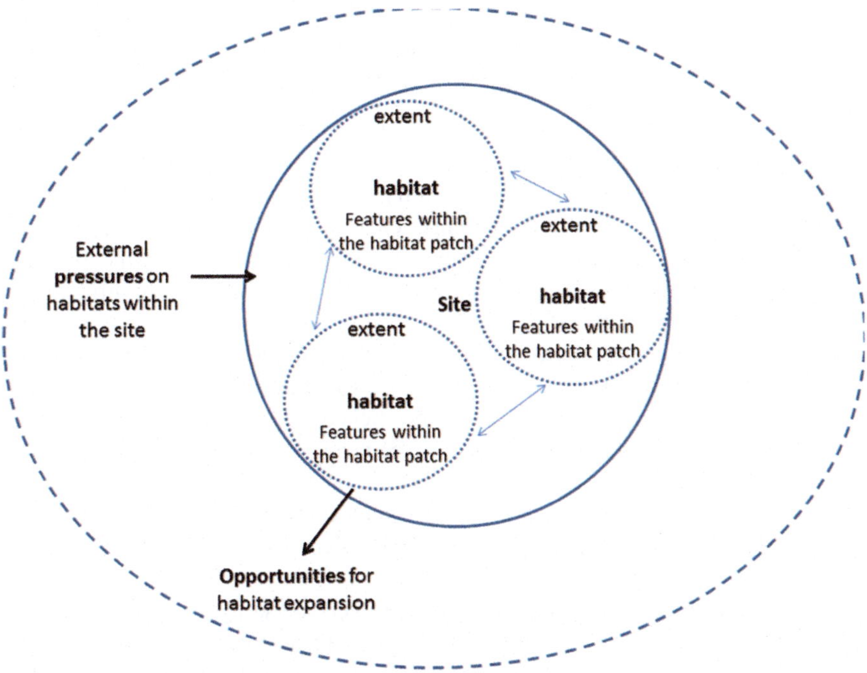

Fig. 9 Processes present around and within a site

Towards a Living Map

The regional map of Norfolk produced by the project provided an overview of habitats in Norfolk at a field/part-field scale level and with knowledge that not all habitats could be identified fully, but areas of search could be delineated and a 'parent class' assigned. Furthermore, there was variation in the accuracy of mapping across the region that was quantifiable.

This knowledge allowed the NBIS to interact with the project team and arrive at a potentially cost-effective way of improving the content of the map through ongoing targeted field work and a citizen science approach (Medcalf et al. 2015. A subsequent assessment and study (Newson et al. in press) enabled the NBIS volunteers to validate and update the map and take forward the concept of a 'living map' in a Defra funded project led by British Trust for Ornithology.

The project assessed the potential for volunteers to assist the process of groundtruthing or validating "Living Maps" using the maps and data produced for Norfolk. It considered the potential sampling methodologies, the technological solutions and the opinions of the volunteer sector, particularly with regards to their interest in contributing to such an exercise, their capabilities and their attitudes towards technological options. A broad spectrum of group leaders spanning charities, councils, leisure groups, recorder networks and conservation agencies were consulted by the

NBIS to identify the volunteer capacity in Norfolk and across the UK. At every stage, the potential for transferability to a UK-wide approach was reviewed.

The project established that:

- Habitat patches were missed from the classification and patches of other habitat mistakenly included;
- Priority habitats can be difficult to identify both in the field by volunteers and from imagery; this is due to the complexity of the gradation of one habitat into another. This is compounded by the existing habitat class names, which despite being pinned to BAP or Annex 1 names as much as possible, were difficult for non-experts to use in the field.
- A stratified sampling strategy for validation would involve both a desk-based and field-based assessment. This is a real issue for habitats of conservation interest because they are often too scarce to be picked up on any random-stratified method. A large number of randomly selected sample squares are needed to provide scientifically robust results in Norfolk.
- Within Norfolk, there is a keen group of volunteers with a range of backgrounds and skills but a limited appetite to add habitat recording to existing activities or carry out desk-based assessments.
- Most volunteers prefer to validate locations very close to where they live, so volunteer sampling strategies need to take this into account.
- A well-designed smart-phone application would appeal to a wide range of users but would be costly to develop, unless rolled-out across the UK with the potential for multi-year assessments to coincide with Living Map updates.
- Training will be critical to participation, along with effective communication of the purpose, outputs and confidence in the product.

Examples of the Use of the Mapped Outputs in Norfolk

Prior to the project, the NBIS held only partial records for selected sites in the county. They had no regional overview of the habitats but held some sub- regional maps and old site maps: the potential for repeating these, however, was limited by resource constraints.

Regional Scale and Sub-regional Mapping

The UK National Planning Policy Framework[7] (NPPF) places requirements on local authorities to map habitats as a means of providing evidence on biodiversity to understand the provision of green infrastructure and ecological networks. In a wider

[7] https://www.gov.uk/government/publications/national-planning-policy-framework--2

context, local authorities, government agencies and conservation charities are working together to prevent biodiversity loss by 2020 and move to net gain in biodiversity as set out in Biodiversity 2020: A Strategy for England's Wildlife and 'Ecosystem Services' objectives (Defra 2011). The regional scale mapping carried out in Norfolk was a significant step towards realising these aspirations as it:

- Increased knowledge of the presence and extent of habitats in the wider countryside;
- Generated habitat maps and data to meet a wide range of landscape scale approaches to biodiversity delivery;
- Generated map layers to support the analysis of ecosystem goods and services, ecological networks, ecological restoration, pollution modeling, species distribution modeling and threat maps for non-native species and plant and animal health issues.

Subsequent activity that has utilised the regional and sub regional mapped outputs and data includes bat habitat suitability modelling by Norfolk Bat Survey and pollution and sediment modelling by the Broadland Catchment Partnership. The Norfolk Wildlife Trust have used the map within their Living Landscape plans, and as a tool to help local parish groups to produce a Phase 1 Map of their parishes: this will contribute to the development of neighbourhood plans linked to the NPPF.

Site Level Mapping

The site level mapping achieved in Norfolk demonstrated the potential for:

- Improving the spatial definition of habitats within some designated sites;
- Producing evidence for management plans for larger sites or discrete areas to inform local planning;
- Identifying threats to habitats and ways of mitigating against and monitoring these (e.g. projects mapping the presence and extent of alien species).

Site managers in Norfolk identified one of the main benefits of the maps being a reduction in survey time: this enables targeted follow-up surveys and increases the time available for planning and site management. Site managers in Breckland, especially at East Wretham Heath, found the maps useful for showing bare ground and scrub, thus assisting with the management of the rabbit population and providing information on the quality and nutrient levels of the Breckland grass-heath. The Himalayan Balsam mapping was exceptionally useful for planning eradication, especially for the large areas of inaccessible broads for which the NBIS had only very limited records of the location of this invasive species.

Conclusion

The Norfolk mapping project demonstrated that the Crick approach was transferable to a lowland situation. It highlighted the importance of understanding the ecology of the area, at a range of spatial scales. Using ecological knowledge, including fieldwork to understand how the habitats are manifested in the imagery, and to document the key ecological features identifiable in EO was an essential step in the process. The constraints of being unable to map certain habitats can be overcome by the NBIS, through the creation of a 'Living Map' programme where targeted field-based assessment decreases the uncertainties in the map. At a site scale, the combined use of satellite and UAV data illustrated that the Crick technique has many potential uses for site managers.

Maps are key learning and engagement tools for local authorities, government agencies, charities and environmental groups. The regional map of Norfolk has provided an overview of habitats in the county at a field/part-field level. The site mapping using UAVs highlighted how useful it is to be able to accurately identify small features on sites such as bare ground, patches of nettles (*Urtica dioica*) and the spread of invasive species. The future uses for EO data are expanding, with habitat mapping a key element, but also with opportunities to inform habitat condition monitoring, assessments of landscape change and to map habitat restoration and green infrastructure.

References

Bell, G., Neal, S., Medcalf, K.: Use of remote sensing to produce a habitat map of Norfolk. Ecol. Inform. **30**, 293–299 (2015)

Blaschke, T., Lang, S., Hay, G.J.: Object Based Image Analysis for Remote Sensing. Springer, Berlin/Heidelberg (2008)

Breyer, J., Pike, S., Medcalf, K., Parker J.: Making Earth Observa-tion Work (MEOW) for UK Biodiversity Monitoring and Surveillance, Phase 4: Testing applications in habitat condition assessment. A report to Defra, prepared by Environment Systems, Ltd (2016)

Burnett, C., Blaschke, T.: A multi-scale segmentation/object rela-tionship modelling methodology for landscape analysis. Ecol. Model. **168**, 233–249 (2003)

Cole, B., McMorrow, J., Evans, M.: Spectral monitoring of moorland plant phenology to identify a temporal window for hyperspectral remote sensing of peatland. ISPRS J. Photogramm. Remote Sens. **90**, 48–58 (2014)

Convention for Biodiversity (CBD): Decision X/2. Strategic Plan for Biodiversity 2011–2020 and the Aichi Biodiversity Targets, Nagoya, Japan, 18–29 October 2010. [online] https://www.cbd.int/decision/cop/default.shtml?id=12268 C and https://www.cbd.int/sp/ (2010)

Defra: Biodiversity 2020: A Strategy for England's Wildlife and Ecosys-tem Services (2011)

Environment Systems: Making Earth Observation Work for UK Biodi-versity Conservation – Phase 2: Crick Framework User manual. JNCC: http://jncc.defra.gov.uk/page-6281 (2012)

Franke, J., Keuck, V., Siegert, F.: Assessment of grassland use intensity by remote sensing to support conservation schemes. J. Nat. Conserv. **20**(3), 125–134 (2012)

Gerard, F.F., Acreman, M.C., Mountford, J.O., Norton, L., Pywell, R.F., Row-land, C., Stratford, C. & Tebbs, E. Earth Observation to Produce Indices of Habitat Condition and Change. CEH Final Report to JNCC. JNCC Ref. C14-0171-0901 (2015)

Gao, B.C.: NDWI – a normalized difference water index for remote sensing of vegetaion liquid water from space. Remote Sens. Environ. **58**(3), 257–266 (1996)

Karl, J.W., Maurer, B.A.: Spatial dependence of predictions from image segmentation: a variogram-based method to determine appropriate scales for producing land management information. Ecol. Inform. **5**, 194–202 (2010)

Kerr, J.T., Ostrovsky, M.: From space to species: ecological applica-tions for remote sensing. Trends Ecol. Evol. **18**, 299–305 (2003)

Lucas, R., Rowlands, A., Brown, A., Keyworth, S., Bunting, P.: Rule-based classification of multi-temporal satellite imagery for habitat and agri-cultural land cover mapping. ISPRS J. Photogramm. Remote Sens. **62**, 165–185 (2007)

Lucas, R., Medcalf, K., Brown, A., Bunting, P., Breyer, J., Clewley, D., Key-worth, S., Blackmore, P.: Updating the phase 1 habitat ma p of Wales, UK, using satellite sensor data. ISPRS J. Photogramm. Remote Sens. **66**(1), 81–102 (2011)

Medcalf K.A., Parker J.A., Turton, N., Finch C.: Making Earth Ob-servation Work for UK Biodiversity Conservation – Phase 1. Report to the JNCC and Defra www.jncc.defra.gov.uk/EO (2011)

Medcalf, K.A., Parker, J.A., Turton, N., Bell, G.: Making Earth Ob-servation Work for UK Biodiversity Conservation – Phase 2. JNCC Report 495 Phase 2, JNCC Peterborough 2014 (2013)

Medcalf, K.A., Parker, J.A., Breyer, J., Turton, N.: Making Earth Ob-servation Work for UK Biodiversity Conservation – Phase 3: Cost Effective Meth-ods to Measure Extent and Condition of Habitats. A Report Produced by Environ-ment Systems Ltd., for Defra and the JNCC. (2015)

Natural England: NCA 76: North West Norfolk Key Facts & Data (2011a)

Natural England: NCA 77: North Norfolk Coast Key Facts & Data (2011b)

Natural England: NCA 80: The Broads Key Facts & Data (2011c)

Natural England: NCA 85: The Brecks (2012)

Newson, S. E., Turvey, D. J., Gillings S. & Neal, S. (in press). Earth Observation Data Integration Pilot Project 5 – Developing community and crowd-sourced validation of 'Living Maps'. A report produced by the British Trust for Ornithology for Defra

Rowland, C., Morton, D.: Developing and Evaluating an Earth Observation-Enabled Ecological Land Cover Time Series System:JNCC://http//jncc.defra.gov.uk/page-6280 (2013)

Tucker, C.J.: Red and photographic infrared linear combinations for monitoring vegetation. Remote Sens. Environ. **8**, 127–150 (1979)

Sub-pixel Mapping of Doñana Shrubland Species

Marcos Jiménez and Ricardo Díaz-Delgado

Abstract Periodically mapping and monitoring the spatial distribution of vegetation at species level increases knowledge about the relationship between ecological processes and ecosystem functioning. In protected natural areas, long-term monitoring of key or invasive species can be applied in a more effective, coherent and consistent way by incorporating regular mapping. To guarantee more reliable monitoring data, these programs should be underpinned by periodic and repeatable measurements. In addition, using standard methodologies will help to ensure that we can make comparisons with other natural areas. In this chapter, we introduce a protocol for mapping the distribution of plant species using airborne imaging spectroscopy. We apply robust and widely used methodologies for acquiring hyperspectral airborne imagery and field spectroscopy, and for processing and analyzing these to generate spatial-explicit distribution maps of plant species. The main aim was to facilitate monitoring programs by supplying periodic maps of plant species that can be used identify major shifts in distribution. The study case focuses on the shrub communities on the stabilized sand dunes of Doñana National Park in south west Spain.

Keywords Plant species mapping protocol • Monitoring shrubland communities • Imaging spectroscopy • Field spectroscopy • INTA-AHS system • Doñana National Park

Introduction

The rate of change in the structure and species composition of ecological communities sometimes occurs in a very dramatic way (Hooper et al. 2005). In order to slow down negative biodiversity trends, conservation decision-makers need knowledge

M. Jiménez (✉)
Remote Sensing Area, National Institute of Aerospace Technology (INTA), Ctra. Ajalvir s/n, Torrejón de Ardoz, 28850 Madrid, Spain
e-mail: jimenezmm@inta.es

R. Díaz-Delgado
Estación Biológica de Doñana (EBD), Consejo Superior de Investigaciones Científicas (CSIC), Sevilla, Spain
e-mail: rdiaz@ebd.csic.es

© Springer International Publishing AG 2017
R. Díaz-Delgado et al. (eds.), *The Roles of Remote Sensing in Nature Conservation*, DOI 10.1007/978-3-319-64332-8_8

about the spatial distribution and temporal dynamics of flora and fauna on a relevant scale (Leutner et al. 2012). In this sense, being able to map and monitor the spatial distribution of vegetation at species level, and identify changes in population in space and time, will increase the knowledge about the relationship between ecological processes and ecosystem functioning (Trochet and Schmeller 2013). Consequently, monitoring species that are introduced or invasive in protected areas can be applied with greater consistency (Schmeller 2008).

Traditional methods for vegetation mapping, such as exhaustive field surveys, are both time consuming and costly (Ustin et al. 2004). Remote sensing is now established as an important tool for researching and monitoring the ecological processes of terrestrial ecosystems, and has the potential to achieve this in an efficient and economical way (Nagendra et al. 2010), particularly in sensitive and inaccessible areas (*e.g.,* mangrove or marshland) (Kamal and Phinn 2011). In the early years of Earth Observation, some vegetation types and communities were mapped using broadband multispectral observations, typically from sensors such as Landsat TM or SPOT (Satellite for Earth Observation). However, with the development of hyperspectral remote sensing, we have the capacity to map at the species level (Ustin et al. 2004). Among hyperspectral techniques, imaging spectroscopy is well adapted for airborne platforms and is reinforced with the application of Remotely Pilot Air Systems (RPAS). However, spaceborne instruments are still in the early stages of development (Schaepman et al. 2009). In this sense, forthcoming space missions like EnMAP (Kaufmann et al. 2008) or PRISMA (Stefano et al. 2013) will present a great stimulus to consolidate this approach.

Although airborne imaging spectroscopy with very high spatial and spectral resolution looks promising in the arena of plant species mapping, operational approaches are lacking because of our limited biophysical understanding of when remotely sensed signatures indicate the presence of unique species within and across ecosystems (Somers and Asner 2012). In this sense, two of the main drawbacks in ecosystems are: high spectral similarity among species with similar ecological adaptations, and, conversely, high 'within species' spectral variability response due to variations in plant constituents (tissues chemistry and structure) (Asner 1998). To improve the mapping efficiency of imaging spectroscopy, the analysis techniques applied to the imagery could be better accomplished if based on ground truth data to help characterise the spectral response of each plant species (Warner 2010). Field spectroscopy is the primary method for registering ground truth data to develop spectral libraries for plants (Manakos et al. 2010). However, to create consistently unique and detectable spectral signatures among species, this spectral library must take into account the spatiotemporal variability of the plants, both throughout the ecosystem and the seasons (Zomer et al. 2009).

Following recommendations of the Rio de Janeiro Convention on Biological Diversity (CBD) in 1992 and European Habitats Directive (Directive 92/43/EEC), natural protected sites should be under continuous observation. Consequently, evaluation is undertaken and reporting required every 6 years, to determine the status of the habitats and species of European importance for nature conservation in a biogeographic region. In this sense, the organizations responsible for the management and conservation of natural protected areas have been given the responsibility to establish long term monitoring programs for terrestrial ecosystems that allow the inference of trends and rates of change (Vaughan et al. 2001). However, these monitoring programs should be informed by periodic and repeatable measurements (Schmeller 2008). Standard methodologies allow better comparison with other maps of natural areas. In the work by Oakley et al. (2003), guidelines for monitoring protocols are outlined, with these highlighting the importance of the collection, management, analysis and reporting of the data.

In this study, we present an approach for establishing a protocol for mapping the distribution of plant species based on airborne imaging spectroscopy. Robust and widely used methodologies for hyperspectral airborne images and field spectroscopy acquisition, processing and analyzing are selected to generate spatially-explicit maps of plant species distribution. The aim behind this work is to facilitate programs to monitor ecological communities based on mapping derived from imaging spectroscopy. These maps will help to interpret shifts in species composition in response to environmental changes induced by climate and land use change and other anthropogenic impacts.

We present a practical case study to demonstrate the usefulness of the program and protocols. In the ecosystem of stabilized sand dunes of Doñana National Park in south west Spain, the shrub communities are a very important habitat for fauna. In-depth knowledge of the spatial distribution of the shrub species is also essential for managing shrubland habitats (Cobo et al. 2002).

Background of Plant Species Mapping Activities Using Imaging Spectroscopy

Before describing the procedure proposed for mapping plant species in this work, we should explain some aspects of the key techniques that underpin the protocols: the characteristics of the airborne imaging spectroscopy, the basis of field spectroscopy data, and the spectral unmixing algorithms applied to the hyperspectral data.

Airborne Imaging Spectroscopy

Airborne remote sensing is characterised by its flexibility in imagery acquisition conditions and continuous maintenance and calibration of the sensors installed. These characteristics offer great advantages for the acquisition of seasonal and diurnal processes (e.g., drought and fire impacts) and can determine the viability of specific research applications. There are several aerial platforms for imaging spectroscopy (e.g., balloons and helicopters) and more recently the Remotely Piloted Air Systems (RPAS), formerly Unmanned Aerial Systems –UAS- (Hruska et al. 2012). However, non-pressurised aircraft are the most widely used platform due to the better combination of autonomy and stability. Airborne imaging spectroscopy operators are usually national institutions involved in aircraft research. In particular, the European Facility for Airborne Research (EUFAR, http://www.eufar.net/) is a European Commission project to integrate 24 of these operators and provide researchers with easy and open access to airborne research facilities.

The first whiskbroom airborne imaging spectrometers developed around 1980 (Green et al. 1998) were manufactured with a full width half maximum (FWHM) of nearly 10 nm in the visible and near infra-red (VNIR) and short wave infra-red (SWIR) wavelength regions. Pushbroom systems offer advantages in robustness, integration time, speed, and spectral/spatial resolution (Schlapfer et al. 2007). Moreover, the pushbroom system increases the levels of signal to noise ratios (SNR) to around 1000:1.

Near-ground flight conditions impose some radiometric and geometric constraints on the hyperspectral imagery: (1) the sensor requires a large (between 40° and 90°) field of view (FOV) to cover as much ground area as possible in each flight line. Consequently, off-nadir pixels tend to increase in pixel size and are subject to radiometric gradients if the scanning plane is not perpendicular to the Sun's principal plane. In addition, rugged terrain enhances these changes of incident sun angle variations and pixel size; (2) to cover a typical local scale study area of 20 × 20 km, several flight lines, which may take 2–3 hours, are needed, to also ensure that Sun angle variations between flight lines are reduced; sun angles variation between flight lines; (3) atmospheric components with more relevance in the radiative transfer optic response are concentrated in the very first kilometers near-ground, thus airborne imagery is also affected by the atmosphere. Likewise, the platform stability is influenced by high-frequency velocity and attitude variations. All these aspects must be corrected and normalized between flight lines to obtain a better georeferenced ground reflectance mosaic of the study area. Algorithms for geometric correction are becoming accurate and can be implemented in a fully automatic way (Biesemans et al. 2007).

A wide variety of studies have been conducted on species-level mapping in different vegetation types, including grasslands (Miao et al. 2006; Möckel et al. 2014), shrublands (Roberts et al. 1998), mangroves (Ustin et al. 2004), marshlands (Silvestri et al. 2003), and forest (Kalacska et al. 2007; Asner et al. 2008).

Field Spectroscopy

Field spectroscopy is the measurement of high-resolution spectral radiance or irradiance in the field to derive the reflectance or emissivity spectral signatures of targets at the Earth's surface under natural environmental conditions. In comparison with airborne or spaceborne imaging spectroscopy, the sensing instrument in the field can remain fixed over the subject of interest for much longer, and the path length between the instrument and the object being measured is thereby reduced (Milton et al. 2009).

The robust and portable field spectroradiometers developed in previous decades have evolved from the non-imaging spectrometers currently used in the laboratory. Fibre optic bundles providing different FOV angles has become widely used. Depending on the application considered, manufacturers offer two kind of spectroradiometer: (1) small and light devices that are designed to work only in the VNIR, with levels of SNR around 250:1; (2) less small and light devices that work across the entire solar spectrum, with refrigerated SWIR detectors and an SNR of around 1000:1. Both instrument types have a VNIR's full FWHM of nearly 3 nm but, for SWIR instruments, the FWHM is nearly 10 nm.

Methods in field spectroscopy are well described by several authors (Salisbury 1998; Goetz 2012). The basic recommendations for better spectra acquisition are to measure under cloudless sky, at high sun zenith angles and with the same illumination and atmospheric conditions for the panel and target. The most widely used acquisition methodology to obtain near-ground reflectance is the *single-beam*, where the same instrument is used to measure both the target and the reference panel spectral radiance. In this case, Spectralon® (Labsphere, North Sutton, NH, USA) has been established as the standard material for panels, due to its high degree of perfect diffuse and lambertian response. Measurements with field spectroradiometers are often hand-held, usually with the sensor head mounted on a pole or yoke to keep it away from the operator's body. In this regard, we should mention the novel carrier-lift system MUFSPEM@MED (Mobile Unit for Field SPEctral Measurements at the MEDiterranean, Manakos et al. 2010), which has the capacity for automated signature acquisition.

Spectral libraries are collections of spectra that characterize the reflectance or emissivity spectral response of Earth's surfaces and materials. Characterizing plant species is always challenging due to variations in vegetation elements, growing states, and phenology (Asner 1998). A large number of studies have focused on the acquisition of plant spectra in the field over different vegetation formations, including shrublands (Manevski et al. 2011), marshlands (Zomer et al. 2009), forest (Somers and Asner 2012) and sub-aquatic environments (Fyfe and Dekker 2001). Nevertheless, there is no standard protocol for the acquisition of a plants reflectance spectral response (Pfitzner et al. 2010), so a universally applied methodology for the collection of field spectra is needed (Manakos et al. 2010).

Spectral Unmixing

When working with mono-specific plant communities, the estimation of plant species cover by remote sensing is better accomplished by sub-pixel algorithms. In remote sensing imagery, the total signal integrated in a pixel is a function of:

- The optic properties of the components inside the Ground Instantaneous Field Of View (GIFOV), and the relationships between them. The components contribution can follow a linear mixture or a non-linear mixture model (Keshava and Mustard 2002);

- The sensor's point spatial function (PSF), which determines if the central parts of the pixel are going to have more relevance than the outer parts, and also the non-negligible part outside of the GIFOV (Schowengerdt 2007);
- Radiance incorporated by the atmosphere, with photons coming outside of the GIFOV by the adjacency effect from surfaces surrounding the pixel (Richter et al. 2006).

Spectral unmixing is the decomposition of a mixed pixel into a collection of distinct spectra (endmembers), and a set of fractional abundances that indicate the proportion of each endmember (Keshava and Mustard 2002). It is a physically based model that transforms radiance or reflectance to physical variables, which are linked to the sub-pixel abundances of endmembers within each pixel. The Linear Spectral Unmixing (LSU) is the most frequently used model because of its simplicity and more direct interpretation, and assumes the pixel spectrum to be a linear combination of a finite number of spectrally distinct endmembers (Keshava and Mustard 2002). The three consecutive procedures for LSU are: (1) reduction of the dimension of the data using Principal Components Analysis (PCA) or Minimum Noise Fraction (MNF), which seeks a minimal representation that sufficiently retains the requisite information for successful unmixing; (2) endmember determination representative of the physical components on the surface. Endmembers can be obtained directly from the image employing statistics to capture variability, such as the Pixel Purity Index (PPI). Similarly, the endmembers could be extracted from spectral libraries derived from laboratory or field spectroscopy; (3) imagery pixel reflectance values are inverted using least square methods to minimise the squared-error and achieve fractional abundances of the components.

Depending on the number of endmembers introduced and constraints applied in the algorithm, different varieties of LSU have been developed. For example, if only a few key endmembers are determined without requiring knowledge of the remaining scene endmembers, Mixture-Tuned Matched Filter (MTMF) (Boardman et al. 1995) is an algorithm that allows false positives to be identified and eliminated from abundance results. Additionally, Multiple Endmember Spectral Mixture Analysis (MESMA) extends LSU by allowing the number and types of endmembers to vary on a per-pixel basis (Roberts et al. 1998).

The need to provide sub-pixel proportions of vegetation components is well reflected in the literature (McGwire et al. 2000). When the endmembers include vegetation, the endmember fraction is considered proportional to the areal abundance of the projected canopy cover. Although the differences in canopy structure and size between species can be very noticeable, which entails non-linear mixture model application, the use of LSU provides statistically significant results (McGwire et al. 2000).

Plant Species Mapping

The protocol for mapping plant species proposed in this work is based on collaboration between an airborne imaging spectroscopy operator, which can be used to acquire and pre-process the hyperspectral imagery, and a user organization (*i.e.*, Natural

Park management unit, University Department), who can be responsible for supplying field data and generating the species cartography.

In order to achieve this mapping, we followed the structured design by Kerekes and Baum (2005) to determine the viability of imaging spectroscopy applications. In Fig. 1, the protocol is schematized and shows the elements that comprise their three different aspects: the procedures and data acquired in the field, the methodologies for acquisition and processing the airborne spectroscopy imagery, and the aspects for cartography generation.

Any kind of digital geographic data should be documented, as much as possible, to ensure that the data producer can characterise the geographic data properly and enable users to apply the data in the most efficient way (International Organization for Standardization (ISO) 2003). A standardized structure for metadata also increases the value of metadata by improving its readability, flexibility and utility for archival processing and usage with software applications (Jiménez et al. 2014). The most relevant organizations publishing standards to define common metadata structures and their hierarchies are the International Organization for Standardization (ISO) and the Open Geospatial Consortium (OGC).

Following the schematic outline shown in Fig. 1, the subsequent sections describe the protocol in depth.

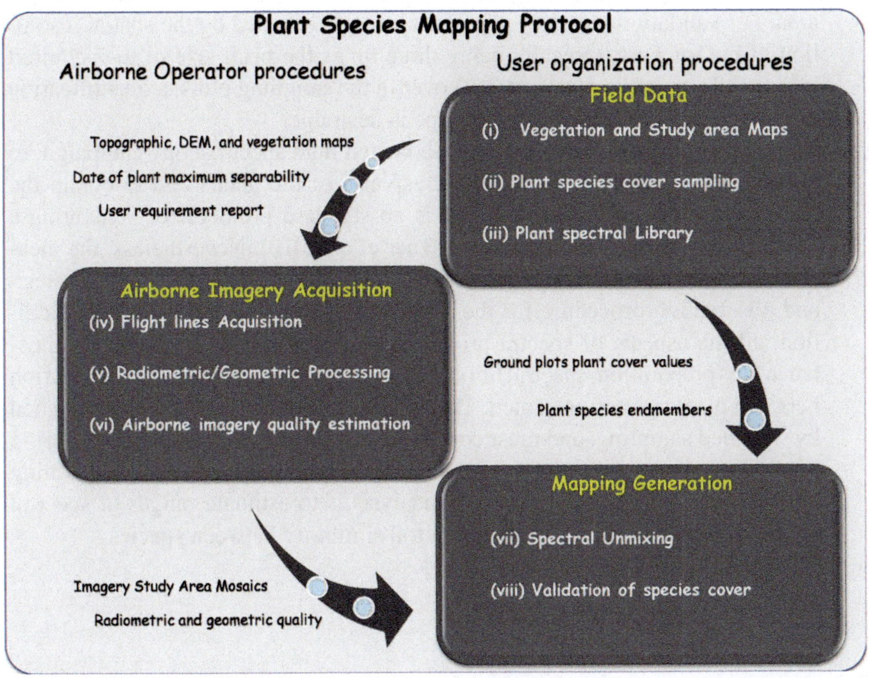

Fig. 1 Plant species mapping protocol based on the use of airborne hyperspectral imagery

Field Data

Field data is needed from the study area describing the plant communities present as auxiliary information for planning the airborne campaign, supporting the imagery analysis, and validating the cartography generated by the process. The user organization plays a key role in the collection and preparation of these field data.

(i) The topographic and plant community maps of the study area are used to plan the hyperspectral flights and provide cover estimates of the plant species. A Digital Elevation Model (DEM) with high spatial resolution is needed to support the processing of airborne imagery. In the case of a natural protected area, the managers often already have this information. As an alternative, information from Spatial Data Infrastructures (SDI) can be obtained and used, although all data cartographies should be in the same projection.

(ii) Field-based cover estimates of plant species are used to validate the plant species maps generated by processing the imagery. The most reliable way to collect plant species cover over heterogeneous ecosystems is by stratified random sampling (Canfield 1941). Moreover, to ensure better results for future programs that monitor species distributions, the sampling plots must be located in as many permanent sites as possible. The number of measurement plots is determined by the number of plant communities, the degree of biodiversity present, and the heterogeneity of environmental factors impacting the survey area. For validation purposes, the plot size is determined by the spatial resolution of the imagery, typically being three times the pixel size of the acquired imagery. To measure plant species cover in the sampling plots, a quantification method independent of vegetation type is desirable.

(iii) A spectral library is generated by a dedicated field spectroscopy campaign, to characterise the spectral reflectance response of the plants and to obtain the endmember for each species. There is no standard protocol for generating a spectral library for plant species (Pfitzner et al. 2010). Nevertheless, the measurement protocol for the spectral library must combine a sampling strategy and observation procedure for the spectra acquisition of the canopy. In addition, all the aspects of spectra processing must comprise the spectral reflectance files preparation, spectral library generation and separability quantification between plant species presented. The sampling strategy is better accomplished by stratified sampling, and must consider obtaining several acquisitions during the phenological cycle to assess the optimum time of the year for separating the species. The aim of separability analysis is to estimate ranges of spectral variability within species and the spectral similarity between species.

Airborne Imaging Spectroscopy

The data required from airborne imaging spectroscopy to finally generate a map of plant species that covers all the study area is a geocoded image with all the bands transformed to ground reflectance. The airborne imaging spectroscopy operator has a key role in delivering this part of the process and must use an accurate radiometric, spectrally and geometrically calibrated airborne imaging spectrometer.

(i) An airborne flight campaign comprises several flight lines designed to cover the study area, constrained by imagery acquisition requisites such as the spatial resolution, flight time and date. In this sense, HYperspectral REmote Sensing in Europe specific Support Actions (HYRESA) establishes a user requisites model that determines the local surveillance area (Reusen et al. 2007). The plant spectral library could help to indicate the best time of the year for maximum separability among species, but it is also important to take into account that higher solar elevation angles correspond with high SNR imagery and a better capacity for discrimination. For mission planning, it is important to remember that the number of flight lines needed to cover the survey area will increase with spatial resolution.

(ii) In general, the operator implements the geometric and radiometric algorithms in a processing and archiving facility (PAF), to integrate an operational workflow that automates the process to transform all the flight lines of the entire campaign. This facility incorporates all the calibrations and auxiliary parameters required. Methods of direct georeferencing rely on high precision position and attitude measurements using an onboard Global Position System and Inertial Navigation System, the bundle adjustment parameters obtained in a geometric calibration flight, and a high-resolution digital elevation model. Radiometric corrections include the calibration coefficients to transform the digital values to at-sensor radiance, and an atmospheric compensation method to obtain ground reflectance. In this sense, the atmospheric compensation methodology could be empirical, such as the Empirical Line Correction (Smith and Milton 1999), or physically-based on radiative transfer models such as MODTRAN (Berk et al. 2006).

(iii) Data quality is an intrinsic property that evaluates the reliability of acquired data (International Organization for Standardization (ISO) 2003). In this sense, airborne spectroscopy imagery must be evaluated against the proposed user requisites. Typically, a georeferenced pixel must be processed with a geolocation error of less than two pixels and the accuracy of reflectance values within 5% (Biesemans et al. 2007). To achieve this quality, we recommend comparison with ground truth data, ground control points (*i.e.*, crossroads) for georeferenced verification, and field spectra of comparable surfaces (*i.e.*, bare soil) for reflectance evaluation. ISO 19157:2013 "*Geographic Information – Data quality*" establishes the principles to describe the quality of geographic data.

Species Mapping Generation

Once the surface reflectance mosaic of the study area has been generated from the airborne imagery and endmembers have been extracted from the spectral libraries, the plant species map can be generated. The user organization plays a key role in the generation of this map.

- In the most widely used commercial remote sensing image processing software, LSU algorithms are implemented by default. The hyperspectral imagery must be prepared by reducing the number of bands by ACP or MNF, which can also be implemented within commercial software. Likewise, endmembers must be selected and prepared by applying the same reduction algorithms as those applied to the imagery. The simple LSU algorithm is restricted to models in which only one spectrum is allowed for each endmember. For this reason, this model does not incorporate the natural variability in scene conditions (i.e., the same material could have different spectral responses). The spectral response of a plant species could be very variable due to the differences in its components and structure. Thus, a spectral unmixing algorithm, such as the MESMA, is more suitable for mapping plant species. MESMA allows the selection of multiple endmembers for each endmember class and incorporates natural variability.
- The outcome of the LSU algorithm is fraction abundances imagery for each introduced endmember; in our case, the fraction cover for each plant species. Accuracy assessment is an integral part of the information extracted from remotely sensed data, since thematic information always contains errors. The accuracy assessment can be performed using the plant cover measurement acquired during the field survey.

Practical Case: Mapping Shrublands Species of Doñana National Park

Study Site

Doñana National Park (DNP) is located on the south-western coast of Spain. It is one of the most important wetlands in Europe (García Novo and Marín Cabrera 2005), and was recognized as a UNESCO World Heritage Site in 1995. The climate at Doñana is Mediterranean sub-humid and has a well-defined seasonality, with mild and wet winters and dry and hot summers. The mean annual precipitation is 550 mm, with rainfall displaying a sharp seasonality, being mostly concentrated between October and March (wet season) and almost absent between June and August (dry season). Three main ecosystems have been traditionally been distinguished in DNP: inland marshes, mobile sand dunes, and stabilised sand dunes. This study was carried out in the stabilised dunes of the Doñana Biological Reserve (DBR), the core area of the DNP (Fig. 2).

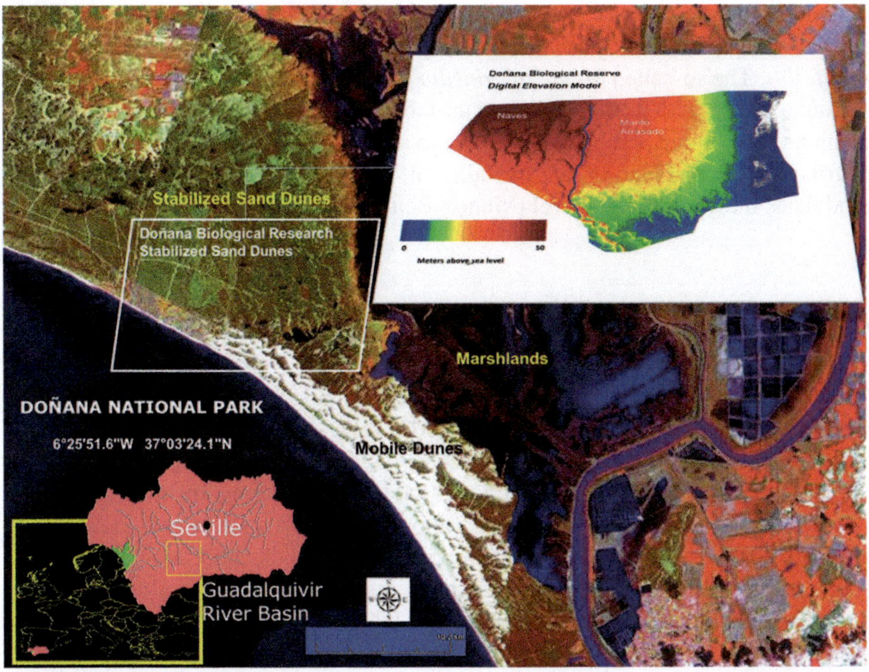

Fig. 2 Overview of Doñana National Park (SW, Spain). *The marked area* is the stabilized dunes ecosystem of Doñana Biological Reserve. *The sub-image* shows Digital Elevation Model of study area

Shrub Communities

Stabilised sands exhibit a rolling topography (see Fig. 2) due to the old dune morphology that has been colonized by vegetation in response to changing groundwater supply. There are three large vegetation zones in the stabilised dunes of DBR: a higher zone dominated by xerophytic shrub (*Naves*), a lower zone dominated by hygrophytic shrub (*Manto Arrasado*), and the transitional grasslands (Vera). The present vegetation is a remnant of the original Juniper woodlands (*Juniperus phoenicea subs turbinata*), stone pine plantations (*Pinus pinea*) and grasslands but mostly a mosaic of three scrubland communities.

Three main types of scrub community are now found on the stabilised sands, depending on the depth of the water table: *Monte Blanco* (Xerophytic sites) occurs on the crests of ancient dunes where the soil water table in summer was deeper than 4 m, and always lies more than 3 m below the soil surface. The community is dominated by *Juniperus phoenicea, Halimium commutatum, Halimium halimifolium, Rosmarinus officinalis*, *Stauracanthus genistoides* and *Cistus libanotis*. On the other hand, *Monte Negro* (Hygrophytic sites) is located in depressions, where the water table in summer rarely lies more than 1 m below the soil surface and where temporary flooding occurs in winter. This plant community is dominated by *Erica scoparia, Erica ciliaris, Calluna vulgaris, Ulex minor, Myrtus communis and Cistus salvifolius*. The so-called *Monte Intermedio* (Mesic sites) is located on the slopes of the dune ridges with a transitional water table depth and no surface flooding. The community is dominated by *Halimium halimifolium* and *Ulex australis*. The spatial distribution of scrubland is determined, at all scales, by the rolling topography that modulates the groundwater level (Muñoz-Reinoso and Novo 2005).

INTA AHS System

The Spanish National Institute for Aerospace Technology (INTA) owns and operates the Airborne Hyperspectral Scanner (AHS) (De Miguel et al. 2014). AHS is an airborne line-scanner imaging spectrometer manufactured by ArgonST (formerly Sensytech Inc.) that covers spectra from 0.45 to 12.8 microns inside atmospheric windows with 80 bands. It is installed onboard the INTA's aircraft (CASA C-212) and is integrated with INS/GPS Applanix POS-AV 414. A calibration and navigation equipment, an auxiliary ground instrumentation and a specific Processing and Archiving Facility (PAF) together form the INTA AHS system. INTA offers this system as a technological service to public institutions or commercial companies, and has performed more than 60 flight campaigns since 2004. The main characteristics of AHS are provided in Table 1.

AHS is considered to be a "generalized" type of sensor that acquires in all atmospheric windows of the optic spectral region. The FWHM in the SWIR region is 15 nm, but 30 nm in the VNIR. As the VNIR region is important for vegetation studies, we carried out a study on the AHS mapping capacity prior to the mapping (Jimenez et al. 2007) showing that the AHS has sufficient radiometric and special resolving power to map Doñana's shrub species.

Table 1 The main characteristics of the AHS sensor

PORT	Spectral coverage (μm)	n° of bands/FWHM (nm)	λ/Δλ (minimum)
Port 1	0.43 > 1.03	20/28	16
Port 2A	1.55 > 1.75	1/200	8
Port 2	2.0 > 2.54	42/13	150
Port 3	3.3 > 5.4	7/300	11
Port 4	8.2 > 12.7	10/400	20

FOV/IFOV: 90°/2.5 mrad
Scan rates: 12.5, 18.75, 25, 35 revolutions per second (pixel 7 to 2 m)
Digitization precision: 12 bits to sample the analog signal, with gain level from ×0.25 to ×10
Two controllable thermal black bodies within the field of view

Doñana Shrub-Species Mapping

The shrub communities that populate the stabilised sand ecosystem are a very important habitat of the DNP. DBR develops important research and conservation programs underpinned by a deep knowledge about both the habitats and the ecosystem. In this sense, the actual location of shrub species is required for monitoring of ecological process and management activities.

The main purpose of the Doñana Biological Station (DBS) is to perform biological research in the DNP and in this case to liaise with INTA to carry out hyperspectral flights over the DNP. Following the guidelines of the protocol illustrated in section "Plant species mapping", the procedure for mapping the dominant species of the Doñana's shrublands with the AHS sensor is as follows:

(i) As the user organization, the DBS provided the auxiliary data required: a digital topography map at 1:25.000 spatial scale, an ecological map at 1:40.000 spatial scale, a habitat map at 1:50.000 spatial scale, and a DEM at 10 m spatial resolution.

(ii) For measuring the cover of plant species, we distributed stratified plots by elevation: first by separating the *Manto Arrasado* and *Naves* zones, and then the upper and lower parts of dunes. The location of the plots was constrained by the local topography and often situated near pathways due to inaccessible areas of very dense vegetation. The size of the plots was 30 m × 30 m and was determined by the resolution of the AHS imagery. The cover of plant species was measured using the line intercept method (Canfield 1941), whereby three parallel transects were sampled inside each plot.

(iii) The spectral reflectance curves of the five dominant species (*E. scoparia, H. halimifolium, U. australis, R. officinalis and S. genistoides)* were measured in several field spectroscopy campaigns, with the primary aim of generating a plant spectral library and extracting the species endmembers. We selected more than 15 individuals for each species (not-randomly) covering the ranges of the Leaf Area Index (LAI measured in m^2/m^2) ranges present. Targets were marked and measured in both seasons. The protocol that we followed is described in Jiménez and Díaz-Delgado (2015). The field spectroradiometer used was the ASD FieldSpec3® (Analytical Spectral Devices, Boulder, CO, USA), which collects energy using a fiber optic with the option of adapting a fore optic lens. It has a spectral range from 350 to 2500 nm with a spectral resolution of 3 nm and a sampling interval of 1.4 nm for the VNIR (350–1000 nm) and 10 m and 2 m for the SWIR-1 (1000–2500 nm) and SWIR-2 (1750–2500 nm) spectral regions respectively. In parallel to spectral measurements, we took hemispherical canopy photographs for each plant with the aim of estimating the LAI. A 180 degree photograph was taken from beneath looking with a NIKON FC-E8 fisheye lens adapted to a Nikon 4000 Coolpix digital camera. We took the photographs shortly before sunset following the protocols for field acquisition outlined by Chen and Cihlar 1995.

(iv) Airborne INTA-AHS flight campaigns were conducted on 28th September 2005 (dry season) and 28th April 2008 (wet season). For the 2005 flight campaign, we acquired two flight lines in a north-east to south-west direction and in the direction of the solar principal plane. The flight altitude was 2743 m above sea level, which gave a pixel size of 6.5 m. For the 2008 flight campaign, we needed three flight lines with the same properties to cover the stabilised sand ecosystem as the solar azimuth was larger. All of the flight lines were acquired using an integrated Applanix POS/AV navigation system, which relies upon a GPS/INS for accurate determination of the instrument position and orientation.

(v) The flight lines were processed with INTA-PAF to generate a mosaic of georefenced ground reflectance over the stabilised sand ecosystem for each date. The sensor radiometric and spectral calibration obtained at INTA facilities encompasses the conversion to digital numbers to at-sensor radiance (in μW/m^2 sr nm units). The imagery was directly georeferenced by PARGE software (Schläpfer and Richter 2002) using the GPS/INS values during the flight, the bundle adjustment parameters calculated in the corresponding year, and the DEM. The atmospheric correction was implemented using the ATCOR-4 software (Richter and Schläpfer 2002), which performs a Look-Up Table (LUT) with the code MODTRAN-5 (radiative transfer model). This compensates for the atmospheric effect in relation to flight altitude, aerosol type, visibility, and water vapor content on a per-pixel basis. Furthermore, ATCOR-4 performs the correction of adjacency effect on a per-pixel basis. Mosaics for the study area can be generated using remote sensing commercial software, including the Exelis Visual Information Solutions (ENVI).

(vi) Ground reflectance mosaics generated for both flight campaigns were evaluated for data quality. Radiometric accuracy is a function of the sensor calibration and atmospheric correction applied. The reflectance values obtained in the INTA-AHS imagery processing were evaluated using ground reflectance data acquired with field spectroscopy in the sand dunes in Doñana. The geometric accuracy was estimated using ground control points extracted from digital cartography.

(vii) A Linear Spectral Unmixing Model was applied to the image mosaics using the endmembers derived in the spectral libraries. MESMA unmixes each pixel using different combinations of potential endmembers and was implemented in the commercial Visualization & Image Processing for Environmental Research (VIPER) software (Roberts et al. 2007). For vegetation mapping studies, the recommended approach is that every pixel in the images can be modeled by a linear combination of three land-cover types (Roberts et al. 1998): photosynthetic vegetation (Veg), non-photosynthetic vegetation (Litter), substrate (soil), and a shade component (Shade) that is typically also present in all pixels. The mixture model that describes Doñana's shrubland is:

$$\rho_{pixel}(\lambda) = \sum \left[F_{veg} \bullet \rho_{veg}(\lambda) + F_{soil} \bullet \rho_{SOIL}(\lambda) + F_{litter} \bullet \rho_{litter}(\lambda) + e(\lambda) \right]$$

where ρ pixel is the reflectance of the pixel, F and ρ are the cover fraction and the reflectance of each endmember, respectively, and e is the error.

(viii) In Fig. 3, the procedure of spectral unmixing is shown schematically. The AHS imagery was dimensionally reduced using MNF, and the spectral reflectance of each pixel was extracted. Identifying a high quality set of reference or image endmembers has been defined as a critical stage of mixture modeling. MESMA incorporates a number of approaches for identifying those spectra that are most representative of a specific class, such as the Endmember Average RMSE (root mean squared error) (EAR) (Roberts et al. 2007): the endmembers that produce the lowest RMSE within a class are selected. In Fig. 3, the groups of representative spectra for each endmember for each date are illustrated.

(ix) The accuracy assessment for plant cover values obtained with airborne imaging spectroscopy was performed by applying a regression analysis on the plant cover fractions in the field plots.

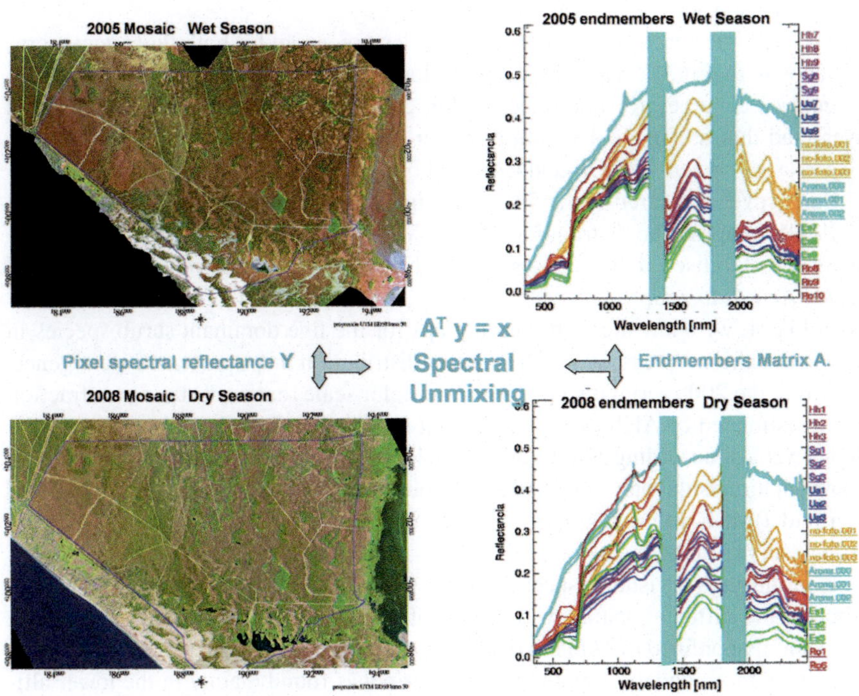

Fig. 3 Linear spectral unmixing elements for Doñana's shrub species mapping with INTA-AHS airborne hyperspectral system

Results

The spectral library for the five dominant shrub species is presented in Jiménez and Díaz-Delgado (2015), and shows the different spectral response for the species in both seasons taking into account LAI range variation along the ecosystem. Looking at the shrub spectra presented in Fig. 3, it is immediately apparent that all species appear to be very similar, as might be expected for a group of species that populate an environment with very similar conditions. Despite the increased variability between the species in the dry season, the comparison of the plant spectral curves between seasons revealed that most marked differences were the reduction in the water absorption bands in the NIR and the increase reflectance in lignin and cellulose absorption bands (centered at 2100 and 2310 nm) for the dry season. These wavebands are important for species discrimination. In Jiménez and Díaz-Delgado (2015), the separability test was applied over spectra in both seasons to determine whether significant differences existed in intra-species variability and the degree of inter-species similarity. T-tests comparing the spectral libraries for both seasons indicated that the dry season had slightly better levels of discrimination than the wet season ($p < 0.05$), where all the species are in the same physiological state. In the estimation of similarity index between species, low and significant values were found for *E. scoparia, H. halimifolium,* and *R. officinalis*, with very high values found between the legume species *U. australis* and *S. genistoides.*

The airborne spectroscopy imagery mosaics processed for both flight campaigns are shown in Fig. 3. The ground reflectance was obtained with less than 5% reflectance accuracy for both dates, when compared to the field spectra acquired over sand dunes. The geolocation error for both imagery mosaics was below two pixels, as estimated with Ground Control Points located in the field using GPS.

Plant cover measured in the 50 ground-truth plots was used to assess the classified AHS images. The distribution of the plots covers all of the altitude variability along the stabilised dunes ecosystem, with plots situated in both the *Naves* and *Manto Arrasado* zones.

In Fig. 4, we show the distribution maps for the five dominant shrub species in the RBD stabilized ecosystem. The species distribution maps presented were generated from the 2008 aerial imagery. The map color scale represents the cover fraction values estimated by AHS, with this ranging from red for lower values to orange for higher values, according to dominant species. The background image for the map representation is the band 15 of AHS. In addition, in Fig. 4, we present the correlation and RMSE values obtained for the AHS imagery cover fractions from the ground survey on both dates.

Even through a visual inspection of the species distribution maps, it is easy to describe the different spatial distribution of species. *E. scoparia* and *U. australis* were mainly confined to *Manto Arrasado* areas whilst *H. halimifolium* and *R. officinalis* were more widely dispersed. *E. scoparia* was found mainly in the lower altitude and wetter areas of the Manto Arrasado, with up to 90 % cover recorded. *H. halimifolium* shows a broad spatial distribution, colonizing both the *Naves* and *Manto Arrasado* areas and attained a fraction cover of up to 80 % in the *Manto*

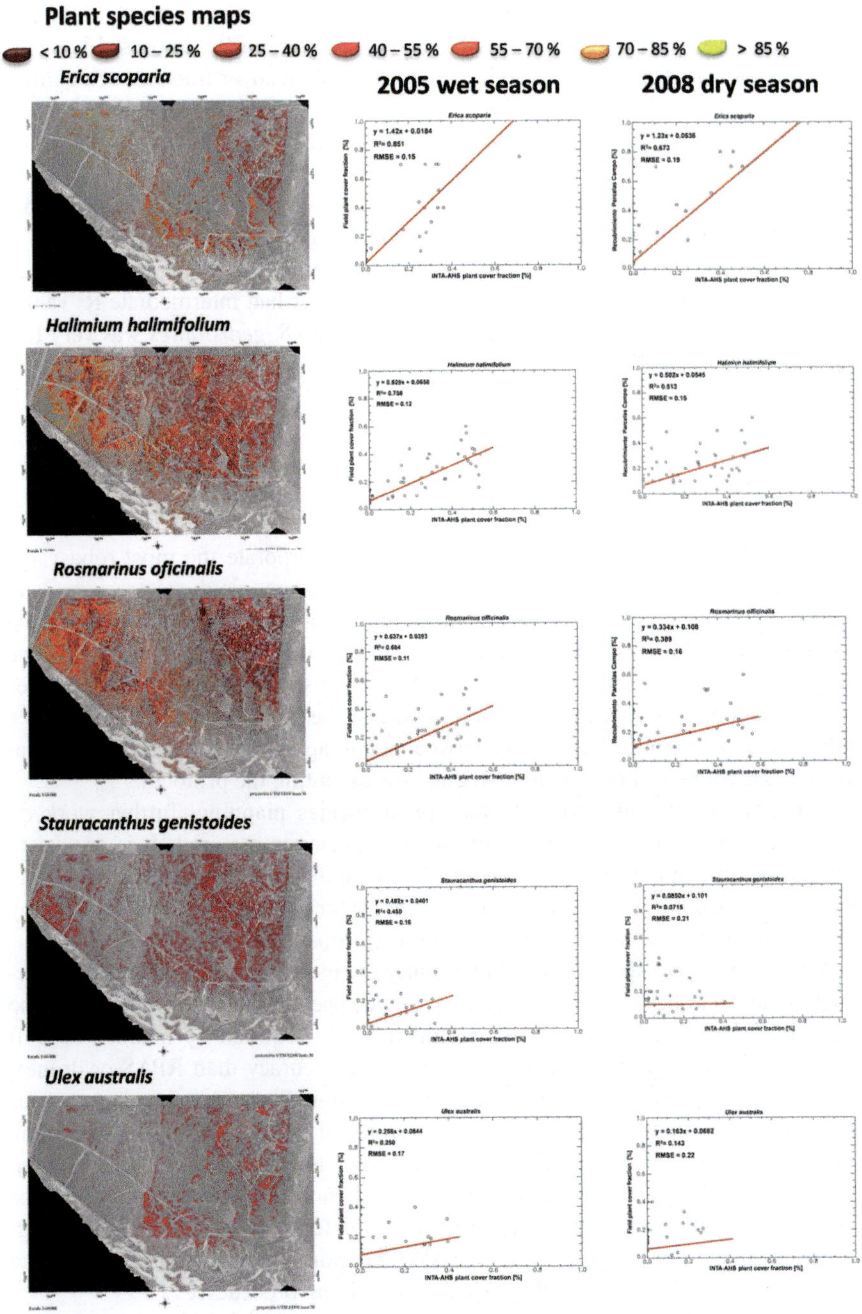

Fig. 4 Doñana Biological Reserve dominant shrub species distribution maps

Arrasado. *R. officinalis* also shows such a broad spatial distribution, but it reached its maximum cover in the *Naves* area. Meanwhile, *S. genistoides* colonised both the *Naves* and *Manto Arrasado* zones but with a lower plant cover fraction, achieving a maximum of 60% in the *Naves*. Finally, *U. australis* was distributed mostly across intermediate altitude areas of the *Manto Arrasado* and reached a maximum fraction cover of 60%.

We used the correlation coefficient and RMSE values between the resulting species fraction cover from the AHS imagery and the ground-truth fraction cover to asses the overall accuracy. There was a strong correlation for *E. scoparia* with high R^2 values ($p < 0.05$). *H. halimifolium* and *R. officinalis* had intermediate R^2 values ($p < 0.05$). Finally, the correlation for *U. australis* and *S. genistoides* was very low and with no statistical significance.

Discussion

The protocol presented in this study intended to incorporate the most robust and widely used procedures in airborne imaging spectroscopy, field spectroscopy, and spectral unmixing to establish a standard protocol for mapping plant species. The spatially-explicit distribution maps of the plant species generated by the airborne imaging spectroscopy increase the knowledge of the spatial spread of each species and its relation with the ecological processes and the perturbation that takes place within the ecosystem. Collaboration between the imaging sensor operator organisation and user organisation was fundamental for executing the protocol.

Although spectral unmixing enables plant species mapping, further work on intra-species variability and similarity among species is required. It is crucial to identify the time of the year with maximum separability among species. There is no standard protocol for developing a spectral library for plant species, but the work by Jiménez and Díaz-Delgado (2015) takes the first steps towards this.

Airborne imaging spectroscopy is currently the most important source of hyperspectral data, and has the best capacity to provide species mapping within and also surrounding protected areas. Hyperspectral imagery acquired by manned aircraft has less uncertainty in radiometric and geometric accuracy than RPAS or drones, which, could represent a future systems for monitoring changes in plant species distribution. In relation to the imaging spectrometer, the image spatial resolution must be adapted to the size and cover for each vegetation type, taking into account the number of flight lines needed to cover the study area. Furthermore, although the VNIR region is the most important spectrum region for vegetation studies, an imaging spectrometer that records within the SWIR region is recommended to enhance plant species discrimination and the spectral unmixing procedure.

Nowadays, MESMA is the most appropriate LSU algorithm to cope with plant species mapping, primarily due to the incorporation of multiple endmembers and algorithms for endmember optimization. However, at present, it is only implemented in the VIPER tool application (Roberts et al. 2007) and not used widely in commercial remote sensing software.

The INTA-AHS hyperspectral system performed well in mapping the dominant shrub species of DNP. The spatial resolution of 2–7 m (see Table 1) was sufficient to deal with the range of canopy sizes (typically 1–2 m) encountered in the shrublands of DNP. The study area was covered with two flight lines in the 2005 campaign and three flight lines in the 2008 campaign, which allowed the generation of a geometrically and radiometrically correct image mosaic for each year.

The plant species distribution maps obtained with the INTA-AHS hypespectral corroborates the knowledge and work of previous expert researchers, in terms of the spatial distribution and plant cover fraction. Furthermore, the spatially-explicit plant species distribution detected gave more insights into the spatial distribution of shrub species that were not possible using traditional vegetation survey methods. For example, *E. scoparia* covered a larger area and the fraction cover was greater in *Naves* than published in previous studies and this was similar to the distribution of *R. officinalis* in the *Manto Arrasado* habitat.

Having the potential to map and monitor the changes in species composition over time would help us to detect, monitor, measure and predict increases or decreases in biological diversity, as well as help to predict the impacts of these changes on ecosystem functions (Trochet and Schmeller 2013). In this work, two flight campaigns were carried out in two different years, which represented an excellent opportunity to show the potential of imaging spectroscopy for mapping and monitoring activities. In Fig. 5, we present a sub-scene of the stabilised dunes at DBR and show

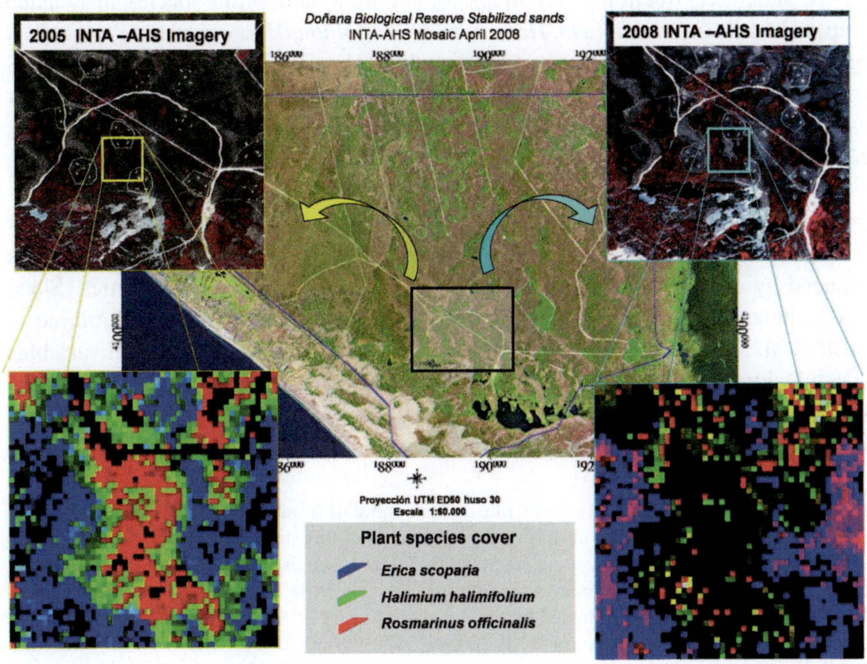

Fig. 5 Monitoring the spatial distribution of the shrubland in Doñana Biological Reserve. INTA-AHS sub-scenes and species map for 2005 a 2008 flight campaigns

the two species maps generated through the two airborne campaigns. A variety of management activities were taking place in the stabilised dune ecosystem as a consequence of the plan for the conservation of Iberian lynx (*Lynx pardinus*), which included clearing selected shrubland regions. Those cleared areas can be seen in the INTA-AHS sub-scenes in Fig. 5. In the 2005 map, *E. scoparia*, *H. halimifolium*, and *R. officinalis* are present in the area, whereas in the 2008 map *R. officinalis* totally disappears and *H. halimifolium* has a reduced presence. This change was not due to clearance activities and instead to a perturbation episode in the winter of 2007. Very low temperatures led to frosts that prevented the species from colonising the crest of the dunes.

Conclusions

In this work, the most widely used procedures in airborne imaging spectroscopy, field spectroscopy for plant endmember generation, and spectral unmixing were selected to establish a standard protocol for plant species mapping. The spatially-explicit distribution of plant species generated by the airborne imaging spectroscopy maps help to obtain a better knowledge of the spatial distribution of species and their relationships with the ecological processes and perturbations that take place on the ecosystem.

We generated fraction cover maps for the dominant shrub species in Doñana National Park (*Erica scoparia, Halimium. halimifolium, Ulex australis, Rosmarinus officinalis* and *Stauracanthus genistoides)* in 2005 and 2008 using the INTA-AHS image data. The comparison between the species cover and the ground survey cover estimates in the images, indicated a correlation that was higher for *E. scoparia, H. halimifolium* and *R. officinalis*, but lower for *U. australis* and *S. genistoides*.

Being able to map and monitor the spatial distribution of vegetation at the species level, and identify changes in population in space and time can help to understand the shifts in species composition in response to environmental changes induced by climate and land use change as well as other human activities. Since 2002, there has been a long term integrated ecological monitoring programme in DBR, with a substantial increase in the monitoring of relevant ecological variables (Díaz-Delgado 2010). The mapping protocols based on airborne and ground-based spectroscopy can be integrated within monitoring programs in the DNS but also in other protected areas and their surrounds.

Acknowledgments The authors would like to thank the staff from Doñana Biological Station and Biological Reserve who helped us during different stages of this study. The authors also recognise the knowledge of the staff of Instituto Nacional de Técnica Aeroespacial (INTA) Remote Sensing Group for performing the airborne campaigns and the geometric correction of the AHS images. The airborne campaigns were funded by the Spanish Ministry of Science and Innovation through the research projects HYDRA (No.CGL2006-02247/BOS) and HYDRA2 (CGL2009-09801/BOS). We appreciate the contribution of Antonio Pou for his helpful direction in the work.

References

Asner, G.P.: Biophysical and biochemical sources of variability in canopy reflectance. Remote Sens. Environ. **64**, 234–253 (1998). doi:10.1016/S0034-4257(98)00014-5

Asner, G.P., Knapp, D.E., Kennedy-Bowdoin, T., et al.: Invasive species detection in Hawaiian rainforests using airborne imaging spectroscopy and LiDAR. Remote Sens. Environ. **112**, 1942–1955 (2008). doi:10.1016/j.rse.2007.11.016

Berk, A., Anderson, G.P., Acharya, P.K., Bernstein, L.S., Muratov, L., Lee, J., Fox, M., Adler-Golden, S.M., Chetwynd, J.H., Hoke, M.L., Lockwood, R.B., Gardner, J.A., Cooley, T.W., Borel, C.C., Lewis, P.E., Shettle, E.P.: MODTRAN 5: Update. In: Proceeding of SPIE 6233, Algorithms and technologies for multispectral, Hyperspectral, and Ultraspectral Imagery XII, 62331F (2006, May 8). doi:10.1117/12.665077

Biesemans, J., Sterckx, S., Knaeps, E., et al.: Image processing workflows for airborne remote sensing. In: Proceedings of the 5th EARSeL Workshop on Imaging Spectroscopy, pp. 1–14. EARSeL, Bruges, Belgium (2007)

Boardman, J., Kruse, F., Green, R.: Mapping target signatures via partial unmixing of {AVIRIS} data, pp. 23–26. JPL Publication, Pasadena (1995)

Canfield, R.H.: Application of the line interception method in sampling range vegetation. J. For. **39**, 34–40 (1941)

Chen, J.M., Cihlar, J.: Quantifying the effect of canopy architecture on optical measurements of leaf area index using two gap size analysis methods. IEEE Trans. Geosci. Remote Sens. **33**, 777–787 (1995). doi:10.1109/36.387593

Cobo, D., Sánchez, E., García, M.P.: Flora y vegetación. In: Parque nacional de doñana, Vicente García Canseco, pp. 109–174. Canseco Editores, S.L., Talavera Reina (2002)

De Miguel, E., Fernandez-Renau, A., Prado, E., et al.: A review of INTA AHS PAF. EARSeL eProceedings, pp. 20–29 (2014)

Díaz-Delgado, R.: An integrated monitoring programme for Doñana natural space: the set-up and implementation. In: Conservation Monitoring in Freshwater Habitats, pp. 325–337. Springer, Dordrecht/Heidelberg/London/New York (2010)

Fyfe, S.K., Dekker, A.G.: Seagrass species: are they spectrally distinct? In: IGARSS 2001. Scanning the Present and Resolving the Future. Proceedings. IEEE 2001 International Geoscience and Remote Sensing Symposium (Cat. No.01CH37217), vol. 6, pp. 2740–2742 (2001)

García Novo, F., Marín Cabrera, C. (eds.): Doñana: agua y biosfera. Confederación Hidrográfica del Guadalquivir, Sevilla (2005)

Goetz, A.F.H.: Making Accurate Field Spectral Reflectance Measurements, vol. 2012. ASD Inc, Boulder (2012)

Green, R.O., Eastwood, M.L., Sarture, C.M., et al.: Imaging spectroscopy and the airborne visible/infrared imaging spectrometer (AVIRIS). Remote Sens. Environ. **65**, 227–248 (1998). doi:10.1016/S0034-4257(98)00064-9

Hooper, D.U., Chapin, F.S., Ewel, J.J., et al.: Effects of biodiversity on ecosystem functioning: a consensus of current knowledge. Ecol. Monogr. **75**, 3–35 (2005). doi:10.1890/04-0922

Hruska, R., Mitchell, J., Anderson, M., Glenn, N.F.: Radiometric and geometric analysis of hyperspectral imagery acquired from an unmanned aerial vehicle. Remote Sens. **4**, 2736–2752 (2012). doi:10.3390/rs4092736

International Organization for Standardization (ISO).: Geographic information—Metadata. International organization for standardization, Geneva (2003)

Jiménez, M., Díaz-Delgado, R.: Towards a standard plant species spectral library protocol for vegetation mapping: a case study in the Shrubland of Doñana National Park. ISPRS Int. J. Geo-Inf. **4**, 2472–2495 (2015). doi:10.3390/ijgi4042472

Jiménez M, Díaz-Delgado R, Vaughan P, et al.: Airborne hyperspectral scanner (AHS) a priori mapping capacity simulation for the Doñana Biological Reserve shrublands. In: Michael Schaepman, Shunlin Liang, Nikée Groot and Mathias Kneubühler (eds.) Proceedings of the

ISPRS Working Group VII/1 Workshop ISPMSRS'07: "Physical Measurements and Signatures in Remote Sensing.", pp. 334–340. Davos, Switzerland (2007)

Jiménez, M., González, M., Amaro, A., Fernández-Renau, A.: Field spectroscopy metadata system based on ISO and OGC standards. ISPRS Int. J. Geo-Inf. **3**, 1003–1022 (2014). doi:10.3390/ijgi3031003

Kalacska, M., Sanchez-Azofeifa, G.A., Rivard, B., et al.: Ecological fingerprinting of ecosystem succession: estimating secondary tropical dry forest structure and diversity using imaging spectroscopy. Remote Sens. Environ. **108**, 82–96 (2007). doi:10.1016/j.rse.2006.11.007

Kamal, M., Phinn, S.: Hyperspectral data for mangrove species mapping: a comparison of pixel-based and object-based approach. Remote Sens. **3**, 2222–2242 (2011). doi:10.3390/rs3102222

Kaufmann, H., Segl, K., Guanter, L., et al.: Environmental Mapping and Analysis Program (EnMAP) – Recent Advances and Status. In: IGARSS 2008–2008 IEEE International Geoscience and Remote Sensing Symposium, pp. IV-109-IV-112 (2008)

Kerekes, J.P., Baum, J.E.: Full-spectrum spectral imaging system analytical model. IEEE Trans. Geosci. Remote Sens. **43**, 571–580 (2005). doi:10.1109/TGRS.2004.841428

Keshava, N., Mustard, J.F.: Spectral unmixing. IEEE Signal Process. Mag. **19**, 44–57 (2002). doi:10.1109/79.974727

Leutner, B.F., Reineking, B., Müller, J., et al.: Modelling Forest α-diversity and floristic composition — on the added value of LiDAR plus hyperspectral remote sensing. Remote Sens. **4**, 2818–2845 (2012). doi:10.3390/rs4092818

Manakos, I., Manevski, K., Petropoulos, G., et al.: Development of a spectral library for mediterranean land cover types (2010)

Manevski, K., Manakos, I., Petropoulos, G.P., Kalaitzidis, C.: Discrimination of common Mediterranean plant species using field spectroradiometry. Int. J. Appl. Earth Obs. Geoinf. **13**, 922–933 (2011). doi:10.1016/j.jag.2011.07.001

McGwire, K., Minor, T., Fenstermaker, L.: Hyperspectral mixture modeling for quantifying sparse vegetation cover in arid environments. Remote Sens. Environ. **72**, 360–374 (2000). doi:10.1016/S0034-4257(99)00112-1

Miao, X., Gong, P., Swope, S., et al.: Estimation of yellow starthistle abundance through CASI-2 hyperspectral imagery using linear spectral mixture models. Remote Sens. Environ. **101**, 329–341 (2006)

Milton, E.J., Schaepman, M.E., Anderson, K., Kneubühler, M., Fox, N.: Progress in field spectroscopy. Remote Sens. Environ. **113**, 92–109 (2009)

Möckel, T., Dalmayne, J., Prentice, H.C., et al.: Classification of grassland successional stages using airborne hyperspectral imagery. Remote Sens. **6**, 7732–7761 (2014). doi:10.3390/rs6087732

Muñoz-Reinoso, J.C., Novo, F.G.: Multiscale control of vegetation patterns: the case of Doñana (SW Spain). Landsc. Ecol. **20**, 51–61 (2005). doi:10.1007/s10980-004-0466-x

Nagendra, H., Rocchini, D., Ghate, R., et al.: Assessing plant diversity in a dry tropical Forest: comparing the utility of Landsat and Ikonos satellite images. Remote Sens. **2**, 478–496 (2010). doi:10.3390/rs2020478

Oakley, K.L., Thomas, L.P., Fancy, S.G.: Guidelines for long-term monitoring protocols. Wildl. Soc. Bull. **31**, 10001003 (2003)

Pfitzner, K., Bollhöfer, A., Esparon, A., et al.: Standardised spectra (400–2500 nm) and associated metadata: An example from northern tropical Australia. In: 2010 IEEE International Geoscience and Remote Sensing Symposium, pp. 2311–2314 (2010)

Reusen, I., Holzwarth, S., Nieke, J., et al.: Towards an improved access to hyperspectral data across Europe (HYRESSA). In: Proceedings of the 5th EARSeL SIG IS workshop on IMAGING SPECTROSCOPY: innovation in environmental research, EARSeL, Bruges, Belgium (2007)

Richter, R., Bachmann, M., Dorigo, W., Muller, A.: Influence of the adjacency effect on ground reflectance measurements. IEEE Geosci. Remote Sens. Lett. **3**, 565–569 (2006). doi:10.1109/LGRS.2006.882146

Richter, R., Schläpfer, D.: Geo-atmospheric processing of airborne imaging spectrometry data. Part 2: atmospheric/topographic correction. Int. J. Remote Sens. **23**, 2631–2649 (2002). doi:10.1080/01431160110115834

Roberts, D.A., Gardner, M., Church, R.M., et al.: Mapping chaparral in the Santa Monica Mountains using multiple endmember spectral mixture models. Remote Sens. Environ. **65**, 267–279 (1998)

Roberts, D.A., Halligan, K., Dennison, P.E.: VIPER Tools user manual (Version1.5). University of California at Santa Barbara, 91P (2007)

Salisbury, J.W.: Spectral Measurements Field Guide. Earth Satellite Corp, Chevy Chase (1998)

Schaepman, M.E., Ustin, S.L., Plaza, A.J., et al.: Earth system science related imaging spectroscopy—an assessment. Remote Sens. Environ. **113**(Supplement 1), S123–S137 (2009). doi:10.1016/j.rse.2009.03.001

Schläpfer, D., Richter, R.: Geo-atmospheric processing of airborne imaging spectrometry data. Part 1: Parametric orthorectification. Int. J. Remote Sens. **23**, 2609–2630 (2002). doi:10.1080/01431160110115825

Schlapfer, D., Nieke, J., Itten, K.I.: Spatial PSF nonuniformity effects in airborne pushbroom imaging spectrometry data. IEEE Trans. Geosci. Remote Sens. **45**, 458–468 (2007). doi:10.1109/TGRS.2006.886182

Schmeller, D.S.: European species and habitat monitoring: where are we now? Biodivers. Conserv. **17**, 3321–3326 (2008). doi:10.1007/s10531-008-9514-1

Schowengerdt, R.A.: Remote Sensing. Models and Methods for Image Processing. Academic, Burlington (2007)

Silvestri, S., Marani, M., Marani, A.: Hyperspectral remote sensing of salt marsh vegetation, morphology and soil topography. Phys. Chem. Earth Parts ABC. **28**, 15–25 (2003). doi:10.1016/S1474-7065(03)00004-4

Smith, G.M., Milton, E.J.: The use of the empirical line method to calibrate remotely sensed data to reflectance. Int. J. Remote Sens. **20**, 2653–2662 (1999). doi:10.1080/014311699211994

Somers, B., Asner, G.P.: Hyperspectral time series analysis of native and invasive species in Hawaiian rainforests. Remote Sens. **4**, 2510–2529 (2012). doi:10.3390/rs4092510

Stefano, P., Angelo, P., Simone, P., et al.: The PRISMA hyperspectral mission: science activities and opportunities for agriculture and land monitoring. In: 2013 IEEE International Geoscience and Remote Sensing Symposium – IGARSS, pp. 4558–4561 (2013)

Trochet, A., Schmeller, D.: Effectiveness of the Natura 2000 network to cover threatened species. Nat. Conserv. **4**, 35–53 (2013). doi:10.3897/natureconservation.4.3626

Ustin, S., Zarco-Tejada, P., Jacquemoud, S., Asner, G.P.: Remote sensing of environment: state of the science and new directions. In: Remote Sensing for Natural Resource Management and Environmental Monitoring, 3rd edn, pp. 679–729. Wiley, Hoboken (2004)

Vaughan, H., Brydges, T., Fenech, A., Lumb, A.: Monitoring long-term ecological changes through the ecological monitoring and assessment network: science-based and policy relevant. Environ. Monit. Assess. **67**, 3–28 (2001). doi:10.1023/A:1006423432114

Warner, T.A.: Remote sensing analysis: from project design to implementation. In: Manual of Geospatial Science and Technology, 2nd edn, pp. 301–318. CRC Press. Taylor & Francis Group, Boca Raton (2010)

Zomer, R.J., Trabucco, A., Ustin, S.L.: Building spectral libraries for wetlands land cover classification and hyperspectral remote sensing. J. Environ. Manag. **90**, 2170–2177 (2009). doi:10.1016/j.jenvman.2007.06.028

Part III
Species-Driven Remote Sensing and New Technologies Studies

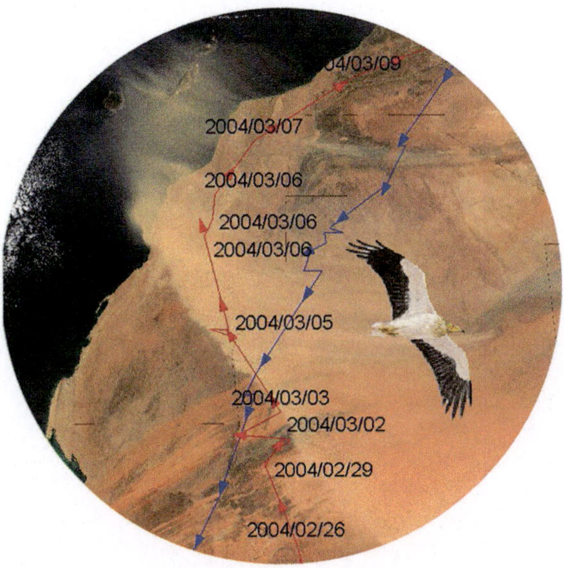

Mapping the Distribution of Understorey *Rhododendron Ponticum* Using Low-Tech Multispectral UAV Derived Imagery

Abigail Sanders

Abstract Invasive species, such as *Rhododendron ponticum*, are an issue of global concern as they are a threat to biodiversity, ecosystem provisioning services and economies and act as a reservoir for emerging pathogen dispersal. *R. ponticum* is a Class A prohibited plant under Schedule 9(2) of the UK wildlife and countryside act, and removal costs in the UK in 2010 were approximately £86 million. This figure is not inclusive of removal costs on private land, an area where *R. ponticum* is spreading rapidly, and which amounts to 74% of the UK woodland. *R. ponticum* is a sporulation host which is tolerant to the highly virulent tree-killing root-fungus pathogen Phytothphera, in particular *P. ramorum* and *P. kernoviae*. These were introduced in 2003 and have now become widespread and are a particular threat to species such as the English Oak (*Quercus robur*). Limited access and high costs of aerial and satellite imagery restrict the progress of research into the spatial distribution patterns of *R. ponticum* and thus their effective removal strategy. Unmanned Aerial Vehicles (UAV) are currently used for conservation and agriculture but are as yet prohibitively expensive. Using commercially available action cams (Sony and Mobius: open source camera adapted for Infrared detection using a Rosco Red Filter) and a DJI Phantom 3 professional quadcopter a multispectral image composed of 7 bands (440–800 nm range) was generated for sites containing invasive rhododendron and other non-target species. Mosaicking efficiency reduced the integrity of images but was rectified through georeferencing and Polynomial (spline) transformation. *R. ponticum* was shown in pairwise comparisons with non-target species to show greatest separability at 540 ($p < 0.05$) and least at 550 ($p > 0.05$) in multivariate analysis. Intra-species difference between two *R. ponticum* populations growing in differing locations was also significant at $p = 0.01$ in the same pairwise comparison. Ivy was not significantly different to larch and *R. ponticum* in all but the 540 and 750 bandwidths ($p > 0.01$). Seven band spectral signatures were all shown to be significantly different at $F(1, 42) = 6.795$, $p < 0.001$. This was confirmed during supervised classifications of the composite spectral checkerboard,

A. Sanders (✉)
Computing, Engineering and Science Department. Glyntaff, Trefforest Campus, University of South Wales, South Glamorgan, UK
e-mail: abigailsanders1234@gmail.com

© Springer International Publishing AG 2017

R. Díaz-Delgado et al. (eds.), *The Roles of Remote Sensing in Nature Conservation*,
DOI 10.1007/978-3-319-64332-8_9

created from samples of the site mosaics which yielded 84% classification accuracy in ERDAS® imagine v2016 (64bit). The 'noisy' mosaic dataset yielded a lower overall classification accuracy of 64 %. The techniques demonstrated show the potential for low-tech, UAV derived, multispectral imagery to aid land management. Spectral separability of rhododendron shows a high level of potential for mapping its distribution using this method. However, more research should be conducted to streamline the process and reduce the potential sources of error.

Keywords Rhododendron • *R.ponticum* • Forestry management • *Phytophthera* • UAV • Invasive species • Multispectral pixel-based classification

Introduction

Invasive species are currently classified, according to the International Union for the Conservation of Nature (IUCN), as the second largest threat to biodiversity worldwide and by the UK Forestry Commission (FC) as a major obstruction to woodland regeneration (Blackburn et al. 2014; Edwards 2006). *Rhododendron ponticum* is one such species, which has become invasive across much of the northern hemisphere; in Europe, the UK and North America (Edwards 2006; Taylor et al. 2013).

Invasive species are traditionally surveyed manually from the ground (Lillesand et al. 2014), with the surveyor undertaking species identification and estimating percentage cover (Brinker and Minnick 2012). Typically, the scope of the exercise would depend on the resources available to the organisation undertaking the survey (Lillesand et al. 2014). Various studies attest to the spatial inaccuracy of manual mapping compared with GIS spatial mapping. However, numerous studies also demonstrate the inability of remote sensing to perform as well as a ground surveyor, in accuracy and breadth of useable information (Powell et al. 2004; Burrough 1986). This study is an example of how remote sensing can enhance the resources available to land managers by providing an alternative means of surveying difficult to access areas to help quantify removal costs. Therefore, one aim of this research was to use low-cost commercially available UAV and cameras to mimic more advanced equipment, which is prohibitively expensive. A second aim was to use the output to identify understorey *R. ponticum* and assess the feasibility and accuracy of this equipment for practical nature conservation purposes.

Species Characteristics and Invasion

R. ponticum has complex global spatial distribution patterns arising from far-reaching radiations and advantageous hybridizations (Milne et al. 2003). The earliest fossilised evidence, found in North America, is carbon dated to around 68

million years ago, during the last stage of the Cretaceous period (Yan et al. 2015; Irving and Hebda 1993). This attests to the durability of this genus to survive and adapt to the changes at the Earth's surface (Milne et al. 2004; Irving and Hebda 1993). *R. ponticum* is an aggressive, allelopathic, evergreen coloniser that threatens biodiversity. Once established, it is difficult and costly to remove (Edwards 2006). *R. ponticum* thrives in milder, wet climatic conditions, favouring acidic ground but capable of colonising very poor soil and cliff faces, growing from 5 to 8 m in height and 4 to 6 m in width to form dense impenetrable areas within seven to ten years (Edwards 2006; FC 2016; Milne and Abbott 2000).

Furthermore, and not insignificantly, *R. ponticum* is a key sporulation host for the emerging virulent tree-killing pathogen *Phytophthora* (*ramorum and kernoviae* species). This is a high concern in the UK due to its rapid spread and diversification, which has enabled it to infect a range of host tree species not previously known to be vulnerable, including the English Oak (*Quercus robur*) (Purse et al. 2013; FC 2016). Infection control dictates that entering known infection sites, as well as areas of surrounding forest, is kept to a minimum; this limits the potential for data collection in these priority areas. (FC 2016).

Remote Sensing Background

Light is measured in waves and the distance between each wave is typically measured in nano-meters (nm). The visible spectrum ranges from 400–700 nm, for human eyes, and can be broken down into groups of colour called 'bands': blue (B), green (G) and red (R) in order of increasing wavelength (nm). The wavelengths just outside the visible spectrum are ultraviolet (UV) light at <400 nm and infrared (IR) at >750 nm. Green plants reflect predominantly in the green band (around 500 nm) and the Near Infrared (NIR, 700 + nm) portion of the spectrum (Heege 2015). Differentiating plants from non-photosynthetic organisms and objects can be achieved using many wavelengths but differentiating between plant groups and species most often requires reference to the visible green and NIR bands green and infrared bands (Heege 2015).

Buschmann et al. (2012) determined, using high resolution 4-band (RGB + NIR) imagery, that leaf reflectance is categorised by three basic parameters: Leaf pigment content, i.e. the absorption of chlorophylls and carotenoids in the pigment protein complexes; leaf tissue structure, i.e. the size of aerial interspaces between cells (influence leaf optical properties); and structure of the leaf surface, e.g. waxes and hairs.

Traditional cameras split the visible spectrum into three bands, as standard: Red, Green, Blue (RGB, 700–400 to respectively). They are designed in this way to mimic our own colour chemicals, present in our eyes, to then produce images that mimic how we, as humans, visualise our surroundings (Pfundel et al. 2008). What we and our traditional cameras. However, we and our traditional cameras cannot see the energy reflected by an object in all wavelengths. Multispectral cameras typically have sensors with numerous discrete bands, greater than just the visible region, such as infrared, whereas hyper-spectral cameras have sensors with continuous sensitivity across the full optical spectrum (350–2500 nm). These types of sensor are available but expensive. Although likely to become a more affordable component of UAV technology in the future, their cost is currently prohibitive. Traditional cameras can be adapted to sense IR radiation by removing or replacing the inbuilt IR filters inside the internal sensor mechanism (Infragram 2016). Therefore, by using multiple traditional cameras (RGB) and IR adapted cameras, and combining the output, we should be able to produce a multispectral image capable of identifying plants such as *R. ponticum*.

Taylor et al. (2013) showed, using lab-based radiometry of more than 500 leaves of *R. ponticum* that the spectral reflectance of leaves from three non-target species; (namely Beech (Fagus sylvatica), Holly (Ilex aquifolium) and Laurel (Prunus laurocerasus)), were significantly different ($p < 0.05$) from *R. ponticum* for all wavelengths except 450 and 460 nm, which yielded no significant difference (Taylor et al. 2013). The bands used for statistical analysis were at 490, 550, 610, 1040 and 1490 nm, based on the absolute reflectance for the key wavelengths shown in previous studies. These reflectance wavelengths relate to the following specific pigments and compounds: Chlorophyll a, phycoerythrin, phycocyanin, oils and cell sugars respectively. The same study also found that the spectral characteristics of *R. ponticum* leaves differ in leaf size and spectral plasticity over four habitat types (garden, oakwood, pinewood and lakeside), with these differences caused by variations in the spectral intensity of specific leaf pigments.

Taylor et al.'s study demonstrates some of the major opportunities and pitfalls for this technology; although differentiation between key non-target species is possible, under controlled conditions, the plasticity of inter species variation could render the information highly site specific. In our study, and contrary to the aims of many UAV studies, site-specific technology would not make this tool less effective if the output is sufficiently accurate to justify the processing time and costs.

The extent to which a remote sensing tool can be used to determine the coverage and distribution of understorey *R. ponticum* is governed by the capacity of a sensor in terms of spatial resolution, spectral sensitivity, extent covered, and temporal frequency. Each remote sensing tool has benefits and limitations associated with these four factors, often where a higher capacity in one factor is attained to the detriment of another attribute, for example; lower spatial resolution for greater coverage in satellite imagery. Monitoring large-scale changes in forest cover, for example, would not require a pixel size of 1 cm so using a lower resolution image would be more appropriate: this highlights that there is no 'one size fits all' model for remotely sensed data. Furthermore, although satellite imagery has the coverage, temporal frequency and the spectral sensitivity to achieve the basic objective of 'identifying understorey *R. ponticum* from surrounding vegetation', the spatial resolution will not be sufficiently high to distinguish individuals between bare branches of the upper canopy. Aerial imagery can overcome this but has a repeatedly high cost for flights and the open-source imagery currently available for our study area lacks the seasonal and temporal frequency (summer 2006 and 2014) to be used to identify understorey *R. ponticum*. UAV imagery does, however, have the potential to provide a platform for overcoming these limiting factors if the low cost can be balanced with sufficient image quality. Rather than a tool to replace ground surveys, we propose that UAV imagery complements the work of ground survey teams, providing a better overview to help map their target habitats.

Study Area

The site was chosen by the Gwaun catchment area Invasive Species Department (IAS) of National Resources Wales (NRW), Pembrokeshire (see Fig. 1). The essential criteria was that there was *Rhododendron* infestation in and around woodland in a site where we could obtain consent to fly without UK Central Aviation Authority (CAA) licencing.

Trecwn Valley is a 400,000 h decommissioned Royal Navy Armaments Depot in north-west Pembrokeshire, Wales (OSGB 1920,2330). There is a traditional herringbone format along the valley giving access to 58 cavern-based storage chambers. These entrances were planted with *R. ponticum* and *Prunus laurocerasus* (Cherry Laurel) to prevent aerial detection: these species were selected because of their fast growth, dense canopy and evergreen cover. Both of these species have now developed into widespread infestations in the area. Critically, an area of Larch within the site was issued with a *Phytophthora* Health Notice in 2014.

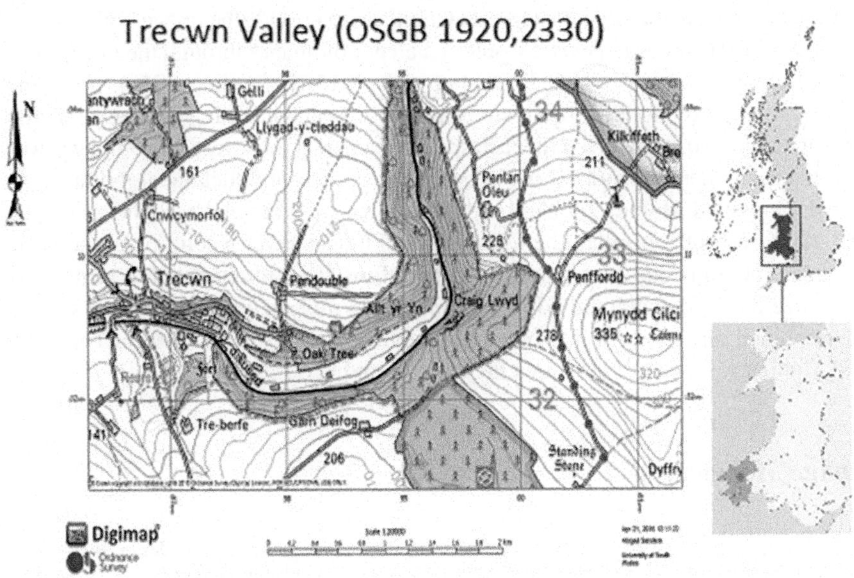

Fig. 1 Trecwn Valley (OSGB 1920,2330), Pembrokeshire, Wales (Digimap 2016)

Methods

Cameras, UAVs and Flight Design

The two RGB cameras were the on-board DJI Phantom 3 professional camera and a Sony ActionCam, which varied in their spectral response curves despite both having Sony sensors. The 4 k (4 Megapixels) resolution DJI camera provided the highest resolution images. NIR spectra were recorded using a Mobius ActionCam with a Rosco RedFire (#19) adapted filter. All three cameras were mounted on the UAV and collected image data simultaneously for each flight. The images collected were treated as one survey, pairing the images from the three cameras together so that environmental conditions were the same for each camera. The cameras took an image every 2 s, or 5 m in any direction (x, y, z), and each flight survey was taken at a constant altitude. It was decided that the altitude for flights should be 5 m above the tallest tree within the flight area. Tree height was measured with an Abney level and altitude ranged from 19–27 m between flights.

Image Pre-processing: Distortion Removal

The on-board DJI camera was gimbal mounted with automatic orthorectification, which ensured that the camera was always perpendicular to the ground. Neither the Sony nor Mobius Action Cams were gimbal-mounted. Both Action-cams were fitted (as standard) with fisheye lenses, which produced a distortion at the edges of an image while compressing a wider field of view into the sensor. This had to be reciprocally distorted to remove the edge misrepresentation and to provide the image with geospatial accuracy. This was achieved by putting batches of images through the image editing software package Adobe© Photoshop, CC 2015 (Photoshop 2016), using the 'batch processing' and 'distortion removal' functions and inputting the degree of distortion required to be removed for each specific camera. Only every 1 in 3 images was kept to ensure there was not excessive overlay to confuse mosaicking algorithms.

The Sony and Mobius cameras had smaller sensor sizes and resolutions (1.9 K and 2.3 K respectively) than the DJI (4 K) and thus a lower resolution at the same altitude. The Sony and Mobius mosaics, therefore, were interpolated to bring their pixel size down to match the DJI resolution of 7.8 mm such that the mosaics could be combined into mosaics into one high-resolution multispectral image. The Sony and Mobius were interpolated to a factor 1:2.1 and 1:1.7 respectively.

Image Mosaicking

DJI phantom output images were automatically georeferenced to WGS 84/ UTM Zone 30 N reference system at a pixel size of 7.8 mm/pixel. DJI georeferenced mosaics were automatically created with high spatial accuracy using Pix4D Mapper v.3.0 (Pix4D 2016). The Sony and Mobius output mosaics were generated without spatial referencing, using the automatic batch-process mosaic tool in Adobe Photoshop CC 2015. All of the mosaics were transferred to ERDAS®Imagine v2016(64bit) (ERDAS IMAGINE 2016) mapping software to be spatially aligned with one another and overlaid.

Mosaic Alignment

Non-georeferenced mosaics were manually tied to the spatially accurate DJI mosaic using visually derived Manual Tie Points (MTPs). The Sony and Mobius mosaics were transformed using 3rd order polynomial spline transformation to conserve the exact overlay of MTPs and effectively mimic a georeferenced Ground Control Point (GCP). Examples of MTPs used in this study are Recognisable Visual Points (RVPs), which stand out in an image. Common RVPs are - manhole covers, forked branches, or obvious tree apexes etc. and were easily seen in the three mosaics to tie images together at these locations. A greater number of MTPs will result in more areas accurately overlapped and thus a more precise overlay. MTPs were used for two reasons; there are unlikely to be many feasible locations, which are clearly visible from the air, for accurate GCPs in a large catchment of woodland, and entering the site poses a risk of spreading the infection.

Multispectral Image Generation

We used the manufacturer's spectral response curves to identify the best wavelengths to extract into bands (Sony 2016; Mobius 2016). A 10 nm bandwidth was extracted around the point of highest quantum efficiency % on the y-axis (or the wavelength at which the most light is absorbed by the sensor/ graphical peak – see Fig. 2). The NIR extracted bandwidth was expanded to 50 nm to include a greater portion of NIR, the most important spectral region for plant identification (Buschmann et al. 2012).

We created a spatial model in ERDAS®Imagine v2016 (64bit) mapping software to extract the bands from the three mosaics and recombine these into a 7 band multispectral mosaic (Fig. 3). Bands were ordered into ascending wavelength in order to appropriately retrieve spectral signatures from the features in the image. The final mosaic had the following discrete bands centred at 440, 450, 540, 550, 600, 610 (all 6 with 10 nm width), and 750 nm (50 nm width).

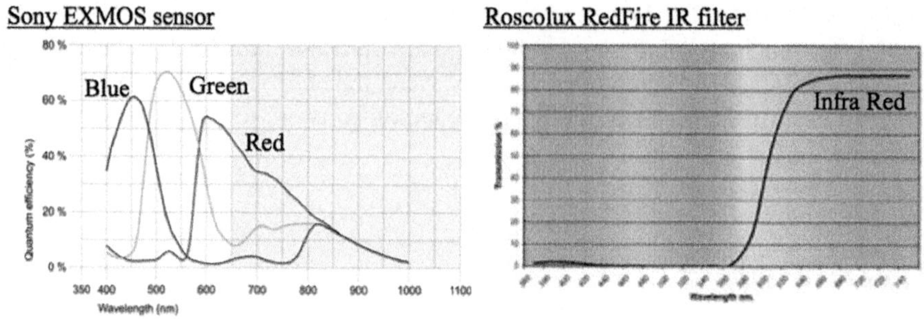

Fig. 2 Manufacturers spectral response curves for; the DJI on-board camera (SONY®EXMOS sensor) and Mobius IR adapted Actioncam with RoscoLux Red fire filter (SONY 2016; Infragram 2016)

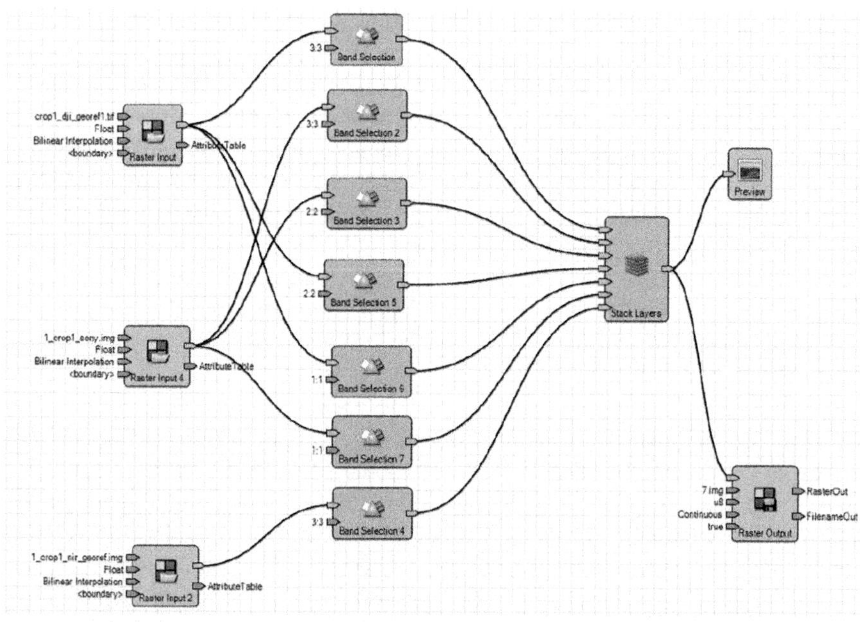

Fig. 3 The Spatial Model used in SpatialModeller, ERDAS®imagine mapping software v2016 (64bit) used to separate bands from the different cameras and re-stack the layers in order of ascending wavelength (ERDAS 2016)

Spectral Checkerboard

When overlapping separate high resolution images of the woodland, there are likely to be areas where the overlapping pictures do not exactly align for all pixels, particularly around object edges. Edge effects are imperfect overlaps of the layers that cause the spectral signature of a pixel found at the edge to show incorrect reflectance spectra. For example, a signature at the edge may have 4 bands showing a leaf

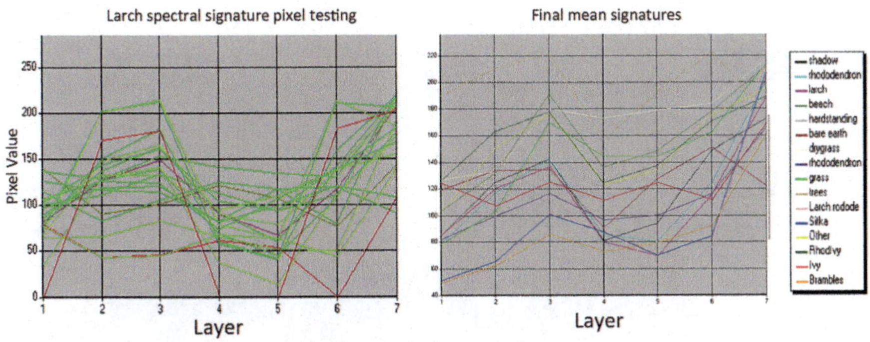

Fig. 4 Spectral signatures of the target species: (*Left*) sample pixels were tested and signatures with any bands showing 0 pixel value removed (*red*). All other signatures (*green*) were averaged to a mean signature (*pink*). The final mean signatures were used for maximum likelihood supervised classification

signature and 3 bands showing the ground signature where they do not perfectly overlap in the mosaic. This signature therefore is not representative of the spectral signature of the plant species. To achieve a more accurate signature for any plant species in the composite mosaic, an average signature across a number of pixels is taken and the mean signature used in classifications. To reduce the likelihood of incorrect pixel signatures being attributed to a species in the signature editor, a spectral checkerboard was created.

The spectral checkerboard was built by taking 100 pixel samples (0.78 m/pixel) from areas of accurate overlap, away from edges and for every target species in the mosaics. Accurate overlap was defined where positive Digital Numbers (DNs) for each band for 20 pixels, chosen at random, were found within each square (Fig. 4). Samples were taken from the mosaics at different locations; 5 samples for target species, 3 samples for non-target species. Target species found at numerous sites, such as *Rhododendron*, had samples taken from all sites because orientation, location and neighbouring vegetative composition has been shown in the literature to cause spectral signature variation (Taylor et al. 2013).

The samples were taken and reprojected to be spatially adjacent in a virtual workspace in ERDAS®imagine v2016 (64bit) (Fig. 5). Removing and recombining samples in this way facilitated the creation of the equivalent of a standardised spectral lab in which to train the classifier to identify individual spectral signatures under controlled conditions.

Pixel-Based Digital Classification and Statistical Analysis

The ERDAS® Imagine v2016 (64bit) signature editor was used on the checkerboard to create a signature file to be used for a maximum likelihood supervised classification. Means of signatures from five randomly selected pixels, within each square,

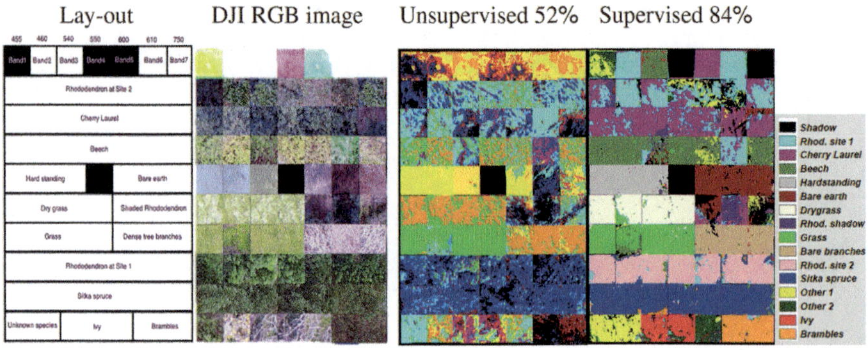

Fig. 5 Checkerboard created in virtual workspace for spectral signature analysis and classification

produced the final signature for each sample. The classification accuracy was calculated using the ERDAS® Imagine Accuracy Assessment feature and 200 randomly distributed points across the chosen 7-band image for both unsupervised and supervised classifications.

We conducted a multivariate analysis between full spectral signatures of samples to determine whether they were significantly different using all 7 bands (spectral separability). Non-target species' signatures were compared to *R. ponticum*, using multivariate analysis, to determine which bands showed the significant variation within samples.

Results

Distortion Effects and Removal

Non-gimbal mounted cameras suffered from vibration wave-distortions and blur on the image outputs, meaning that many pictures were unsuitable for mosaicking. Identifying errors in the images and removing the distortion was time consuming: gimbal mounting and the removal of fisheye lenses would fully resolve these issues.

Spectral Analysis

R. ponticum was shown in inter-species pairwise comparisons to show the greatest spectral separability at 540 ($p < 0.05$) and least at 550 ($p > 0.05$) using multivariate analysis. Intra-species difference between two populations growing in different locations was also significant ($p = 0.01$) in the same pairwise comparison. Ivy was not significantly different to Larch and *R. ponticum* in all but the 540 and 750 bandwidths ($p > 0.01$). The MANOVA between species' full spectral signatures were shown to all be universally significantly different at $F(1, 42) = 6.795$, $p < 0.001$. This result confirms that the analysis using 7 band signatures is sufficient to distinguish species in principle.

Digital Classification

The spectral analysis checkerboard, as a tool in itself, provided a platform for standardising comparison of samples from multiple areas and was a valuable output from this study as a stand-alone utensil (Fig. 5). Both supervised and unsupervised classifications of the checkerboard displayed a clear visual distinction between many of the samples, which was confirmed by the overall accuracy assessment: 52% unsupervised and 84% supervised (Fig. 5).

When a classification was conducted on 7-band test sections of the site the accuracy assessment yielded 62% overall classification accuracy for 100 pixel samples using the ERDAS® imagine internal accuracy assessment feature. Cherry Laurel, Beech and Ivy were all incorrectly identified as *R. ponticum* or *vice versa*. Beech foliage would not be present at the optimal time of survey (midwinter) and Cherry Laurel is an additional reservoir host for Phytophthora so both were considered as target species. Ivy, however, was a non-target species of concern as it would be expected to be present in most natural and semi-natural woodlands across the UK and would confound results (Fig. 5). Sample sizes for the checkerboard were greater than the coverage of Ivy (0.78 m) and this could have contributed to errors.

Discussion

Feasability Study

This study demonstrates the feasibility of creating multispectral mosaics from commonly sourced equipment. The separate bands of the intra-band variation were significantly different. Variability in significance between 540 nm and 550 nm showed the importance of this region for the identification of *Rhododendron* from other species, which confirmed other results in the literature (Taylor et al. 2013). This information enables future studies to be targeted for more rapid optimisation of tools for identification: a greater bandwidth in the 550 nm, for example, could have made this region incorrectly significant due to overlap into the 540 nm bandwidth (Fig. 6).

The accuracy levels seen in both the checkerboard and the mosaic demonstrate the ability of a pixel-based supervised classification of 7-band composites to identify groups of species, including the *R. ponticum* and Cherry Laurel group, both of which are reservoir hosts of *Phytophthora*.

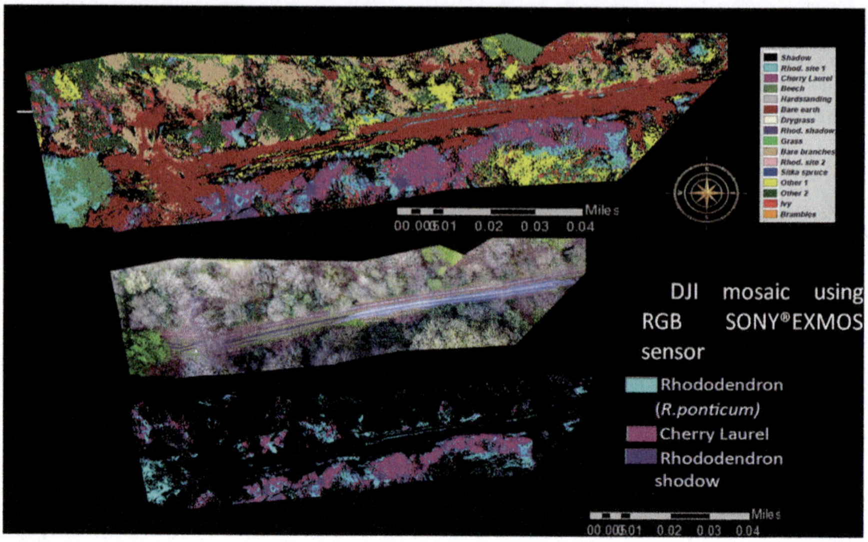

Fig. 6 Final supervised classification maps of test site for broadleaved woodland at Trecwn Valley, Pembrokeshire. These mosaics were created in ERDASimagine. (*top*) This mosaic was classified using an supervised classification based on the spectral signature set created from the checker-board. (*middle*) This mosaic shows the RGB bands from the DJI camera output. (*bottom*) Isolated in this image are the averaged signatures for; *R. ponticum*, Cherry laurel and the shadow areas for *R. ponticum*. This image highlights the signature similarity with Beech, which would not be present at the optimal time of survey

This was conducted using ArcMap and ERDASimagine, which are both high cost programmes, so this cannot reflect the feasibility of using open source software, such as QGIS, though this software is rapidly advancing to meet the needs of users globally. For practical application, even at 62% accuracy, distribution and temporal changes could be reliably identified provided that Ivy could be reliably separated. Application of Object Based Image Analysis (OBIA), based on leaf morphology and positioning, has the potential to separate both Ivy and Laurel from *R. ponticum*: this has increased vegetation classification accuracy in the literature (Hernando et al. 2012).

Replicate flights, at other sites and at different times in the year, would quantify the temporal and inter-site feasibility of either sharing site spectral data or conserving site-specific datasets for achieving the highest accuracy.

Limitations and Cost-Benefit Analysis

UAVs cannot be flown in winds >25 m/s nor in rain which limits their use during winter months in the UK: this coincides with the optimal time for *R. ponticum* mapping (Anderson and Gaston 2013). Although aerial flights have a higher tolerance for unfavourable weather conditions, they are also restricted, and charges may still apply if the job is outsourced and a flight is cancelled due to the weather (Whitehead and Hugenholtz 2014). Owning a UAV enables conservation managers to produce their own imagery as regularly as their schedule and weather permits, without costs beyond capital expenses, training, and electricity for charging.

General limitations include the availability of sites for consensual flight; liability and insurance cover is needed for any moving vehicle used for charity or business purposes, which covers UAV use for nature conservation (CAA 2017). The only directly relevant EU-wide regulation, at the time of writing, is that any Small Unmanned Aircraft (SUA) weighing 150 kg or less must have adequate insurance cover (CAA 2017). Permissions on private land are at the discretion of the land-owner. This study, for example, was flown without certifications on private land. The Civil Aviation Authority (CAA), which controls UK airspace, provides the relevant rules and regulations within the Civil Aviation Publication (CAP) 393 Air Navigation Order on topics such as negligence and privacy (articles 138 and 167 respectively) (CAA 2017). The CAA provides training for professionals, which many common and large-scale land managing bodies require for permission to fly (CAA 2017; National Trust 2013). Commercial businesses are required to have met the requirements of the UKCAA's CAP722: Chapter 4: Civil UAS Remote Pilot Competency to undertake commercial flights (CAA 2017). The cost of a professional course to gain a licence for flights over common land is between £900–£1300, a one off cost for training and examination, and £112–240 for commercial permissions, which must be renewed annually at a reduced cost (CAA 2017).

The increasing demand and rapid development of drone technology means owning a drone with a high resolution gimbal mounted camera can cost £800 upwards and can be flown using any smartphone (PCMag 2016). At an altitude of < 30 m, a UAV can produce a resolution of 0.1–10 cm depending on the camera sensor size (Lillesand et al. 2014). The cost of hiring a professional to collect drone imagery for a day can be £500–1000. However, an UAV is roughly the same cost (Wood et al. 2015). Small aircraft and helicopters are often used for surveying large-scale forestry, particularly *Phytophthora* sites, and allow an on-board surveyor to 'spot' areas of interest by eye However, these flights do not yet automatically include accurate spatial mapping (FC 2016). There is no literature on the actual accuracy of these types of survey as there is no alternative. The ability of many quad copters to hover, remain in a stationary position, and use GPS tracking systems allows rapid investigation of points of interest. This was demonstrated during this study, where indicators of *Phytophthora* infection (dead upper branches) could be clearly seen in real time and from stored images. Furthermore, as UAVs use electricity, the use of a drone can reduce the CO_2 emissions of any organisation with a sustainability portfolio that relies on aerial surveys.

Future Directions for Nature Conservation

The spectral information generated using this technique can be carried forward into future spectral analysis using multi and hyper-spectral tools by targeting and disseminating spectral signatures of species. Critically, this method can be used to inform management decisions now, and remains relevant through technological advances.

Of importance is finding novel, useful UAV procedures within nature conservation that will drive the industry to develop to fill this niche. Developing protocols for using UAVs in nature conservation is central for shaping policy on UAV use in the UK. It is also highly relevant within conservation organisations, particularly in the light of changing legislative frameworks.

References

Anderson, K., Gaston, K.J.: Lightweight unmanned aerial vehicles will revolutionize spatial ecology. Front. Ecol. Environ. **11**(3), 138–146 (2013)

Blackburn, T.M., Essl, F., Evans, T., Hulme, P.E., Jeschke, J.M., Kühn, I., Kumschick, S., Marková, Z., Mrugała, A., Nentwig, W., Pergl, J.: A unified classification of alien species based on the magnitude of their environmental impacts. PLoS Biol. **12**(5), e1001850 (2014)

Brinker, R.C., Minnick, R.: The Surveying Handbook. Springer Science & Business Media (2012)

Burrough, P.A.: Principles of geographical information systems for land resources assessment (1986)

Buschmann, C., Lenk, S., Lichtenthaler, H.K.: Reflectance spectra and images of green leaves with different tissue structure and chlorophyll content. Israel J. Plant Sci. **60**(1–2), 49–64 (2012)

CAA Civil Aviation Authority: Unmanned Aircraft System Operations in UK Airspace – Guidance. [online] Available at: https://publicapps.caa.co.uk/docs/33/CAP%20722%20Sixth%20Edition%20March%202015.pdf. Accessed 26 Aug 2016 (2017)

Digimap.: – OS MasterMap Topography Layer [TIFF geospatial data], Scale 1:20000, Tile(s): Trewcn Valley, Updated: Apr 2016, Ordnance Survey, Using: EDINA Digimap Ordnance Survey Service, + URL + , Downloaded: September 2015. (2016)

Edwards, C.: Managing and controlling invasive rhododendron. Forestry Commission Practice Guide, Forestry Commission, Edinburgh. pp. i–iv + 1–36 (2006)

ERDAS IMAGINE: USA: Hexagon Geospatial (2016)

FC, Forestry Commission. Control of Invasive Rhodoendron Ponticum (Forest Research). [online] Forestry.gov.uk. Available at: https://www.forestry.gov.uk/fr/infd-6tvh7x (2006). Accessed 4 Mar 2016

Heege, H.J.: Precision in Crop Farming. Springer (2015)

Hernando, A., Arroyo, L.A., Velázquez, J., Tejera, R.: Objects-based image analysis for mapping Natura 2000 habitats to improve forest management. Photogramm. Eng. Remote Sens. **78**(9), 991–999 (2012)

Infragram P.. Public Lab Open Source Software: Near-Infrared Camera. [online] Publiclab.org. Available at: https://publiclab.org/wiki/near-infrared-camera (2016). Accessed 9 Feb 2016

Irving, F., Hebda, R.: Concerning the origin and distribution of Rhododendrons. Am. Rhododendron Soc. **47**(3) (1993)

Lillesand, T., Kiefer, R.W., Chipman, J.: Remote Sensing and Image Interpretation. Wiley, New York (2014)

Milne, R.I.: Phylogeny and biogeography of Rhododendron subsection Pontica, a group with a tertiary relict distribution. Mol. Phylogenet. Evol. **33**(2), 389–401 (2004)

Milne, R.I., Abbott, R.J.: Origin and evolution of invasive naturalized material of Rhododendron ponticum L. in the British Isles. Mol. Ecol. **9**(5), 541–556 (2000)

Milne, R.I., Terzioglu, S., Abbott, R.J.: A hybrid zone dominated by fertile F1s: maintenance of species barriers in Rhododendron. Mol. Ecol. **12**(10), 2719–2729 (2003)

Mobius. Mobius [online] Mobius Actioncam. Available at: Accessed 24 Apr 2016

National Trust. Tackling invasive rhododendron in the south Lakes. [online] Available at: https://www.nationaltrust.org.uk/wray-castle/features/tackling-invasive-rhododendron-in-the-southlakes (2013). Accessed 19 May 2016

Nel, J.L., Le Maitre, D.C., Nel, D.C., Reyers, B., Archibald, S., Van Wilgen, B.W., Forsyth, G.G., Theron, A.K., O'Farrell, P.J., Kahinda, J.M.M., Engelbrecht, F.A.: Natural hazards in a changing world: a case for ecosystem-based management. PLoS One. **9**(5), e95942 (2014)

PCMag.: Drones. [online] PCMag UK – Consumer Electronics Reviews, Ratings & Comparisons|Robotics & Automation – Products – Drones. Available at: http://uk.pcmag.com/drones-1 (2016). Accessed 19 Aug. 2016

Pfündel, E.E., Agati, G., Cerovic, G.Z.: Optical properties of plant surfaces. Biology of the plant cuticle. **3**, 216–239 (2008)

Photoshop. USA: Adobe (2016)

Pix4D Mapper pro: Switzerland: Pix4D (2016)

Powell, R.L., Matzke, N., De Souza, C., Clark, M., Numata, I., Hess, L.L., Roberts, D.: Sources of error in accuracy assessment of thematic land-cover maps in the Brazilian Amazon. Remote Sens. Environ. **90**(2), 221–234 (2004)

Purse, B.V., Graeser, P., Searle, K., Edwards, C., Harris, C.: Challenges in predicting invasive reservoir hosts of emerging pathogens: mapping Rhododendron ponticum as a foliar host for Phytophthora ramorum and Phytophthora kernoviae in the UK. Biol. Invasions. **15**(3), 529–545 (2013)

Smith, C.S., Byers, J.E., Reichard, S., Randall, J.M., Parker, I.M., Lonsdale, W.M., Atkinson, I.A.E., Seastedt, T.R., Williamson, M., Chornesky, E., Hayes, D.: Directing research to reduce the impacts of nonindigenous species. Conserv. Biol. **16**(3), 630–640 (2002)

Sony: Sony AS20 Action Cam. [online] Available at: https://www.sony.co.uk/electronics/action-cam/hdr-as20/specifications. Accessed 30 May 2016

Taylor, S.L., Hill, R.A., Edwards, C.: Characterising invasive non-native Rhododendron ponticum spectra signatures with spectroradiometry in the laboratory and field: Potential for remote mapping. ISPRS J. Photogramm. Remote Sens. **81**, 70–81 (2013)

Williams, F., Eschen, R., Harris, A., Djeddour, D., Pratt, C., Shaw, R.S., Varia, S., Lamontagne-Godwin, J., Thomas, S.E. and Murphy, S.T., 2010. The Economic Cost of Invasive Non-native Species on Great Britain. CABI Proj No VM10066, pp. 1–99.

Whitehead, K., Hugenholtz, C.H.: Remote sensing of the environment with small unmanned aircraft systems (UASs), part 1: a review of progress and challenges 1. J. Unmanned Veh. Syst. **2**(3), 69–85 (2014)

Wood, C.M., Norton, L.R., Rowland, C.S.: What are the costs and benefits of using aerial photography to survey habitats in 1km squares? NERC Centre for Ecology & Hydrology, 23pp., Lancaster (2015)

Yan, L.J., Liu, J., Möller, M., Zhang, L., Zhang, X.M., Li, D.Z., Gao, L.M.: DNA barcoding of Rhododendron (Ericaceae), the largest Chinese plant genus in biodiversity hotspots of the Himalaya–Hengduan Mountains. Mol. Ecol. Resour. **15**(4), 932–944 (2015)

The Potential of UAV Derived Image Features for Discriminating Savannah Tree Species

J. Oldeland, A. Große-Stoltenberg, L. Naftal, and B.J. Strohbach

Abstract Mapping tree species at the single-tree level is an active field of research linking ecology and remote sensing. However, the discrimination of tree species requires the selection of the relevant spectral features derived from imagery. We can extract an extensive number of image parameters even from images with a low spectral resolution, such as Red-Green-Blue (RGB) or near-infrared (NIR) images. Hence, identifying the most relevant image parameters for tree species discrimination is still an issue. We generated 42 parameters from very high resolution images acquired by Unmanned Aerial Vehicles (UAV), such as chromatic coordinates, spectral indices, texture measures and a canopy height model (CHM). The aim of this study was to compare the relevance of these components for classifying savannah tree species. We obtained very high (5 cm) pixel resolution RGB-NIR imagery with a delta-wing UAV in a thorn bush savannah landscape in central Namibia in April 2016. Simultaneously, we gathered ground truth data on the location of 478 individual trees and large shrubs belonging to 16 species. We then used a Random Forest classifier on single and combined thematic sets of image data, e.g. RGB, NIR, texture and in combination with CHM. The best average overall accuracy was 0.77 and the best Cohen´s Kappa value was 0.63 for a combination of RGB imagery and the CHM. Our results are comparable to other studies using hyperspectral data and LiDAR information. We further found that the abundance of the tree species is crucial for successful mapping, with only species with a high abundance being classified satisfactorily. Diverse ecosystems such as savannahs could therefore be a challenge for future tree mapping projects. Nevertheless, this study indicates that UAV-borne RGB imagery seems promising for detailed mapping of tree species.

J. Oldeland (✉)
Biodiversity, Ecology and Evolution of Plants, Biocentre Klein Flottbek and Botanical Garden, University of Hamburg, Ohnhorststr. 18, Hamburg 22609, Germany
e-mail: jens.oldeland@uni-hamburg.de

A. Große-Stoltenberg
Applied Landscape Ecology and Ecological Planning Institute of Landscape Ecology, Heisenbergstraße 2, Münster 48149, Germany
e-mail: ags@uni-muenster.de

L. Naftal • B.J. Strohbach
Faculty of Natural Resources and Spatial Sciences, Namibia University of Science and Technology, P/Bag, 13388, Windhoek, Namibia
e-mail: wexly4one@gmail.com; bstrohbach@nust.na

© Springer International Publishing AG 2017

R. Díaz-Delgado et al. (eds.), *The Roles of Remote Sensing in Nature Conservation*,
DOI 10.1007/978-3-319-64332-8_10

Keywords Biodiversity monitoring • Drone • GLCM • LAS • Namibia • NDVI • Point cloud • Spectral discrimination

Introduction

Classifying and mapping individual trees is increasingly applied in forestry, urban management and nature conservation. According to a review by Fassnacht et al. (2016), since the year 2000 and particularly from 2010 onwards, there has been a dramatic increase in the number of studies that compare the suitability of different datasets, classifiers, and sensor platforms. However, most studies on the classification of tree species use expensive technology to capture data, e.g. hyperspectral or LiDAR sensors, with only a few studies applying relatively cheap solutions such as UAVs carrying consumer-grade cameras that provide Red-Green-Blue (RGB) or Near-Infrared (NIR) imagery. Furthermore, most studies to date have focussed on temperate or boreal forest ecosystems while Savannah ecosystems, which are relatively rich in tree species, remain understudied.

This chapter evaluates the suitability of image parameters derived from low-cost, UAV-borne, consumer-grade cameras for classifying tree species in a savannah ecosystem. In particular, we aim to test (a) whether savannah tree species can be discriminated successfully with very high resolution UAV imagery, (b) whether RGB or NIR spectral indices perform better, and (c) if a canopy height model can significantly improve the classification. Finally, we discuss the role of a species abundance for it´s potential to be accurately mapped.

Background

Mapping the distribution of tree species using remote sensing means producing a vector or raster layer that contains the information on locations of tree species either at the stand-level or single-stem level. These maps or data sets are valuable in nature conservation, particularly in a biodiversity monitoring context. However, until recently, the most commonly used image data for mapping tree species were from hyperspectral and LiDAR sensors (Fassnacht et al. 2016) which are costly, difficult to preprocess, and require expert knowledge in their analysis. The recent advent of drones, also called Unmanned Aerial Vehicles (UAVs), provides new tools and the opportunity to obtain more spatial detail for tree species mapping, and the use of UAVs for mapping tree species is becoming increasingly popular (Singh et al. 2015; Lisein et al. 2015). UAVs have several advantages over satellite or airborne data. They are extremely flexible in usage, can be scheduled in a very short time interval (e.g. daily or weekly), are easily carried to diverse locations and, unlike satellites, are not limited by clouded skies. The main drawbacks are the limited spatial coverage and a restricted carrying capacity which renders UAVs unsuitable for heavy hyperspectral or LiDAR sensors and thus being restricted to consumer-grade cameras or small multispectral cameras. Only a few studies have evaluated RGB spectral indices for species discrimination, e.g. Rasmussen et al. (2013), who tested the potential of RGB indices for site-specific weed management, Dvořák et al. (2015) used RGB imagery for invasive plant detection, and Rasmussen et al. (2016) studied the performance of RGB indices to measure barley biomass. Hence, the question remains open as to whether very high spatial resolution imagery taken by a standard UAV can successfully discriminate tree species. If so, nature conservation could make use of a very flexible image acquisition platform for monitoring small areas, i.e. covering several square kilometres with a small number of flights. Furthermore, the question how consumer-grade cameras with RGB or a NIR-filter perform in such a task needs to be addressed. Is NIR really necessary or are observations in RGB sufficient?

Most of the studies reviewed by Fassnacht et al. (2016) had two things in common: they used hyperspectral imagery in combination with LiDAR data and were undertaken in temperate or boreal forests. Bunting and Lucas (2006) and Lucas et al. (2008) established the use of CASI and HYMAP hyperspectral data for discriminating tree species in open woodlands and forests in Queensland, Australia, confirming that differences in the mean spectra from crown objects increased the accuracy of discrimination. However, only a few studies set out to classify savannah tree species in southern Africa (Naidoo et al. 2012; Cho et al. 2012; Colgan et al. 2012). These studies classified between six and 15 tree species. Cho et al. (2015) also tested the suitability of very high resolution satellite imagery for this purpose, but only used three out of ten dominant canopy species. It seems that the abundance of a tree species also contributes to its capability for being mapped precisely. We are of the opinion that this issue has not been sufficiently highlighted in the literature (but see comments in Fassnacht et al. 2016).

Study Area

The study was part of the Biodiversity Observatory S05 of the BIOTA Africa project (www.biota-africa.org), which is a cross-country biodiversity monitoring project with a standardized monitoring approach performing monitoring in southern, western, and northern Africa (Jürgens et al. 2012). The observatory is located on the cattle farm Erichsfelde (coordinates: 16.935° E 21.597° S) in central Namibia. The Biodiversity Observatory spans 1 km^2 and is divided into 100 ha from which 20 were selected in the year 2001 for permanent annual monitoring of vegetation and animal diversity. The vegetation monitoring was undertaken within plots of 20 × 50 m, which were situated at the mid-point of a selected hectare. The vegetation consisted of typical Thornbush savanna *sensu* Giess (1998), dominated by *Acacia mellifera* subsp. *detinens* and *Boscia albitrunca*. Other *Acacia* species also occurred, in particular *A. hebeclada* subsp. *hebeclada, A. tortilis, A. reficiens* and *A.*

Fig. 1 Landscape perspective of the thorn bush savannah vegetation on the private cattle-farm Erichsfelde. The tree layer consists mainly *Acacia mellifera* subsp. *detinens* with one larger *Acacia tortilis* in the back. Small shrubs and a dense grass layer leaving some open soil patches characterize the landscape. Picture taken 05. April 2016, by L. Naftal

karroo. The vegetation structure is a typical semi-closed bushland, with a low grass layer, and shrubs of up to 4 m high (Fig. 1). These are interspersed by few trees, between 4 and 8 m in height. Erichsfelde is a private cattle farm of about 13,000 ha that is used extensively for meat production (cattle grazing). Game species, including Oryx and Kudu are also present. The Observatory is not excluded from regular land-use.

Methods

UAV Imagery Acquisition

On 21.03.2016, we acquired an image mosaic with 5 cm ground resolution for the whole Biodiversity Observatory S05, i.e. approximately 1 km^2 (Fig. 2). We covered the area in two flights with an eBee 3 drone (costs ca. 30.000€, SenseFly 2015, Cheseaux-Lausanne, Switzerland). The settings for the flight missions were 70% longitudinal and 60% lateral overlap, with a flying height of 115 m above take-off point. Each image had a width of 160 m and a length of 120 m. The first flight was conducted using a modified Canon S110 where the blue filter was replaced by a NIR filter, recording at 850 nm. This camera also recorded a green band (550 nm) and a red band (625 nm). The second flight was performed using a regular RGB camera (Canon S110), recording at 450 nm (blue), 520 nm (green) and 660 nm (red) (SenseFly 2014). The flights to collect the images took about 30 min each and took place on the same day between 10h45 and 12h00. The first flight (NIR) yielded 411 single images; the second flight (RGB) 358 images. We then mosaicked the image sets into two single orthomosaics using the PiX4D software. We did not use ground

Fig. 2 UAV imagery of the BIOTA Observatory S05 Otjiamongombe / Erichsfelde. (**a**) 1 km^2 RGB image with a 5 cm resolution acquired on 21.03.2016.(**b**) a subset focused on hectare 46 from the centre of the observatory (**c**) photo of hectare 46 from the year 2003

control points for the mosaicking procedure, so we needed to adjust the imagery by shifting the NIR image manually by five pixels on the x-axis and three pixels on the y-axis to ensure proper matching with RGB imagery. We used QGIS v.2.16 (QGIS Development Team 2016) to shift the image.

Ground Truth Data

After image processing, subsections of the orthomosaics corresponding to each of 20 permanent monitoring plots were taken to the field. Within each plot, all trees and shrubs in the image were compared to each tree and shrub in the vegetation plot. Then, for each single-stem individual, an outline for each individual was drawn onto the image. This was necessary as the crowns of species were commonly overlapping where multiple stemmed individuals occurred. In total, 16 species were recorded and only living individuals were considered. However, we excluded seven of these species from further analysis because they occurred with an abundance of less than ten individuals within the samples (Table 1). We then converted our paper drawings into polygons per species in QGIS based on the RGB orthomosaic. Although

Table 1 Sampled tree and shrub species with observed and relative abundance

Nr.	Species name	Family	Short	Individuals	Rel. %
1	*Acacia mellifera*	Fabaceae	AM	174	0.38
2	*Grewia flava*	Malvaceae	GF	109	0.24
3	*Acacia tortilis*	Fabaceae	AT	45	0.10
4	*Lycium eenii*	Solanaceae	LA	35	0.08
5	*Acacia reficiens*	Fabaceae	AR	17	0.04
6	*Dichrostachys cinerea*	Fabaceae	DS	16	0.03
7	*Acacia hereroensis*	Fabaceae	AHR	14	0.03
8	*Boscia albitrunca*	Capparaceae	BA	13	0.03
9	*Acacia hebeclada*	Fabaceae	AH	12	0.03
10	*Phaeoptilum spinosum*	Nyctaginaceae	PS	8	0.02
11	*Acacia luederitzii*	Fabaceae	AL	7	0.02
12	*Leucosphaera bainesii*	Amaranthaceae	LB	4	0.01
13	*Ziziphus mucronata*	Rhamnaceae	ZZ	3	0.01
14	*Commiphora africana*	Burseraceae	CA	3	0.01
15	*Acacia fleckii*	Fabaceae	AF	2	0.00
16	*Acacia erubescens*	Fabaceae	AE	1	0.00

Short = species name abbreviation, Rel.% = relative abundance in percent. Taxonomy follows *A Checklist of Namibian Indigenous and Naturalised Plants* (Klaassen and Kwembeya 2013) as Namibian standard. Updated names for the genus *Acacia* can be found in Kyalangalilwa et al. (2013). Species with less than ten observed individuals were not considered in this analysis, as these do not provide sufficient ground truth information

several software solutions exist that promise to segment tree crowns from point clouds and RGB imagery, such as the TIDA algorithm (Culvenor 2002), these algorithms are difficult to handle and come with their own error. As the number of polygons required was small, we preferred the manual approach of delineating tree crowns. In total, we drew 478 polygons to build the species dataset. The abundance of the observed species is described in Table 1. Note that this includes only living individuals.

RGB and NIR Spectral Indices

As the aim of this study was to identify the most suitable predictors for species discrimination, we derived an extensive set of parameters from both the RGB and NIR imagery. Based on the RGB imagery, we calculated the chromatic coordinates (Woebbecke et al. 1995; Meyer and Neto 2008) as well as the excessive Red (exR) and Green (exG) indices (Table 2). Recent studies had found these to be the most suitable to discriminate crop species (Woebbecke et al. 1995; Meyer and Neto 2008) or to predict vegetation parameters based solely on RGB imagery (Zhang et al. 2010; Schirrmann et al. 2016; Vergara-Díaz et al. 2016). In addition, we calculated the normalised green-red difference index (NGRDI) and the exG–exR parameter.

Table 2 Overview of all image parameters extracted for crown polygons

Image parameter	Short	Data	Formular	References
Normalized Difference Vegetation Index	NDVI	NIR	(NIR-RED)/ (NIR+RED)	Tucker (1979)
Thiam's Transformed Vegetation Index	TTVI	NIR	sqrt(ABS(NDVI + 0.5))	Thiam (1998)
Transformed Soil Adjusted Vegetation Index	TSAVI	NIR	a(NIR-a) (R-b) / R + aNIR -ab	Baret et al. (1989)
Perpendicular Vegetation Index 1	PVI84	NIR	(bNIR-R) + a / (sqrt(b^2+1))	Perry and Lautenschlager (1984)
Perpendicular Vegetation Index 3	PVI94	NIR	aNIR-bRED	Qi et al. (1994)
Normalized Green-Red Difference Index	NGRDI	NIR	(G- R) / (G + R)	Rasmussen et al. (2016)
excessive Redness	exR	RGB	1.4*chrR-chrG	Meyer and Neto (2008)
excessive Greeness	exG	RGB	2*chrG-chrR-chrB	Meyer and Neto (2008)
exG-exR	exG-exR	RGB	exG-exR	Meyer and Neto (2008)
excessive Greeness 2	exG2	RGB	(2*G-R-B) / (G+R+B)	Rasmussen et al. (2016)
chromatic coordinate R	chrR	RGB	R* / R* + G* + B*	Meyer and Neto (2008)
chromatic coordinate G	chrG	RGB	G* / R* + G* + B*	Meyer and Neto (2008)
chromatic coordinate B	chrB	RGB	B* / R* + G* + B*	Meyer and Neto (2008)
Energy	Energy	Texture	$\sum_{i,j} g(i,j)^2$	Haralick et al. (1973)
Entropy	Entropy	Texture	$-\sum_{i,j} g(i,j)\log_2 g(i,j)$	Haralick et al. (1973)
Correlation	CorrL	Texture	$\sum_{i,j} \frac{(i-\mu)(j-\mu)g(i,j)}{\sigma^2}$	Haralick et al. (1973)
Inverse Distance Moment	IDM	Texture	$\sum_{i,j} \frac{1}{1+(i-j)^2} g(i,j)$	Haralick et al. (1973)
Inertia	Inertia	Texture	$\sum_{i,j} (i-j)^2 g(i,j)$	Haralick et al. (1973)
Cluster Shade	ClusSha	Texture	$\sum_{i,j} ((i-\mu)+(j-\mu))^3 g(i,j)$	Haralick et al. (1973)
Cluster Prominence	ClustPro	Texture	$\sum_{i,j} ((i-\mu)+(j-\mu))^4 g(i,j)$	Haralick et al. (1973)
Haralick's Correlation	HarrCorr	Texture	$\frac{\sum i,j(i,j)g(i,j)-\mu_i^2}{\sigma_t^2}$	McInerney and Kempeneers (2015)

a=intercept, b=slope of soil line, R* = normalized Red channel. μ = window average, σ = window variance, g(i,j) = function for pixel pair i and j

A recent, simplified version of the excessive greenness index (Rasmussen et al. 2016) was also calculated, in this study called exG2.

To exploit the additional NIR information, we calculated slope and distance based vegetation indices (Silleos et al. 2006). Slope based vegetation indices make use of the difference in the slope of the red and NIR channel; the famous Normalized Difference Vegetation Index (NDVI, Tucker 1979) belongs here. The SAVI is a modified NDVI which adjusts for potential effects of bare soil (Huete 1988). Thiam's vegetation index improves on the NDVI by multiplying the absolute NIR and Red band values with their square root (Thiam 1998). Distance based vegetation indices make use of the concept of the so-called "soil-line" (Silleos et al. 2006). The distances refer to the distance of samples in the two dimensional red-NIR spectral space to the soil line, that describes the lower boundary of pixels in this space, usually aligning across a clearly visible axis. To determine the soil line parameters required for the calculation of the distance based vegetation indices, a set of $n = 100$ bare soil pixels were selected, stratified by the hectare grid of the Biodiversity Observatory, and the NIR and red values were extracted. Based on these values, a linear regression ($R^2 = 0.89$, $p < 0.001$) was used to estimate the intercept and the slope of the soil line. The linear regression parameters intercept (alpha = −227.29) and slope (beta=1.877) were used to calculate the Perpendicular Vegetation Index III (Qi et al. 1994; Silleos et al. 2006). All indices were calculated and image manipulations were performed with the open source software SAGA-GIS (Conrad et al. 2015). For all individual tree crown polygons, we calculated values for the mean and standard deviations using the zonal statistics tool in SAGA-GIS.

Image Texture

Richards (2013) suggested that the texture of an image can be described as smooth, rough or repetitive in terms of the spatial arrangement of grey values. In terms of canopy cover this would describe whether tree crowns consist of repeating patterns of shadow and greenness or whether the canopy is closed and thus equal in colour. Often texture measures will improve remote sensing classifiers (Krefis et al. 2011). As our main interest was to discriminate between tree species canopies, the greenness (exG) of the canopy seemed to be a good parameter for a texture analysis. We used the Orfeo Toolbox v.5.6.1 (McInerney and Kempeneers 2015), a free open source software for remote sensing image analysis, to calculate eight different types of simple image texture measures. Haralick's grey level occurrence matrix (GLCM, Haralick et al. 1973), which is a standard for describing image texture, was the basis for calculating all of the texture measures. We choose a constant window size of 5×5 pixels and an offset of 1 for x and y. The number of grey levels was set to 16. We then loaded the calculated image texture measures into SAGA-GIS and extracted the texture as average and standard deviation for each individual tree crown canopy polygon.

Canopy Height Model

We generated a canopy height model (CHM) based on the overlapping single image tiles. We used the software Postflight Terra 3D Vs4.0.104 (SenseFly 2015) to generate dense point clouds as *.las-files. Then we imported the *.las point cloud files into the LAStools software (Isenburg 2016) to generate buffered tiles. Ground points representing bare ground were identified visually. Based on these ground points, we calculated the height (height normalization) for all non-ground points of all tiles. These tiles were then mosaicked in SAGA-GIS using a b-spline interpolation with feathering to create a seamless normalized Digital Surface Model (nDSM). This nDSM describes the maximum heights of the point cloud. Next, we generated a Digital Terrain Model (DTM), that describes the minimum heights of the point cloud. Finally, the CHM was generated by subtracting the DTM from the DSM, which gave values in the range of −0.11 to 1.89 m. The lower range was adjusted to zero. Average canopy height and its standard deviation were extracted for each canopy polygon.

Random Forest Classification

The Random Forest algorithm (Breiman Breimanx) is now a common standard non-parametric classifier with high performance as was found by many comparative studies in a remote sensing context (Pal 2005; Duro et al. 2012; Qian et al. 2014). Random Forest makes use of the concept of classification and regression trees (CART) but combines them with ensemble modelling and bagging. Random Forest is a non-parametric classifier that creates thousands of single decision trees and averages their results. Each decision tree is a subsample of the whole dataset. The split for each tree node is determined by the Gini criterion, which measures the entropy of the dataset. The best split is that parameter value that leads to the largest decrease in the Gini criterion. When the classifier is applied to the test dataset, the final class label is then based on the majority vote of all constructed decision trees (Immitzer et al. 2012).

A Random Forest classifier was used to predict species labels, with this achieved by first dividing the dataset into training and testing polygons, with an 80:20 ratio per class. To establish if any single set of parameters were sufficient alone, the dataset was split into a RGB, a NIR, a texture and a complete dataset (ALL). For quantifying the importance of the CHM, we added these values to each parameter dataset. Before classification, all parameters with Pearson correlations higher than 0.75 were deleted to ensure that multicollinearity issues would not be an issue. Only two texture parameters where omitted because of multicollinearity, the Cluster Prominence and Haralick's correlation. The latter was correlated with "correlation (corrL)" and the first with "inverse distance moment (IDM)". Finally, in order to test the effect of species abundance on the classification results, the species data were divided into three datasets, considering (a) all species with an abundance larger than 10, (b) frequent species with an abundance larger than 30 and (c) infrequent species with abundances between 30 and 10 individuals (Table 2). These species subsets were all tested in combination with all parameter subsets, leading to a final number of 24 single datasets.

For each single dataset, a classifier was produced and its accuracy was verified using the test dataset. For accuracy assessment, confusion matrices were generated and the following accuracy measures were derived: Overall accuracy (OA), confidence limits for OA based on cross-validation, and Cohen´s Kappa which takes class imbalance into account (Kuhn and Johnson 2013). As a null-model for the overall accuracy, we calculated the No-Information Rate (Kuhn and Johnson 2013), which is simply defined as the proportion of the largest class expressed as a percentage. A one-sided test of equal proportions was then conducted to provide a p-value for the null-model.

The relevance of the single predictors was assessed by calculating their variable importance. Variable importance describes the relationship between each parameter and the outcome of the classification or regression procedure. It is measured as the loss in performance when the respective parameter is not considered. Variable importance was measured for all parameters in the three species subsets in order to identify consistently important predictors across all predictors considered.

Random Forests were run with 5000 trials. The parameter *mtry* was set to 1/3 of the number of variables considered. The parameter *mtry* describes the number of parameters that are included in each single decision tree. In addition, a repeated cross validation was implemented using a tenfold cross validation with five repetitions to be able to achieve standard errors and confidence intervals for the overall accuracy. Classification was performed in the free and open source software R (R Core Team 2016) using the packages caret (Kuhn et al. 2016), randomForest (Liaw and Wiener 2002) and e1071 (Meyer et al. 2015).

Results

Only frequent species constantly exhibited significant p-values (Table 3, Fig. 3) meaning that OA was higher than the respective null-model. When using all species or only infrequent species, this was not the case. The species subsets "All" and "Infrequent" had always low Kappa and OA values except for infrequent species with the ALL and ALL+CHM (Table 3).

The highest OA and Kappa values were obtained for the combined RGB and CHM dataset for the frequent species, with an OA value of 0.77 and a Kappa value of 0.63. Globally, the "ALL" model was ranked second by Kappa for frequent species. However, "ALL" is much more complex (42 parameters) than RGB+CHM (16 parameters). Thus, the simpler RGB solution can be regarded as much more informative and easier to reproduce as fewer parameters have to be derived from the imagery (Table 3). The lower quality of the infrequent species dataset was also evi-

Table 3 Accuracy measures of 24 random forest classification models with different combinations of species and UAV imagery products

Species	Dataset	Kappa	OA	OA_{Lower}	OA_{Upper}	OA_{Null}	p-value
All	ALL	0.33	0.54	0.45	0.63	0.57	0.739
	ALLCHM	0.34	0.54	0.45	0.63	0.56	0.677
	NIR	0.22	0.46	0.37	0.55	0.51	0.880
	NIRCHM	0.25	0.48	0.39	0.58	0.53	0.841
	RGB	0.36	0.56	0.47	0.65	0.57	0.609
	RGBCHM	0.39	0.58	0.49	0.67	0.57	0.465
	TEXT	0.31	0.53	0.43	0.62	0.56	0.794
	TEXTCHM	0.28	0.51	0.42	0.60	0.56	0.882
Freq	ALL	0.52	0.70	0.60	0.79	0.54	**>0.001**
	ALLCHM	0.49	0.67	0.57	0.76	0.52	**>0.01**
	NIR	0.43	0.64	0.54	0.73	0.53	**>0.05**
	NIRCHM	0.49	0.68	0.58	0.77	0.55	**>0.01**
	RGB	0.41	0.62	0.52	0.72	0.46	**>0.001**
	RGBCHM	0.63	0.77	0.67	0.85	0.53	**>0.001**
	TEXT	0.45	0.67	0.57	0.76	0.64	0.306
	TEXTCHM	0.49	0.70	0.60	0.79	0.62	0.062
Infreq	ALL	0.53	0.63	0.38	0.84	0.32	**>0.01**
	ALLCHM	0.47	0.58	0.34	0.80	0.32	**>0.05**
	NIR	0.27	0.42	0.20	0.67	0.37	0.399
	NIRCHM	0.14	0.32	0.13	0.57	0.26	0.383
	RGB	0.34	0.47	0.24	0.71	0.32	0.111
	RGBCHM	0.27	0.42	0.20	0.67	0.32	0.226
	TEXT	0.26	0.42	0.20	0.67	0.32	0.226
	TEXTCHM	0.27	0.42	0.20	0.67	0.26	0.100

OA = Overall Accuracy (%), OANull= Null model, p-value describes whether OA is significantly different from OANull. Kappa is Cohen's unweighted Kappa

dent by the large confidence intervals (Fig. 3). These are much smaller in the case for ALL and frequent species. This effect was attributed to the small number of samples within the infrequent species (72 samples spread across five classes, see Table 1). Models that used texture (TEXT) alone or a combination of texture and CHM (TEXTCHM) were never significant in any of the species sets (Table 3). Best results for texture models were found for the frequent species with an OA of 0.70 and a Kappa value of 0.49.

The inclusion of the CHM led to an increase in model quality for 7 out of 12 image parameter pairs (Table 3). The largest increase in the Kappa (0.22) was found for the RGB – RGB+CHM pair in the frequent species dataset. However, the second largest change was a decrease of 0.13 for the NIR – NIR+CHM in the infrequent species dataset (Table 3). Except for these two values, the average increase in Kappa was zero. Hence, we did not find that the CHM contributed additional information.

In the variable importance analysis (Fig. 4), none of the NIR-derived image parameters occurred in the top ten parameters. The RGB indices exG2, exR, exG

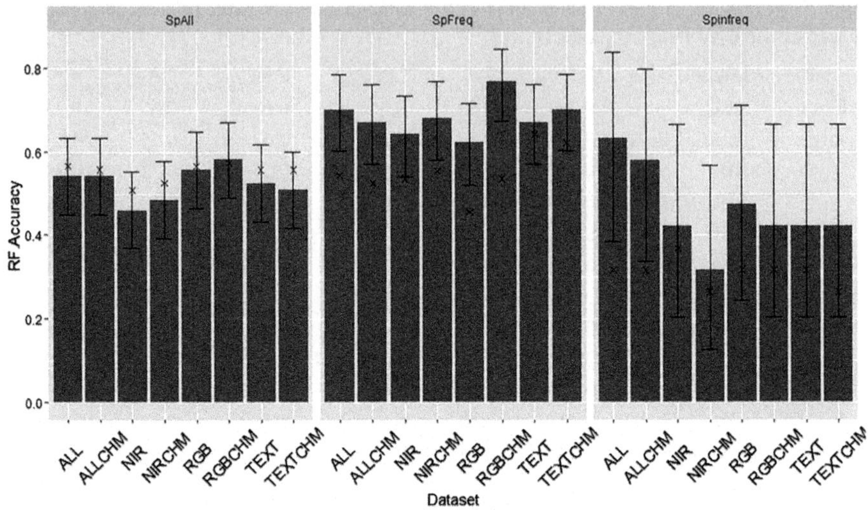

Fig. 3 Average overall accuracy with confidence limits based on tenfold cross-validation with five repetitions. The *stars* denote the overall accuracy derived by a null-model. *Stars* within confidence limits signify models that were not significantly better than the null-model and thus do not provide credible results. Note the very high range of confidence limits for the infrequent species data set (Spinfreq)

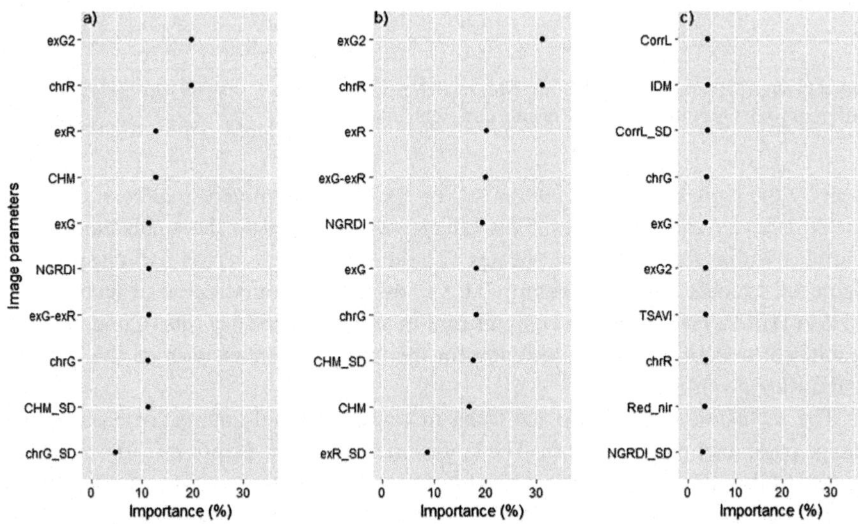

Fig. 4 Variable importance of the ten most important image parameters for the three species subsets, i.e. all species (**a**), frequent species (**b**) and infrequent species (**c**). For full names of image parameters see Table 2

and NGRDI occurred in the All and Frequent species subsets. Highest importance values were achieved by the RGB indices exG2 and chrR both with 32% and 20% for Frequent and All species subsets respectively. In the case of the infrequent species, two texture measures appear in the ten most important variables. However, their variable importance did not reach values higher than five percent, rendering all parameters for the infrequent species redundant.

In summary, the frequent species dataset together with RGB+CHM image parameters provided the highest accuracy in the discrimination of tree species, used the lowest number of predictors and provided the smallest confidence intervals.

Discussion

We evaluated the relevance of different parameters derived from very high resolution RGB-NIR imagery for the discrimination of savannah tree species. We could confirm the commonly found pattern that information based on the visual part of the spectrum is important for discriminating tree species (Fassnacht et al. 2016). In particular, we found that RGB-based spectral indices (Meyer and Neto 2008; Rasmussen et al. 2016) and simple chromatic coordinates (Woebbecke et al. 1995) in combination with a canopy height model (CHM) achieved the best results. By contrast, the performance of the NIR image parameters was weak and deteriorated when combined with the CHM. Our overall accuracy of 77% (on average, maximum was 0.83) is comparable to the results of the recent review of Fassnacht (see Fig. 3 in Fassnacht et al. (2016)) who analysed 129 case studies on tree species mapping.

Most case studies have used a combination of hyperspectral and/or LiDAR for tree species mapping and have typically achieved overall accuracies of between 75 and 90%. Of these, three were carried out in southern Africa, all in Kruger National Park, and all used hyperspectral image data and height information derived from LiDAR sensors. Naidoo et al. (2012) achieved 82% with four hyperspectral indices (including NDVI) and height information; whilst Cho et al. (2012) also used hyperspectral data but resampled them to seven World View 2 multispectral bands and combined them with LiDAR based height information. They achieved OA values of between 63 and 81%. Colgan et al. (2012) used LiDAR-based height information and BRDF corrected reflectance values for the VIS-NIR region. The bidirectional reflectance distribution function (BRDF) is a function describing the change in reflectance values due to view angle and sun position during image assessment. The BRDF correction improved the hyperspectral information and thus led to OAs of between 70 and 78%. Hence, our UAV based tree species discrimination approach, requiring only a ≈100$ RGB camera, performed equally well, when compared to the technically more sophisticated, and also much more expensive, hyperspectral and LiDAR sensors.

VIS-NIR imagery is by far the most commonly used data source for generating spectral indices. Pure RGB based spectral information has been used less often in remote sensing studies that have focused on the discrimination and mapping species.

In their review on tree species mapping, Fassnacht et al. (2016) also state that the VIS region (350–650 nm) contains the most often selected features for tree species mapping, without mentioning the relation to RGB. A recent study comparing spectral indices derived from RGB and NIR camera images to multispectral imagery (Rasmussen et al. 2016) showed that cheaper RGB / NIR cameras are equal in performance for mapping barley biomass in agricultural fields. Another study (Fischer et al. 2012) compared an NDVI calculated from high spectral resolution field spectrometer (i.e. ASD Field Spec 3) with an NDVI derived from a modified Olympus consumer-grade camera for mapping the spatial variability of NDVI in biotic soil crusts; they found strong correlations with R^2 of 0.91.

In this study, we found that RGB bands were an important predictor but the least important was texture. This is surprising as we thought that texture would have a high potential for describing crown properties related to shadow patterning or variation in greenness. Fassnacht et al. (2016) list several studies that applied texture to improve tree species classification by 10–15%. However, using texture also creates problems that make its use seem clumsy and time consuming. First of all, the idea of texture is a multiscale problem. The relevant scale (i.e., the size of the window in which a texture value is calculated) has to be identified empirically. Therefore, different window sizes have to be compared with these usually being 3×3, 5×5, 7×7, 9×9 and so on. When UAV data volumes are large (>1 GB), this quickly becomes unwieldy. Also, with very high resolution images, larger window sizes are needed but these slow processing times. Other options such as the offset and the number of grey levels considered, provide further opportunities to optimize the results; yet leading to a seemingly endless endeavour in finding the right parameter settings. Second, different species might require different window sizes. This seems logical but is difficult to realize technically. Third, the large number of available texture measures makes it difficult to select those that are optimal or most appropriate. This is complicated by the typically high correlation between the different texture measures. In our study, all parameters were kept stable (i.e., a window size of 5×5, offset of 1×1 and 16 grey levels). Not testing different settings might explain the poor performance of the texture parameters. We also used only a small fraction of the available texture measures. Other texture measures related to tree crown shape and size could have been considered (Fassnacht et al. 2016). The Orfeo Toolbox provides around 40 different texture measures in total. Hence, texture measures derived from UAV imagery require more studies on selecting and optimising the best measures and optimal window sizes for tree species discrimination or tree crown analyses.

Tree species mapping through remote sensing data can become an efficient tool in biodiversity monitoring. However, the nature of biodiversity is that communities under study almost always consist of common and rare species (Magurran and McGill 2011). The occurrence of rare tree species (rare equal to less than ten individuals overall) severely affected the potential to classify the whole tree species pool. Out of 16 species, seven species were considered as too rare to be used in the classification. Another five infrequent species, i.e. with less than 30 tree crowns for training and testing, led to poor classification results due to the small amount of training data. However, the infrequent species were impossible to classify correctly, as can be seen from the non-significant or low quality models (Table 2, Fig. 3). Thus out of 16 species, only the four frequent species could be classified to acceptable levels of accuracy. The pattern that only frequent species can be mapped with sufficient accuracy is confirmed by many other studies (Naidoo et al. 2012; Immitzer et al. 2012; Cho et al. 2012; Baldeck et al. 2015). The review of Fassnacht et al. (2016) reports the number of species that were classified in the analysed studies ranging from two to seventeen with an average of five or six. This finding has important implications for future biodiversity monitoring that should be based on tree species mapping leading to a complete census. Mapping rare tree and shrub species becomes a challenge when too few individuals can be found for training and testing a classifier. Thus, in future studies, more emphasis should be put on high quality ground truth data gathering an equal number of ground truth tree locations per species. In the study of Colgan et al. (2012), three species made up 30% of the landscape while the category "other" also had 20% of all occurring tree crowns. Thus, rare species can make a large fraction of tree crowns in a savannah but are represented by a small number of individuals per species. Common trees however, bear different challenges. For example, a high genus – species ratio (i.e. where many species of the same genus occur as in the genus *Acacia* or *Combretum*) means these species are sometimes lumped together into a single tree category at genus level (Naidoo et al. 2012). The species abundances in (semi-)natural ecosystems are much more complex than in temperate forests and will require special considerations for an operative tree species mapping based on remote sensing imagery.

Although our study was successful in discriminating selected savanna tree species with a UAV-borne RGB camera, the limitations of UAVs in comparison to airborne or satellite-borne sensors requires discussion. In our case, the largest obstacle was the mismatch in the co-registration of NIR and RGB imagery, which had to be corrected manually. Better results could be achieved when using multispectral cameras or even lightweight hyperspectral cameras. The spatial mismatch could have been the reason why the averaged tree crown parameters were worse for NIR than for RGB. Digitization of the tree crowns was also undertaken manually using only the RGB imagery. Hence, it is possible that the NIR imagery parameters contained a higher shadow fraction or parts of neighbouring tree crowns. Although manual digitization seems straightforward, it is also error prone and could be avoided by using specifically designed algorithms or software packages, e.g. TIDA (Culvenor, 2002), JSEG (Kang et al. 2016) or ITCsegment (Dalponte and Coomes 2016). Other serious problems connected to light and shadowing effects that can occur when using UAV imagery are discussed by Rasmussen et al. (2016). In our study, the different flight directions during the drone overflight affected the brightness pattern. Rasmussen et al. (2016) also mentioned that BRDF effects affect the outcome of a study when not taken into consideration. These issues, co-registration and changing light conditions (including BRDF effects), seem to compromise the utility of UAV imagery. Ground control points should be essential for proper image co-registration, however, often these require expensive differential-GPS equipment. Further improvement of technical equipment or standardized procedures for UAV image acquisition should bring remedies in the future.

Conclusions

In this study, we evaluated the relevance of RGB and NIR image products, derived from UAV images, for discriminating tree species in a Namibian savannah. We found that data acquired in the NIR wavelength region only were not sufficient or even necessary, although this conclusion might have been incorrectly drawn because of co-registration problems between the NIR and RGB imagery. Permanently marked, well-surveyed ground control points therefore need to be planned for future image acquisition campaigns. Nevertheless, the OAs achieved with RGB data and CHM were comparable to other studies that used more expensive hyperspectral data and LiDAR instruments. This indicated that UAVs have a high potential for future tree species mapping tasks if areas less than 1 km^2 are to be monitored. However, the number of species that can be mapped or discriminated seems independent of the sensor type. The assumption is that hyperspectral data theoretically can outperform RGB-NIR data when a large number of species are present. However, for the process of training a classifier, such as a Random Forest, the number of training polygons needs to be at least 30 in order to achieve sufficient and acceptable accuracies. This seems not feasible when rare species (i.e., less than ten individuals per square kilometre) are present. Hence, a significant future challenge is the task of mapping species with low abundances.

Practical Application for Nature Conservation

This chapter dealt with the application of UAV-borne consumer grade cameras for discriminating savannah tree species and has several important messages for practical applications in nature conservation. Firstly, the delta-wing UAV that we employed, the eBee system (SenseFly 2015), is capable of capturing an area of 1 km^2 during a single flight when the desired resolution is a 5 cm pixel size or greater. Smaller pixel sizes, e.g. 2 cm, can only be achieved in several flights (four to five). However, this this also doubles the disk space required for storing the imagery. Affordable quadcopter systems cannot usually cover 1 km^2 in a single flight. Secondly, we showed that tree species discrimination based solely on RGB + Canopy Height is possible, suggesting that a second flight with a NIR camera is potentially unnecessary. However, we need more studies comparing RGB based spectral indices to NIR based spectral indices in order to see whether RGB can replace NIR indices in the future. Finally, we found that ground truthing should take the abundance or frequency of the species into consideration. We suggest using a minimum of 50 individuals per species for training purposes in order to be successfully mapped. Species from the same genus, e.g. the different *Acacia* species in our study, often share similar spectral properties and thus are very difficult to distinguish. One alternative is to map these at the genus level, if it is not the users demand to produce species specific map. In conclusion, we have shown that using UAVs to map the individual stems of tree species could be a cheap and very flexible tool for nature conservation in the near future.

Acknowledgments Our gratitude to the Pommersche Farmereigesellschaft and their staff for allowing us to work on the farm Erichsfelde. The work was financially supported by the SASSCAL initiative, with funding by the German Federal Ministry of Education and Research; BMBF Funding Nr: 01LG1201M.

References

Baldeck, C.A., Asner, G.P., Martin, R.E., et al.: Operational tree species mapping in a diverse tropical forest with airborne imaging spectroscopy. PLoS One. **10**, e0118403 (2015). doi:10.1371/journal.pone.0118403

Baret, F., Guyot, G., Major, D.: TSAVI: a vegetation index which minimizes soil brightness effects on LAI and APAR estimation. In: 12th Canadian Symposium on Remote Sensing and IGARSS'90, p. 4, Vancouver, Canada, 10–14 July 1989 (1989)

Breiman, L.: Random forests. Mach. Learn. **45**, 5–32 (2001). doi:10.1023/A:1010933404324

Bunting, P., Lucas, R.: The delineation of tree crowns in Australian mixed species forests using hyperspectral Compact Airborne Spectrographic Imager (CASI) data. Remote Sens. Environ. **101**, 230–248 (2006). doi:10.1016/j.rse.2005.12.015

Cho, M.A., Mathieu, R., Asner, G.P., et al.: Mapping tree species composition in South African savannas using an integrated airborne spectral and LiDAR system. Remote Sens. Environ. **125**, 214–226 (2012). doi:10.1016/j.rse.2012.07.010

Cho, M.A., Malahlela, O., Ramoelo, A.: Assessing the utility WorldView-2 imagery for tree species mapping in South African subtropical humid forest and the conservation implications: Dukuduku forest patch as case study. Int. J. Appl. Earth Obs. Geoinf. **38**, 349–357 (2015). doi:10.1016/j.jag.2015.01.015

Colgan, M.S., Baldeck, C.A., Féret, J.-B., Asner, G.P.: Mapping savanna tree species at ecosystem scales using support vector machine classification and BRDF correction on airborne hyperspectral and LiDAR data. Remote Sens. **4**, 3462–3480 (2012). doi:10.3390/rs4113462

Conrad, O., Bechtel, B., Bock, M., et al.: System for automated geoscientific analyses (SAGA) v. 2.1.4. Geosci. Model Dev. **8**, 1991–2007 (2015). doi:10.5194/gmd-8-1991-2015

Culvenor, D.S.: TIDA: an algorithm for the delineation of tree crowns in high spatial resolution remotely sensed imagery. Comput. Geosci. **28**, 33–44 (2002). doi:10.1016/S0098-3004(00)00110-2

Dalponte, M., Coomes, D.A.: Tree-centric mapping of forest carbon density from airborne laser scanning and hyperspectral data. Methods Ecol. Evol. **7**, 1236–1245 (2016). doi:10.1111/2041-210X.12575

Duro, D.C., Franklin, S.E., Dubé, M.G.: A comparison of pixel-based and object-based image analysis with selected machine learning algorithms for the classification of agricultural landscapes using SPOT-5 HRG imagery. Remote Sens. Environ. **118**, 259–272 (2012). doi:10.1016/j.rse.2011.11.020

Dvořák, P.J., Müllerová J., Bartaloš, T., Brůna J.: Unmanned aerial vehicles for alien plant species detection and monitoring. ISPRS – international archives of the photogrammetry, remote sensing and spatial information sciences XL-1/W4: 83–90 (2015). doi:10.5194/isprsarchives-XL-1-W4-83-2015

Fassnacht, F.E., Latifi, H., Stereńczak, K., et al.: Review of studies on tree species classification from remotely sensed data. Remote Sens. Environ. **186**, 64–87 (2016). doi:10.1016/j.rse.2016.08.013

Fischer, T., Veste, M., Eisele, A., et al.: Small scale spatial heterogeneity of Normalized Difference Vegetation Indices (NDVIs) and hot spots of photosynthesis in biological soil crusts. Flora – Morphol. Distrib. Funct. Ecol. Plants. **207**, 159–167 (2012). doi:10.1016/j.flora.2012.01.001

Giess, W.: A preliminary vegetation map of Namibia. Dinteria. **4**, 1–112 (1998)

Haralick, R.M., Shanmugam, K., Dinstein, I.: Textural features for image classification. IEEE Trans. Syst. Man Cybern. **3**, 610–621 (1973)

Huete, A.R.: A soil-adjusted vegetation index (SAVI). Remote Sens. Environ. **25**, 295–309 (1988). doi:10.1016/0034-4257(88)90106-X

Immitzer, M., Atzberger, C., Koukal, T.: Tree species classification with random forest using very high spatial resolution 8-Band WorldView-2 satellite data. Remote Sens. **4**, 2661–2693 (2012). doi:10.3390/rs4092661

Isenburg, M.: LAStools, efficient LiDAR processing software. rapidlasso GmbH (2016)

Jürgens, N., Schmiedel, U., Haarmeyer, D.H., et al.: The BIOTA biodiversity observatories in Africa—a standardized framework for large-scale environmental monitoring. Environ. Monit. Assess. **184**, 655–678 (2012). doi:10.1007/s10661-011-1993-y

Kang, J., Wang, L., Chen, F., Niu, Z.: Identifying tree crown areas in undulating eucalyptus plantations using JSEG multi-scale segmentation and unmanned aerial vehicle near-infrared imagery. Int. J. Remote Sens. **38**, 1–17 (2016). doi:10.1080/01431161.2016.1253900

Klaassen, E.S., Kwembeya, E.G.: A checklist of Namibian indigenous and naturalised plants. 2013. Occasional Contributions No 5, National Botanical Research Institute, Windhoek, Namibia (2013).

Krefis, A.C., Schwarz, N.G., Nkrumah, B., et al.: Spatial analysis of land cover determinants of malaria incidence in the Ashanti Region, Ghana. PLoS One. **6**, e17905 (2011). doi:10.1371/journal.pone.0017905

Kuhn, M., Johnson, K.: Applied Predictive Modeling. Springer, New York (2013)

Kuhn, M.K., Weston, S., Williams, A., et al.: Caret: Classification and Regression Training. R package version 6.0-70. https://CRAN.Rproject.org/package=caret (2016)

Kyalangalilwa, B., Boatwright, J.S., Daru, B.H., et al.: Phylogenetic position and revised classification of Acacia s.l. (Fabaceae: Mimosoideae) in Africa, including new combinations in Vachellia and Senegalia. Bot. J. Linn. Soc. **172**, 500–523 (2013). doi:10.1111/boj.12047

Liaw, A., Wiener, M.: Classification and regression by random Forest. R News. **2**, 18–22 (2002)

Lisein, J., Michez, A., Claessens, H., Lejeune, P.: Discrimination of deciduous tree species from time series of unmanned aerial system imagery. PLoS One. **10**, e0141006 (2015). doi:10.1371/journal.pone.0141006

Lucas, R., Bunting, P., Paterson, M., Chisholm, L.: Classification of Australian forest communities using aerial photography, CASI and HyMap data. Remote Sens. Environ. **112**, 2088–2103 (2008). doi:10.1016/j.rse.2007.10.011

Magurran, A.E., McGill, B.J.: Biological Diversity: Frontiers in Measurement and Assessment, 1st edn. Oxford University Press, Oxford/New York (2011).

McInerney, D., Kempeneers, P.: Orfeo toolbox. In: Open Source Geospatial Tools, pp. 199–217. Springer International Publishing, Basel (2015).

Meyer, G.E., Neto, J.C.: Verification of color vegetation indices for automated crop imaging applications. Comput. Electron. Agric. **63**, 282–293 (2008). doi:10.1016/j.compag.2008.03.009

Meyer, D., Dimitriadou, E., Hornik, K., et al.: e1071: Misc Functions of the Department of Statistics, Probability Theory Group (Formerly: E1071), TU Wien. R package version 1.6-7 https://CRAN.R-project.org/package=e1071 (2015)

Naidoo, L., Cho, M.A., Mathieu, R., Asner, G.: Classification of savanna tree species, in the Greater Kruger National Park region, by integrating hyperspectral and LiDAR data in a Random Forest data mining environment. ISPRS J. Photogramm. Remote Sens. **69**, 167–179 (2012). doi:10.1016/j.isprsjprs.2012.03.005

Pal, M.: Random forest classifier for remote sensing classification. Int. J. Remote Sens. **26**, 217–222 (2005). doi:10.1080/01431160412331269698

Perry, C.R., Lautenschlager, L.F.: Functional equivalence of spectral vegetation indices. Remote Sens. Environ. **14**, 169–182 (1984). doi:10.1016/0034-4257(84)90013-0

QGIS Development team. QGIS Geographic Information System. Open Source Geospatial Foundation (2016)

Qi, J., Chehbouni, A., Huete, A.R., Kerr, Y.H., Sorooshian, S.: A modified soil adjusted vegetation index. Remote Sens. Environ. **48**, 119–126 (1994). doi:10.1016/0034-4257(94)90134-1

Qian, Y., Zhou, W., Yan, J., et al.: Comparing machine learning classifiers for object-based land cover classification using very high resolution imagery. Remote Sens. **7**, 153–168 (2014). doi:10.3390/rs70100153

R Core Team: R: A Language and Environment for Statistical Computing. R Foundation for Statistical Computing, Vienna (2016)

Rasmussen, J., Nielsen, J., Garcia-Ruiz, F., Christensen, S., Streibig, J.C.: Potential uses of small unmanned aircraft systems (UAS) in weed research. Weed Res. **53**, 242–248 (2013). doi:10.1111/wre.12026

Rasmussen, J., Ntakos, G., Nielsen, J., et al.: Are vegetation indices derived from consumer-grade cameras mounted on UAVs sufficiently reliable for assessing experimental plots? Eur. J. Agron. **74**, 75–92 (2016). doi:10.1016/j.eja.2015.11.026

Richards, J.A.: Remote Sensing Digital Image Analysis: An Introduction, 5th edn. Springer, Berlin (2013)

Schirrmann, M., Giebel, A., Gleiniger, F., et al.: Monitoring agronomic parameters of winter wheat crops with low-cost UAV imagery. Remote Sens. **8**, 706 (2016). doi:10.3390/rs8090706

SenseFly. User Manual: S110 RGB/NIR /RE camera. SenseFly Ltd., Lausanne, Switzerland (2014)

SenseFly. eBee Sensefly: Extended User MANUAL eBee and eBee Ag. Revision 17, June 2015. SenseFly Ltd., Lausanne, Switzerland (2015)

Silleos, N.G., Alexandridis, T.K., Gitas, I.Z., Perakis, K.: Vegetation indices: advances made in biomass estimation and vegetation monitoring in the Last 30 years. Geocarto Int. **21**, 21–28 (2006). doi:10.1080/10106040608542399

Singh, M., Evans, D., Tan, B.S., Nin, C.S.: Mapping and characterizing selected canopy tree species at the Angkor World Heritage Site in Cambodia using aerial data. PLoS One. **10**, e0121558 (2015). doi:10.1371/journal.pone.0121558

Thiam A.K.: Geographic Information Systems and Remote Sensing Methods for Assessing and Monitoring Land Degradation in the Sahel Region: The Case of Southern Mauritania (1998)

Tucker, C.J.: Red and photographic infrared linear combinations for monitoring vegetation. Remote Sens. Environ. **8**, 127–150 (1979)

Vergara-Díaz, O., Zaman-Allah, M.A., Masuka, B., et al.: A novel remote sensing approach for prediction of maize yield under different conditions of nitrogen fertilization. Front. Plant Sci. (2016). doi:10.3389/fpls.2016.00666

Woebbecke, D.M., Meyer, G.E., Von Bargen, K., Mortensen, D.A.: Color Indices for Weed Identification Under Various Soil, Residue, and Lighting Conditions. Trans. ASAE. **38**, 259–269 (1995). doi:10.13031/2013.27838

Zhang, F., Zaman, Q.U., Percival, D.C., Schumann, A.W.: Detecting bare spots in wild blueberry fields using digital color photography. Appl. Eng. Agric. **26**, 723–728 (2010)

A Toolbox for Remotely Monitoring Large Carnivores in Sweden

Michael Schneider and Holger Dettki

Abstract Five species of large carnivore are the main focus of interest in Sweden when it comes to the conservation of biological diversity, ecosystem completeness, and the traditional herding of reindeer (*Rangifer tarandus*) by indigenous Sámi people. Successful work with the wolf (*Canis lupus*), brown bear (*Ursus arctos*), lynx (*Lynx lynx*), wolverine (*Gulo gulo*) and golden eagle (*Aquila chrysaetos*) necessitates good knowledge of their numbers, distribution and population dynamics as well as their effects on prey species and the reindeer herding economy. However, large carnivores are relatively few, elusive, wide ranging and secretive, and therefore notoriously hard to observe. Hence, collecting standardized data of sufficient quality and quantity is a challenge for both research and management. In this chapter, we describe how this challenge is being met in Sweden.

We define remote sensing as *observing and measuring from a distance*, and different approaches for remote sensing of carnivores are used in Sweden. These include non-invasive methods such as DNA-sampling and snow tracking, partly invasive methods such as automatic cameras, and highly invasive methods such as tagging with biotelemetry sensors. Attitude surveys are used to monitor public opinions about carnivores and their management. We also present infrastructure solutions (Rovbase, UC-WRAM) for handling the wealth of data that are acquired through remote sensing of carnivores in Sweden.

Keywords Large carnivores • Brown bear • Wolf • Wolverine • Lynx • Golden eagle • Sweden • Norway • Multi-species system • Multi-method monitoring • Data management

M. Schneider (✉)
Västerbotten County Administration, Umeå, Sweden
e-mail: michael.schneider@lansstyrelsen.se

H. Dettki
Umeå Center for Wireless Remote Animal Monitoring (UC-WRAM),
Department of Wildlife, Fish, and Environmental Studies,
Swedish University of Agricultural Sciences, Umeå, Sweden
e-mail: holger.dettki@slu.se

© Springer International Publishing AG 2017

R. Díaz-Delgado et al. (eds.), *The Roles of Remote Sensing in Nature Conservation*,
DOI 10.1007/978-3-319-64332-8_11

Introduction

In many areas of research and conservation, remote sensing is seen as a tool for finding appropriate habitat and to define its extent and quality in order to make predictions on species occurrence and population size (Neumann et al. 2015). When it comes to carnivores, the opposite is often true, and remote sensing results on individual distribution, movement and activity are used to infer distribution, extent and quality of the species' habitat (e.g. Lande et al. 2003; Dahl et al. 2015).

Five species of large carnivore are the focus of interest in Sweden with respect to the conservation of biological diversity, ecosystem completeness, and the traditional herding of reindeer (*Rangifer tarandus*) by indigenous Sámi people (Schneider 2006a). To undertake successful work with wolf (*Canis lupus*), brown bear (*Ursus arctos*), lynx (*Lynx lynx*), wolverine (*Gulo gulo*) and golden eagle (*Aquila chrysaetos*) requires good knowledge of their numbers, distribution and population dynamics as well as their effects on prey species and the reindeer herding economy. However, large carnivores are relatively few, elusive, wide ranging and secretive, and therefore notoriously hard to observe. Catching and handling them may also be difficult and dangerous. Collecting standardized data of sufficient quality and quantity is therefore a challenge for both research and management.

We define remote sensing as *observing and measuring from a distance*, where distance can occur both in space and time. Several approaches are used to remotely sense large carnivores in Sweden. These include non-invasive methods such as DNA-sampling and snow tracking, partly invasive methods such as automatic cameras, and highly invasive methods such as tagging with biotelemetry sensors. In the following sections, we describe the system that is used for the surveillance and monitoring of the five carnivore species in Sweden. As carnivore management aims at a population-level management of the species in Scandinavia (Linnell et al. 2008), a similar system is used in Norway.

A Compensation System for Reindeer

In the northern half of Sweden, as in the northern parts of Norway and Finland, year-round free-ranging, semi-domestic reindeer are herded by Sámi pastoralists. During the twentieth century, the size of the reindeer population in Sweden has been cyclic on a 30 year basis and it has been fluctuating between ca 150,000 and 300,000 animals. In recent years, the population size has been about 250,000 animals (Sametinget and Naturvårdsverket 2013).

All five species of large carnivore in Sweden predate on reindeer to some extent. An estimated 19,500–72,500 reindeer are killed by large carnivores each year (Sametinget and Naturvårdsverket 2013). Predation on reindeer by large carnivores can have a negative effect on the livelihood of herders (Hobbs et al. 2012), and compensation payments are made to mitigate these effects. Yearly payments from the Swedish state to the reindeer herders averaged about 5 million Euros annually for the period 2010–2014 (statistics from the Sami Parliament, Sametinget). As dead reindeer are hard to find in the woods and mountains and it is difficult for reindeer herders to prove their losses so, in 1996, a new compensation system was launched. Since then, County Administrations (i.e. the regional governments that are responsible for the management of large carnivores, among other things (Sandström and Lindvall 2006)) have the obligation to count carnivores instead, and compensation is payed based on the number of carnivores and their potential effects on reindeer, rather than on proven losses (Schneider 2012).

The Swedish compensation system has been evaluated scientifically and it has been found to be effective in supporting the conservation of large carnivores and in helping to mitigate their negative effects on reindeer herding (Zabel et al. 2010; Persson et al. 2015). The Swedish system is an example of conservation performance payments, which are a type of payment for environmental services (Zabel and Holm-Müller 2008; Zabel and Engel n.d.). In this case, reindeer herders are paid to conserve carnivores, instead of hunting them in retaliation or as precaution to protect reindeer. Performance payments are tied to quantitatively measureable indicators of conservation outcomes, here being the number of carnivores that are present on reindeer grazing grounds. In order to collect unbiased measures of carnivore numbers, County Administrations (as independent parties) do the surveillance.

The approaches to carnivore management in Sweden and Norway are quite different, but the two countries share common populations of the species. Consequently, Sweden and Norway recently launched a joint system for the monitoring of the trans-border populations of brown bear, wolf, wolverine and lynx. A process for developing a joint system for golden eagles started in 2015 and should be operational from 2019. The advantages of a joint system are many: comparable results between countries and between years, regular estimations of carnivore population sizes at regional, national and Scandinavian levels, updated maps of species distribution, involvement of stakeholders in carnivore management, and a well-informed public due to great media interest in carnivore issues.

Each County Administration is responsible for being well-informed about carnivore numbers and distribution, and about problems caused by the predators. For this purpose, specially trained wildlife rangers are employed who mostly work in the field. In co-operation with reindeer herders, hunters and ornithologists, they survey the populations of carnivores. Large carnivores are surveyed according to the rules and guidelines issued by the Swedish Environmental Protection Agency and the Sami Parliament (links to relevant publications (in Swedish) are listed at www.naturvardsverket.se).

Research, Monitoring and Management

Humans have always been interested in large carnivores, primarily because these animals can be dangerous to them, but also because they can be a threat to livestock, or because the strength, agility and beauty of predators fascinates and inspires human fantasy. In former times, a good knowledge of carnivores was important to be able to exterminate them; today, we need good knowledge in order to be able to protect these species, which is demanded by EU-legislation.

Research on large carnivores has a long tradition in Scandinavia. The era of effective and scientifically sound projects began in 1984, when the Scandinavian Brown Bear Research Project was started. This became one of the most successful long-term research projects on carnivores worldwide (www.bearproject.info). A Swedish research project on lynx started in 1993 and, since 2005, there has been formalized cooperation with Norway within Scandlynx, the Scandinavian Lynx Project (http://scandlynx.nina.no/). The Swedish Wolverine Project was started in 1993 with the aim of providing a sound knowledge of wolverine ecology to facilitate science-based management and conservation of wolverines in Scandinavia (www.wolverineproject.se). The Scandinavian Wolf Research Project SKANDULV was established in 1999 (www.slu.se/skandulv/) and has since then increased our understanding of wolf ecology in Scandinavia tremendously. The youngest member in this suite of projects is the Swedish Golden Eagle Research Project (www.goldeneaglesweden.com). This was started in 2010 to investigate the effects of wind farms on golden eagles, but since then, the scope of the project has been broadened.

The management of large carnivores in Sweden is, as far as possible, based on scientific knowledge. Much of this knowledge originates in the five research projects mentioned above. The management authorities ask for specific information and, at least in part, fund research to produce it. The research projects conduct basic as well as applied research. Applied research often is reactive, answering the questions that management has put forward. Basic research is more pro-active, producing information and methods that management might find useful in the future. With respect to surveillance, research findings can enhance survey performance by making possible higher spatial and temporal resolutions, by increasing accuracy, or by testing and suggesting entirely novel methods. Then, the monitoring system uses the results from surveillance to inform management, in order to reach the adaptive, continuously improving management system that is the goal (see Fig. 1).

Methods for Surveillance

Several different methods are used for the surveillance of large carnivores in Sweden (Table 1). Many of them focus on family groups, i.e. females with young. As family groups are often more easily found than single individuals, they are more easily counted, and they can be used as an indicator of the dynamics of the whole population in a given area.

Snow-tracking during winter is the main method used to follow the lynx population and to track wolves; both packs and dispersing individuals. Also, for these two species, we use biotelemetry sensor tags to GPS-track individual animals to some extent (e.g. Liberg et al. 2012). Wolverines are surveyed by looking for and visiting

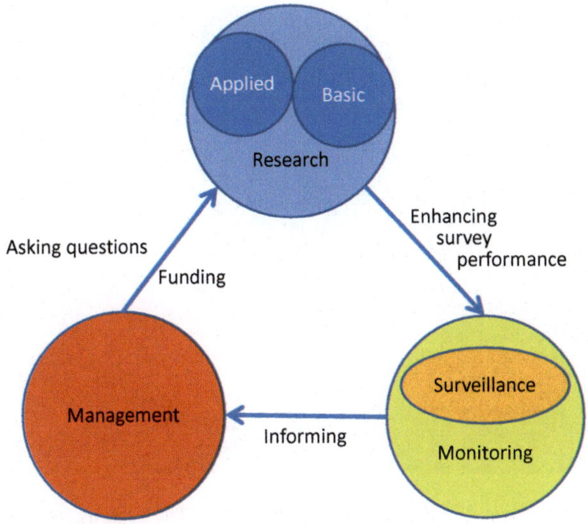

Fig. 1 A schematic view of the interplay of research, surveillance and monitoring, and management with respect to large carnivores in Sweden

Table 1 Many different methods are used during the surveillance of large carnivores in Sweden today

	Brown bear	Wolf	Wolverine	Lynx	Golden eagle
Snow tracking		XXX	XX	XXX	
Nest or den survey			XX		XXX
DNA	XXX	XX	XX	X	X
Biotelemetry tags	X	X	X	X	X
Automatic cameras			XXX	XX	
Aerial surveys			X	X	X
Reports from moose hunters	XX	X		X	
Reports from the public		X		X	
Attitude surveys	X	X	X	X	

X = used to some extent; XX = important method; XXX = very important method

den sites during spring. If encountered during tracking, hair, blood and excrement of the three species are sampled for DNA-analysis. The bear population is censused by direct observations during the moose (*Alces alces*) hunting season in autumn and by DNA-analyses of droppings. The population of Golden eagles is tracked by annual nest surveys, though DNA-analysis is increasingly used for this species too. Also biotelemetry backpacks with GPS-sensors are used to track the movement of individual birds in near real-time. Aerial surveys are conducted to some extent. Finally, attitude surveys are conducted regularly to measure and to track changes in attitudes towards large carnivores and large carnivore management (Schneider 2012).

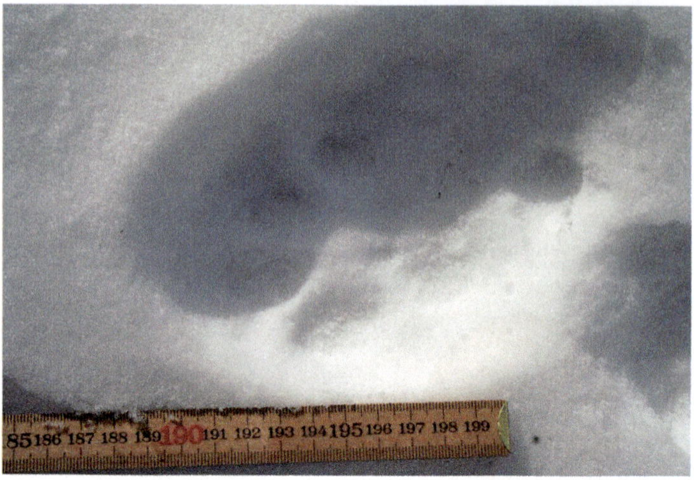

Fig. 2 A typical print from the right front paw of a lynx. The scale is in centimeters (Photo: Michael Schneider, April 2012)

Snow Tracking

Large carnivores move over huge areas almost every day and therefore, in winter, they leave tracks in the snow. Tracks reveal where carnivores move, what they do and how many they are. Often, it is also possible to determine the sex of an animal and, sometimes, its age (Aronson 2011).

Snow tracking is intensively used when surveying wolves, wolverines and lynx in northern Sweden. Bears hibernate and, therefore, they leave very few tracks in the snow. A pre-requisite for the method is a lasting snow cover, which makes it less suitable in more southerly parts of Sweden or in warmer countries.

Snow tracking is central for the monitoring of the lynx population (Fig. 2). The survey season for lynx lasts from October until the end of February, if snow cover is sufficient. Experienced and well-trained wildlife rangers from County Administrations search for tracks of lynx, using snow mobiles or even helicopters under certain circumstances. Also, reports from reindeer herders, hunters or the general public are collected and followed up in the field.

In areas with reproducing lynx populations, only family groups are counted, i.e. females with their young from last spring, which remain together until the onset of the following mating season in March. Neighboring family groups are separated in the field by tracking, or later on by applying spatio-temporal distance rules, which are based on results from research on the mobility and home range usage of lynx (Gervasi et al. 2013). From the number of family groups found during the survey, the total number of individuals in the population can be inferred (Andrén et al. 2002). In areas where lynx do not reproduce, single individuals are also tracked and it is determined if the species occurs regularly or temporarily only. Among other things, this information is important for the compensation system, where reindeer herders get paid for the occurrence of large carnivores on their grazing grounds.

In central Sweden, where most of the Swedish wolves occur, snow tracking is an important method for keeping track of the wolf population. Every winter, County Administrations use this method to count the number of wolf packs and, if possible, the number of individuals in each pack, and to determine if reproduction occurred in any given territory the previous spring. The number of newly established territorial pairs is also determined (Liberg et al. 2012).

Nest and Den Surveys

Reproducing females are also the center of interest during wolverine surveys, which last from March until the snow disappears in May to July, depending on the area. Wolverines use dens in the snow to give birth to their young. Females can use the same den site year after year, and good den sites may be used by consecutive females under a long period of time. Wolverine surveys have been conducted since 1996 in northern Sweden and the County Administrations have accumulated a good knowledge of important sites. This makes it possible to target these sites and to focus survey activities there. The objectives of survey activities at natal dens are to find out initially if the site is occupied and, if so, where the den is located, if there are young in the den, and how many. At the same time, the female should not be disturbed. Snow tracking, direct observation from a distance, collection of samples for DNA-analysis and automatic cameras are used at den sites to accomplish this (Schneider 2006b). Sites where it is unclear if reproduction has occurred can even be visited after snowmelt, when the wolverines have left, to look for signs of reproduction (juvenile hair, bite marks, prey remains, well-used trails etc.). Often, helicopters are used to effectively reach these usually remote places.

Even more so than wolverines, golden eagles can use the same place year after year for rearing their young. Sites can be used by eagles for several decades and ornithologists accumulate knowledge of where nests are, how often they are used, and what quality a given site has. Each territory, in some cases with multiple nest sites, of golden eagles has to be visited every year to find out if birds are present, if they breed, and if breeding is successful, i.e. if the young survive until they leave the nest (Ekenstedt and Schneider 2008). Survey activities include direct observation of displaying birds from a distance in the beginning of the breeding season, closer-range nest observation during the season, as well as banding of young eagles and taking samples for DNA-analysis at the end of the season. In remote areas in northern Sweden, helicopters can be used as an effective means to survey known eagle territories. Several nests can be visited within a single day and the view from above renders information about the number and age of the young.

Sampling and Analysing DNA

DNA-based methods are of ever-increasing importance for large carnivore surveys. For the brown bear, it is the most important method in Sweden for counting the numbers of individuals that are encountered in a given area, which is then used to estimate the total population size (Kindberg 2010). Brown bear DNA surveys are usually undertaken intermittently in the counties. Management authorities provide sampling material and ask volunteers, most of which are hunters, to collect samples of bear droppings that they find in the autumn and to send them in (Fig. 3). The results include regular estimates of bear population size at regional and national levels, updated maps of bear distribution, involvement of stakeholders in bear management and a well-informed public due to a multitude of media reports when scat sampling is started and when the outcomes are presented (Schneider 2006c).

DNA-based methods are also important for the wolf (Liberg et al. 2012; Åkesson 2017). The Scandinavian wolf population is small and it is affected by inbreeding. Therefore, it is possible, and necessary for wolf management, to keep track of the degree of inbreeding in the different wolf packs. Also, regular sampling of material (droppings, urine, blood etc.) makes it possible to identify immigrant individuals and their offspring and to adapt wolf management in order to protect these important individuals (e.g. Åkesson and Svensson 2015). This is especially important in northern Sweden, where many wolves are killed because of their possible negative impact on reindeer herding.

The importance of DNA-techniques has also increased during recent years with respect to wolverine surveillance. Accumulated results from several years make it possible to reconstruct the home ranges of individuals as well as the age of the animals.

Fig. 3 Droppings from a brown bear, containing DNA which is used to survey bear populations (Photo: Michael Schneider, June 2006)

They also make it possible to look at the spatial dynamics of the population and the dispersal of individuals and, in certain cases, to infer reproductive success where it could not been proven with other methods.

DNA-Data and Wolf Monitoring: A Case Study

The genetic situation of the wolf population in Scandinavia is an important issue for the management of the species in Sweden. The population is inbred and in great need of immigrating wolves from Finland and Russia, which diversify the Scandinavian gene pool once they reproduce. Therefore, the collection of samples for DNA-analysis is an important part in the monitoring of the wolf population in Sweden. However, DNA-information supports wolf monitoring in several ways.

Samples are collected by professional wildlife trackers mostly during snow tracking in winter, when they follow wolf packs, territorial pairs or, in the reindeer herding area, even single animals. Samples are also taken at rendez-vous sites of family groups and in places where wolves have attacked livestock. Trackers collect faeces, urine, blood, hair and saliva when they encounter such material and send samples of it to Grimsö Research Station in central Sweden, which is part of the Swedish University of Agricultural Sciences. There, samples are sorted according to their priority; emergency samples (e.g. samples from the reindeer herding area or from possible immigrants) are analysed within five working days, while the analysis of samples of normal priority takes some weeks, and some low-priority samples are not processed at all. Currently, more than 2000 wolf samples are collected and about 1200 of them are analysed each year (e.g. Åkesson 2017). This figure includes even tissue samples from wolves that are found dead or that have been killed by anthropogenic causes (traffic, hunting, illegal killing). On average, about 50 dead wolves per year are examined by the National Veterinary Institute in Sweden (e.g. Meijer and Ågren 2017).

In the first step of the analysis, DNA is extracted from the sample and purified with different methods, depending on the material provided in the sample. The next step is a polymerase chain reaction (PCR) process where the DNA is amplified. In total, 30 autosomal microsatellites are looked at, plus two markers for sex determination and five markers on mitochondrial DNA. The different alleles of the microsatellite markers are then separated and visualised by electrophoresis. Different genotypes (wolf individuals) are defined by their individual variation of their alleles of the different markers.

Currently, a new method is being tested at the laboratory in Grimsö. Instead of microsatellites, this method uses markers for single nucleotid polymorphism (SNP-markers). The method is supposed to be faster, cheaper and less prone to analytical error. It has successfully been used for bears in Sweden and is currently being developed for golden eagles as well.

(continued)

The objective of the analyses is to determine species, origin and sex of each sample and to find out if the individual already is known from other samples. These DNA-results are important as they help managers to assess the status of stationary wolves, to identify territorial individuals, to separate neighbouring territories and to confirm reproduction. They tell managers where individual wolves come from and they enable county administrations to follow the route of migrating wolves when their DNA is found in different places. Genetically important wolves that have immigrated from the east can be identified and subsequently protected from legal hunting or illegal activities. DNA-results have also made it possible to construct an almost complete pedigree of the Scandinavian wolf population, which gives us information on the degree of inbreeding in all wolf territories and in the population as a whole, telling us how severe the genetic situation is and how many immigrating wolves from the east are needed.

Currently, DNA-sampling is being intensified in order to get better background information on the structure of the wolf population. This information is needed to test a recently developed new conversion factor that can be used to translate the number of wolf packs (which are counted during the annual survey) to the total number of wolf individuals (single wolves are surveyed in northern Sweden only) in the population. Information on the number of individuals is important for understanding the population dynamics of the species, for setting hunting quotas, for comparing population size with objectives that have been set, for reporting to the European Union within the framework of the EU Habitats Directive, and for the communication between managers, media and the public.

With respect to golden eagles, methods for surveillance based on DNA are currently being tested and new methods have been developed. In the future, DNA-based results will give us insights into the turnover of individuals at breeding sites and parenthood of the birds present, as well as the size of the home range and shape of the territory. All of this will contribute to a better understanding of the golden eagle population size and dynamics as well as habitat quality and the frequency of illegal killing.

Tagging with Biotelemetry Sensors

The use of biotelemetry sensors mounted on tags, e.g. neck collars or backpacks, sometimes also referred to as 'radio tagging', is not a method that is primarily used for the purpose of surveilling large carnivores. However, the biotelemetry tags incorporating GPS-sensors for tracking during research projects can produce results of the utmost importance for both the surveillance and management of large carnivores.

Fig. 4 The wolf Bullmarksvargen, who was the catalyst for the development of Vargwebben. See text for further explanations (Photo: Michael Schneider, March 2007)

Results from GPS-tracked lynx have very much increased our understanding of home range sizes and the animals' movement within their ranges in different parts of Scandinavia. These results are reflected in the spatio-temporal distance rules that are applied when the number of family groups is determined during annual lynx surveys (Gervasi et al. 2013).

Results from GPS-telemetry collars on wolves are regularly used to determine which parts of the reindeer herding area these wolves have visited and for how long they stayed in each part (Fig. 4). This is important information for the compensation system. Also, an early warning system for hunters and reindeer herders ('Vargwebben') has been developed (Schneider 2008a), where near-real-time GPS-locations from collars on wolves are automatically transferred to a site on the Internet (http://webmap.slu.se/website/vargwebb). On this site, each 10×10 km^2 which contains the latest position of one or more collared wolves is highlighted on the map, so that people can avoid the actual area or so that herders can take care of their livestock when a wolf approaches. In order to avoid illegal killing, the location of the 10×10 km^2 containing the wolves is provided rather than the actual location of the animals. The date and time of the observation is also provided. This system, of course, depends on the presence of wolves with radio collars. In recent years, the number of tagged wolves has been decreasing, due to a shortage of funding for wolf research: currently there are very few wolves with radio collars.

A project on the predation of reindeer calves by brown bears has produced results that are important for a future compensation system that better reflects the damage done by bears. A sophisticated system consisting of biotelemetry collars

Fig. 5 A female wolverine moving her cub out of the den (Photo: One of the County Administration's automatic cameras in Västerbotten County, March 2015)

on bears and female reindeer that communicated with each other made this possible (Karlsson et al. 2012).

GPS-equipped biotelemetry backpacks on golden eagles have also given important insights into home range sizes, home range use over time, dispersal of juveniles, and adult seasonal migrations. They also have supplied information on the birds' susceptibility to wind power plants and the mortality of eagles in general (Singh et al. 2017).

Automatic Cameras

Today, automatic cameras are used widely, both in research and wildlife management as well as by the public. In wolverine surveillance, automatic cameras are employed by County Administrations in Sweden to determine the occurrence and number of young animals at the dens in a relatively non-invasive way (Fig. 5). These cameras are also used to take photographs of the bellies of wolverines at feedings stations, in order to identify lactating females, which are the focus of the surveys.

In the surveillance of golden eagles, automatic cameras are used to get background information regarding the development of the plumage of young birds and to aid age determination. Cameras have also given interesting insights into prey choice and predation rates of the birds. They have also contributed to the understanding of breeding success and illegal activities of humans at nest sites.

Winters are getting increasingly warmer in Scandinavia and the snow cover is becoming more unpredictable. Therefore, in many areas, snow tracking can no longer be used as the main method for surveying lynx. Currently, a method is being developed where automatic cameras are used instead (Odden 2015). Whole family groups can be captured by the cameras and, due to individual color patterns, individuals can be identified on photographs. In this way, the number of family groups and total population size can be inferred from photographs.

Aerial Surveys

When conditions are right in northern Sweden (sunny, calm weather and lots of snow on the ground), helicopters are sometimes used to search for tracks of lynx and wolverines. The results from these aerial surveys are then used as a starting point for a more thorough, ground-based work. Helicopters are also used as an effective means of transportation, when remote den sites of wolverines have to be visited after the snowmelt and when the animals have left. These visits have to be made when it is unclear at the end of the survey season if reproduction has occurred at a site.

In remote areas in northern Sweden where golden eagles are abundant but roads are scarce, helicopters can be used as an effective means to survey known eagle territories. However, new territories and nests cannot effectively be searched for from helicopters, which is why airborne surveys have to be augmented by ground-based activities. In Finland, the use of helicopters is the main method for surveying golden eagles and in 2016 almost 500 territories were visited in this way (Ollila 2016).

There have been ideas of using drones in the surveillance of both golden eagles and wolverines. However, the current legislation surrounding the usage of drones, and filming and photographing from the air, is very restrictive in Sweden. Furthermore, territorial golden eagles can show aggressive behavior towards approaching flying objects, which may result in collisions (Ahlgren 2015). Therefore, currently drones are not used when surveying large carnivores in Sweden.

Reports from Moose Hunters

Every autumn, thousands of hunters spend days and weeks in the Swedish forests to participate in moose hunting. Around 100,000 moose are killed in Sweden during the hunting season each year. For many people, moose hunting is an important recreational activity, and a means of acquiring most of the meat for their domestic cooking during the year to come. Most of the hunters are organized in hunting parties, i.e. groups of people hunting together in an area which they own or lease (Schneider 2012).

Hunting parties are asked to keep track of bear sightings during moose hunting and to report their results to the Swedish Hunters' Association (Kindberg et al. 2009). There, the results are compiled for the whole country and an annual index of bear sightings per observation effort is published for every bear county. The 'sightability' of bears is different in different areas and at different times. Therefore, this bear observation index in itself does not tell us very much about the exact number of bears in the woods, but it indicates the dynamics of the bear population in a given area over time. Together with results from dropping surveys, trends from observation indices can be used to estimate bear population size in the counties in different years (Kindberg 2010).

Reports from the Public

Citizen science, the collection of scientific data by members of the public on a voluntary basis, has many advantages. Data collection is relatively cheap, interested individuals have the possibility to participate in what they feel is important, and science and the public get a common knowledge base. It is a goal of the large carnivore management system in both Norway and Sweden to include the public in the survey of large carnivores.

In Sweden, the importance of the general public as rapporteurs increases from the north to the south. In the north, there are fewer reports from the public as fewer people live there. However, as there is snow cover for longer in the north it is possible for professional trackers to work effectively and successfully. Furthermore, the compensation system for reindeer herders demands that any observation of large carnivores has to be checked and verified by County Administration field personal in order for compensation payments to be made. Also in the north, public reports of lynx family groups and of wolves play an important part in the surveillance of these species.

The public, at least the part interested in hunting, is an important player when it comes to the surveillance of brown bears. Bear dropping surveys depend largely on samples collected by the public and bear observation indices are similarly based on bear sightings reported by the public.

Most golden eagle surveillance is undertaken by ornithologists. It is debatable if these people still can be looked upon as "the public", as many of them are specially trained and have huge experience in the field. However, most of their work is carried out on a voluntary basis.

In order to make it easier for the public to report their sightings of carnivores, an internet site has been launched where observations can be reported (www.skandobs.se). Also, an application has been developed for mobile phones, which can be used for reporting directly in the field.

Attitude Surveys

Among other things, human attitudes depend on the levels of predator damage to dogs, livestock, reindeer and game animals, on actual or perceived threats to humans, and on local levels of involvement during decision making. Many people have strong feelings towards large carnivores. These feelings may be positive or negative and may encompass anything from hatred to love, from deep fear to enthusiastic sympathy. Often, it is not the predators *per se* that are the problem, more an underlying conflict between a central administrative institution and the countryside community, a conflict that may exist at several scales. All of these feelings and attitudes have to be taken seriously when managing large carnivores (Schneider 2008b).

A multitude of attitude surveys regarding large carnivores has been conducted in Sweden. Since 2004, these surveys are carried out at a large scale every 5 years, encompassing the northern half of the country and treating all species of large carnivore except golden eagles (Ericsson and Sandström 2005; Ericsson et al. 2010). Although the results from these surveys show that an overwhelming majority of the people is supportive of both carnivores and their current management, many of those who live closest to the carnivores have a more negative attitude. Over time, people's support for large carnivores and large carnivore management has been decreasing. This is true especially for wolves and bears (Sandström and Ericsson 2009; Sandström et al. 2014). These are important findings for the management system, as negative attitudes may be implemented in an illegal killing, which not only threatens carnivore survival but also threatens the fulfillment of regional and national objectives as well as international treaties and directives.

Handling Loads of Data

The Scandinavian Carnivore Database Rovbase

Surveillance data on large carnivores are collected by many different players in many places and in many different ways. In Sweden, the data collectors include wildlife rangers and ornithologists in the field, indoor-personnel at County Administrations, DNA-laboratories, the Swedish Veterinary Institute, and the Swedish Wildlife Damage Centre. The internet-based database 'Rovbase' is used to help smooth the reporting of all these data.

Rovbase was developed by the Norwegian Environment Agency (Miljødirektoratet). Initially it was restricted to Norway, whereas today Rovbase is used in both Norway and Sweden and increasingly in Denmark. Currently (February 2017), Rovbase has about 200,000 records from Sweden, and approximately 20,000 new data points are added each year. Rovbase is also used to store, to visualize, and to make large carnivore data available for inspection, for completion if necessary, and for further analysis.

As parts of the data in Rovbase are classified information (nests of golden eagles, dens of wolverines etc.), access to the database is restricted. Only personnel actively involved with the monitoring or management of large carnivores is granted access. However, to satisfy the broader interest of the public, an open version of the database (www.rovbase.se) is available for anyone interested in carnivores. This open version of Rovbase is also used to give feedback to people who have reported observations of carnivores or who have participated in bear dropping surveys and collected fecal samples (Schneider 2015).

Bear observations that are made by moose hunters in the autumn are not reported in Rovbase: the Swedish Hunters' Association has its own routines and methods for processing and publishing these data. Similarly, Rovbase is not used to store data from carnivore research. Rovbase data can, however, be used for subsequent research.

The Wireless Remote Animal Monitoring Database E-Infrastructure

Data from research projects and data from large carnivores that are acquired by using biotelemetry sensors are not stored in Rovbase. The wealth of data that originates from these sources is handled in another e-infrastructure solution, the Wireless Remote Animal Monitoring (WRAM), which is an instrument for automatic reception, long-time storage, sharing and analyzing of biotelemetry sensor data from animals (Dettki et al. 2014). It has been developed and operated as the national Swedish e-infrastructure for biotelemetry data from both research and environmental monitoring programs by the Umeå Center for Wireless Remote Animal Monitoring (UC-WRAM) at the Swedish University of Agricultural Sciences (SLU) in Umeå.

During the last 40 years, in Sweden and other countries, radio collars, fish tags, and other telemetry equipment have been used to track fish and wildlife. The equipment used most often relied on traditional radio-telemetry techniques using 'Very High Frequency' (VHF) or 'Ultra High Frequency' (UHF) radio collars. These techniques are labor intensive and result in a relatively low numbers of positions per individual.

In recent years, Global Positioning System (GPS) telemetry collars and automated telemetry equipment have been used to an ever increasing extent. Due to the automated fashion of data gathering and storage as well as improved battery lifetime of the equipment, the average cost per position is shrinking, resulting in large datasets being collected per animal studied. Furthermore, additional biosensors can be integrated in the same collar, measuring, for example, activity of the animal, body temperature, heart beat or proximity to other sensors.

Due to the huge amount of data available, single research groups are often no longer able to analyze their data in a timely fashion and they want to share their data with similar research projects to obtain synergy effects and to open up for new fields of research. Thus, the WRAM database e-infrastructure has been developed to enable present and future national and international cooperation by connecting together WRAMs own database with similar national and international data repositories for biotelemetry data, as e.g. German-US 'Movebank', Italian 'EuroDeer', or Norwegian 'Dyreposisjoner'. The objective is to provide an open database network which is independent of location or the platform used, and which is used to store, share, secure and analyze data from wireless remote animal monitoring. The database is accessible to all participating researchers and research groups and features a web portal used to select, visualize, and access raw data, together with simple spatial analysis tools and statistical tools. Currently (May 2017) the database contains approximately 186 million records from real-time biotelemetry sensors and is used to date by 38 user groups from eight countries, monitoring more than 24 species and 2892 individual animals (www.slu.se/wram). One recent user is the Sami Parliament, who decided to use WRAM for handling biotelemetry data collected from reindeer in Sweden.

Geodata and Their Integration with Tracking Data

In Sweden, most geodata are easily available to anyone who wants to use them. These open geodata may be accessed, used and published free of charge and they are available in a machine-readable interface. Lantmäteriet (www.lantmateriet.se), the National Land Survey of Sweden, carries out aerial photography and airborne laser scanning as well as land surveys. They also ensure that companies, authorities and the public have access to the information, e.g. in the form of maps and images. When it comes to carnivore monitoring, useful information on e.g. topography, elevation and vegetation cover can be downloaded from their site (Lantmäteriet 2017).

The satellite database Saccess contains satellite images of Sweden that cover every decade from the 1970s onwards, and every year since 2007. The nationwide data sets consist of optical multispectral data from satellites with a spatial resolution of 10–30 m (except MSS data from 1970 that have 80-meter resolution). The information is easily accessible via the Internet and is available free of charge, due to special government funding. Different satellites and different sensors have been used through the decades. For 2015 e.g., satellite information is available based on IRS-P6 (© *ANTRIX, SI, Euromap Neustrelitz*) and Landsat 8 (*USGS/NASA Landsat, Processing Metria AB)* (Lantmäteriet 2017).

Users of Rovbase usually do not have to care very much about geodata, as these are automatically included when maps are produced. The database contains cached maps (images) that are activated depending on the level of zooming. These images are supplied by Metria (www.metria.se) and they build on vector-data from Lantmäteriet, who also deliver the Swedish aerial photographs that are available in Rovbase. The map of Scandinavia in Rovbase is delivered by Statens Kartverk (www.statenskartverk.no), the Norwegian Mapping Authority, which also supplies maps and aerial photos for Norwegian users.

The Wireless Remote Animal Monitoring (WRAM) database e-infrastructure includes a web portal which can be used to select, visualize, and access raw data, together with simple spatial analysis tools and statistical tools. Currently, this portal uses geodata information from Lantmäteriet for its background maps, as e.g. in Vargwebben, but a transition to Google Maps is currently contemplated. In many cases, users of WRAM will apply their own GIS-solution within their research projects for further analyses. How an integration of GPS tracking data and remotely acquired environmental data can be accomplished in order to look at habitat usage, has been presented by e.g. Rauset et al. (2012) for lynx and wolverine and by Nellemann et al. (2007) for brown bears.

Summarizing Conclusions

Sweden has an elaborate system for the surveillance, monitoring and management of large carnivores, in part because it is demanded by the compensation system for reindeer herding. As Sweden has five different species of carnivores to deal with in landscapes which differ in climate, topography, vegetation, human population density and infrastructure, a whole battery of different methods has to be used. Cooperation with Norway is essential, as large carnivore populations are cross-boundary and Scandinavian rather than Swedish or Norwegian. The surveillance system is expensive and costs about 5 million Euros in Sweden annually, but the results obtained are of high quality. Tight cooperation between management and research contributes to the up-dated quality and effectiveness of the system.

Acknowledgements We are grateful to Peter Jaxgård for providing information on Rovbase, to Linda Backlund for directing us to surveillance photos of wolverines, and to Edith Marie Pasquier and Clive Hurford for commenting on the manuscript. We are also grateful to all those people who work with carnivore surveillance in Scandinavia.

References

Ahlgren, C.-G. (ed.): The Golden Eagle in Sweden 2015. Kungsörnen, 2015, 11–19 (2015) (in Swedish with English summary)

Åkesson, M.: Technical report on genetic analyses of wolves in Sweden 2016. Swedish University of Agricultural Sciences (2017) (in Swedish)

Åkesson, M., Svensson, L.: Compilation of the pedigree of the Scandinavian wolf population until 2014. Swedish University of Agricultural Sciences (2015) (in Swedish)

Andrén, H., et al.: Estimating total lynx *Lynx lynx* population size from censuses of family groups. Wildl. Biol. **8**, 299–306 (2002)

Aronson, Å.: Tracks and tracking of large carnivores. Swedish Wildlife Damage Center, Swedish University of Agricultural Sciences, Riddarhyttan (2011) (In Swedish)

Dahl, E.L., Nilsen, E.B, Brøseth, H., Tovmo, M.: Estimating the number of breeding pairs of golden eagle based on known occurrence in Norway for the period 2010–2014. NINA Report 1158. 23 pp (2015) (in Norwegian with English summary)

Dettki, H., Brode, M., Clegg, I., Giles, T., Hallgren, J.: Wireless Remote Animal Monitoring (WRAM) – A new international database e-infrastructure for management and sharing of telemetry sensor data from fish and wildlife. In: Ames, D.P., Quinn, N.W.T., Rizzoli, A.E. (eds.) Proceedings of the 7th International Congress on Environmental Modelling and Software, June 15–19, pp. 122–128. San Diego, California (2014)

Ekenstedt, J., Schneider, M.: The Golden Eagle (*Aquila chrysaetos*) in the North Calotte Area 1990-2007. Report No. 55, The North Calotte Council (2008)

Ericsson, G., Sandström, C.: Partial report on the attitudes of Swedes towards carnivore politics and management. FjällMistra-report nr. 10 (2005) (in Swedish)

Ericsson, G., Sandström, C., Kindberg, J., Støen, O.-G.: On the fear of Swedes of large carnivores, moose and wild boar. Report 2010:1, Department of Wildlife, Fish, and Environmental Studies, Swedish University of Agricultural Sciences (2010) (In Swedish)

Gervasi, V., Odden, J., Linnell, J. D. C., Persson, J., Andrén, H., Brøseth, H.: Re-evaluation of distance criteria for classification of lynx family groups in Scandinavia. NINA report 965 (2013) (In Norwegian with English summary)

Hobbs, N.T., Andrén, H., Persson, J., Aronsson, M., Chapron, G.: Native predators reduce harvest of reindeer by Sami pastoralists. Ecol. Appl. **22**, 1640–1654 (2012)

Karlsson, J., et al.: Bear predation upon reindeer and potential effects of three preventive measures. – Report 2012:6, Swedish Wildlife Damage Center, Swedish University of Agricultural Sciences, Riddarhyttan (2012) (in Swedish)

Kindberg, J.: Monitoring and management of the Swedish Brown Bear (*Ursus arctos*) population. Acta Universitatis Agriculturae Sueciae 2010, 58 (2010)

Kindberg, J., Ericsson, G., Swenson, J.E.: Monitoring rare or elusive large mammals using effort-corrected voluntary observers. Biol. Conserv. **142**, 159–165 (2009)

Lande, U.S., Linnell, J.D.C., Herfindal, I., Salvatori, V., Broseth, H., Andersen, R., Odden, J., Andrén, H., Karlsson, J., Willebrand, T., Persson, J., Landa, A., May, R., Dahle, B., Swenson, J.: Potential habitats for large carnivores in Scandinavia: GIS-analyses on an eco-regional level. NINA Fagrapport 64, Norwegian Institute for Nature research (2003) (in Norwegian)

Lantmäteriet.: www.lantmateriet.se. Accessed on 25 February 2017 (2017)

Liberg, O., Aronson, Å., Sand, H., Wabakken, P., Maartmann, E., Svensson, L., Åkesson, M.: Monitoring of wolves in Scandinavia. Hystrix. **23**, 29–34 (2012)

Linnell, J., Salvatori, V., Boitani, L.: Guidelines for population level management plans for large carnivores in Europe. Contract nr. 070501/2005/424162/MAR/B2. Final version 1st July 2008 (2008)

Meijer, T., Ågren, E.: Large carnivores 2016. Report, National Veterinary Institute (2017) (in Swedish)

Nellemann, C., Støen, O.-G., Kindberg, J., Swenson, J.E., Vistnes, I., Ericsson, G., Katajisto, J., Kaltenborn, B.P., Martin, J., Ordiz, A.: Terrain use by an expanding brown bear population in relation to age, recreational resorts and human settlements. Biol. Conserv. **138**, 157–165 (2007)

Neumann, W., Martinuzzi, S., Estes, A.B., Pidgeon, A.M., Dettki, H., Ericsson, G., Radeloff, V.C.: Opportunities for the application of advanced remotely-sensed data in ecological studies of terrestrial animal movement. Mov Ecol. **3**(8), 1–13 (2015). doi:10.1186/s40462-015-0036-7

Odden, J.: Using camera traps to monitor lynx – a pilot study in three areas of southeastern Norway – NINA Report 1216 (2015) (in Norwegian with English summary)

Ollila, T.: The Golden Eagle in Finland 2016. Kungsörnen. 24–25 (2016) (in Swedish with English summary)

Persson, J., Rauset, G.R., Chapron, G.: Paying for an endangered predator leads to population recovery. Conserv. Lett. **8**(5), 345–350 (2015). doi:10.1111/conl.12171

Rauset, G.R., Mattisson, J., Andrén, H., Chapron, G., Persson, J.: When species' ranges meet: assessing differences in habitat selection between sympatric large carnivores. Oecologia, published online 14 December 2012. doi:10.1007/s00442-012-2546-y (2012)

Sametinget & Naturvårdsverket: A management tool for the occurrence of large carnivores based on tolerance levels for reindeer herding. Report 6555, Naturvårdsverket (2013) (in Swedish)

Sandström, C., Ericsson, G.: On Swedes' attitudes towards carnivores and carnivore politics. Report 2009:1, Department of Wildlife, Fish, and Environmental Studies, Swedish University of Agricultural Sciences (2009) (in Swedish)

Sandström, C., Lindvall, A.: Regional management of carnivores in Västerbotten and Norrbotten – about similarities and differences from a co-management point of view. FjällMistra-report nr. 18 (2006) (in Swedish)

Sandström, C., Ericsson, G., Dressel, S., Eriksson, M., Kvastegård, E.: Attitudes towards carnivores and carnivore politics. Report 2014:1, Department of Wildlife, Fish, and Environmental Studies, Swedish University of Agricultural Sciences (2014) (in Swedish)

Schneider, M.: Management of large carnivores in Västerbotten County. An introduction to the management plans for Brown Bear, Wolverine, Lynx, Wolf and Golden Eagle 2006–2010. Meddelande 5–2006. Västerbotten County Administration, Umeå (2006a) (in Swedish with English summary)

Schneider, M.: Monitoring the Wolverine *Gulo gulo* in Västerbotten County. In: Hurford, C., Schneider, M. (eds.) Monitoring Nature Conservation in Cultural Habitats, pp. 215–230. Springer, Dordrecht (2006b)

Schneider, M.: Monitoring the Brown Bear *Ursus arctos* in Västerbotten County. In: Hurford, C., Schneider, M. (eds.) Monitoring Nature Conservation in Cultural Habitats, pp. 195–214. Springer, Dordrecht (2006c)

Schneider, M.: The wolf Bullmarksvargen September 2006 until March 2008. Report, Västerbotten County Administration, Umeå (2008a) (in Swedish)

Schneider, M.: Managing large carnivores in Northern Sweden. BarentsWatch 2008, English edition: Large carnivores in the Barents region, pp. 10–11 (2008b)

Schneider, M.: Managing large carnivores in Västerbotten County. Meddelande 19–2012, Västerbotten County Administration, Umeå (2012)

Schneider, M.: The bear dropping survey of 2014 in Västerbotten County. Report, Västerbotten County Administration, Umeå (2015) (In Swedish)

Singh, N., Hipkiss, T., Ecke, F., Hörnfeldt, B.: The importance of home ranges, habitat choice and movements of golden eagles for wind power establishment. Report 6734, Naturvårdsverket (2017) (in Swedish)

Zabel, A., Engel, S.: Conservation performance payments. Policy Brief, PEPE, Swiss Federal Institute of Technology Zurich (without year) (n.d.)

Zabel, A., Holm-Müller, K.: Conservation performance payments for carnivore conservation in Sweden. Conserv. Biol. **22**, 247–251 (2008)

Zabel, A., Bostedt, G., Engel, S.: Outcomes and Determinants of Success of a Performance Payment Scheme for Carnivore Conservation. Working paper 7, CERE (Center for Environmental and Resource Economics), Umeå, Sweden (2010)

Coupling Field Sampling with Earth Observation Increases Understanding of Tiger Movement and Behaviour

Sonali Ghosh and Richard Lucas

Abstract Tigers (*Panthera tigris*) are critically endangered worldwide, with the geographical range of native populations reduced to less than 7% in the last hundred years. Currently, there are only 6 subspecies residing in thirteen range countries. The main causes of their decline are habitat loss and fragmentation, prey depletion and poaching for the illegal wildlife trade. Addressing these issues has been severely constrained by inadequate spatial information on tiger distributions, movements, habitat preferences and behaviour. Focusing on the Indo-Bhutan Manas Tiger Conservation Landscape (IBMTCL) in the Indian subcontinent (India and Bhutan), this study sought to demonstrate the use of field-based camera traps for providing ground-based observations of tigers and other large cats, namely leopards (*Panthera pardus)* and clouded leopards (*Neofelis nebulosa*), and their prey. These observations were also coupled with classifications of land cover, elevation data and fire observations from Landsat TM and the Terra ASTER and MODIS respectively. The study indicated a large but variable range for individual tigers, a preference of tigers for forest cover and proximity to burned areas (attributed to greater access to prey), and a spatial separation from populations of other large cats. The study illustrates how earth observation data can provide some of the elements needed to better understand how large cats utilise the landscapes they inhabit thereby contributing to efforts aimed at long-term conservation of these endanged species across their range.

Keywords C12 • Tiger • *Panthera tigris* • Carnivores • Royal Manas National Park • Manas National Park • Camera trap • Earth observation • Fire • Topography • Land cover

S. Ghosh (✉)
Wildlife Institute of India, Chandrabani, P.O.Box.18, Dehradun 248001, Uttarakhand, India
e-mail: ghoshsonali@gmail.com

R. Lucas
Centre for Ecosystem Science, School of Biological, Earth and Environmental Sciences, The University of New South Wales, High Street, Kensington (Sydney), NSW 2052, Australia
e-mail: richard.lucas@unsw.edu.au

© Springer International Publishing AG 2017

R. Díaz-Delgado et al. (eds.), *The Roles of Remote Sensing in Nature Conservation*,
DOI 10.1007/978-3-319-64332-8_12

Introduction

The tiger (*Panthera tigris*) is the largest and one of the most charismatic and endangered cat species in the world. It is a flagship icon for wildlife conservation, being at the apex of the food chain in the majority of the eco-regions in Asia (Sunquist et al. 1999; Jhala et al. 2008, 2015). The tiger has been held in mythical reverence across the world, is worshipped in certain faiths and religions and is the national animal for India, Bangladesh, Nepal, Malaysia and both North and South Korea. Until the early nineteenth century, nine subspecies (Luo et al. 2004) of tiger were found across Asia, from the Caucasus and the Caspian Sea to Siberia and Indonesia. Three of these subspecies are now extinct; the Bali Tiger (*Panthera tigris balica*), Javan Tiger (*P.t.sondaica*) and Caspian Tiger (*P.t.virgata*). The remaining six occur as fragmented populations in the Russian Far East, northeast China and the Asian subcontinent.

India supports about 70% of the World's tiger population and one of the key areas of importance is the Indo-Bhutan Manas Tiger Conservation Landscape (IBMTCL), which offers a contiguous protected landscape of over 6000 sq km and constitutes the core area of Manas National Park (NP) in India and Royal Manas NP in Bhutan. Between November 2010 and February 2011, a large camera trap survey was conducted in Manas NP and the southern section of Royal Manaus NP by organisations in both India and Bhutan. This survey was undertaken to locate, identify and track the movements of individual Royal Bengal tigers (*Panthera tigris tigris*) and two other sympatric species, the leopard (*Panthera pardus*) and clouded leopard (*Neofelis nebulosa*). Landsat-derived classifications of land cover, historical MODIS fire records and spatial information on human development of the landscape were then used in combination with the survey results to provide a better understanding of niche separation as well as influences on the movement and range of these large cat species over the period of observation. Through this approach, the potential contribution of integrated field campaigns and earth observation to supporting management of Tiger Conservation Landscapes (TCLs) was assessed.

Background

Why Are Tigers Endangered?

Whilst once widespread, the range of tigers has been reduced to a mere 7% (less than 1,184,911 km^2) of their former geographical range, and approximately 55% of the inhabited landscape is essentially non-tiger habitat (Sanderson et al. 2006). A general compilation of national estimates suggests a global population of approximately 4240 tigers remaining in the wild (Goodrich et al. 2015). The causes for their decline has largely been habitat loss and poaching and the pressures continue to escalate (Sanderson et al. 2006; Dinerstein et al. 2007). They are also vulnerable to extinction because of their low densities (relative to other mammals, including their prey species) and relatively low recruitment rates (where few animals raise offspring that survive to join the breeding population) (Kerley et al. 2003). Tigers also require large undisturbed landscapes with ample prey to raise young and maintain long term genetic and demographic viability (Seidensticker and McDougal 1993; Karanth and Sunquist 1995; Carbone et al. 1999; Jhala et al. 2008). However, such areas are contracting or becoming increasingly fragmented because of high anthropogenic pressures on natural resources. Whilst India has the world's largest population of wild tigers, it has also suffered the highest range contraction for the species. Poaching is mainly for the supply of organs, bones and hides that are smuggled for use in traditional medicine, mainly into China and south-east Asia (Gratwicke et al. 2008; Nowell 2009).

Efforts to save the tiger started rather late. For example, Project Tiger was initiated in India in 1973 (GOI 2005). However, the strategies used were not time-tested and were also often weak on scientific rigor, partly because of a general lack of baseline information on tiger distributions *(ibid).* Conservation strategies for tigers had previously focused only on protecting the species in pristine wilderness areas and were based on the premise that estimates of tiger densities derived from sign-based indices (e.g., pugmark census techniques) were adequate to provide information on the survival status of the species. However, detecting tigers through traditional field survey methods is a challenging task. Tigers require sufficient numbers of large-bodied prey and therefore roam across large that may exceed 100s of kilometres. They also occur in low densities and often as lone individuals, particularly in closed forests (Gittleman and Harvey 1982; Carbone et al. 2001), are cryptic by nature and are rarely seen. As a consequence, records relating to tiger distributions and their use of habitats have been inadequate and this has severely impeded the formulation and implementation of appropriate conservation measures (Blake and Hedges 2004; Sanderson et al. 2006).

For these reasons, there has been an increasing and now urgent need to devise rapid yet rigorous population analysis and viability methods with spatial and temporal observations that are repeatable across the varied landscapes that tigers occupy (Linkie et al. 2006). It is also widely accepted that a landscape approach to tiger conservation that takes into account the distribution and condition of habitat also fares better than those that focus only on protecting the species. Although breeding populations are currently found in eight Asian range states, there are few regular empirical studies to indicate trends on a countrywide basis (Goodrich et al. 2015). Sanderson et al. (2006) put forward the concept of establishing Tiger Conservation Landscapes (TCLs) and estimating tiger populations across their entire range using multiple methods; from expert opinion to land cover (e.g., vegetation) and change assessments based on satellite sensor data. The study also highlighted that 76 TCLs held the major tiger subpopulations in the world. An additional 543 fragments were also identified, but they were considered too small to support long-term viable populations. 491 areas were designated as Tiger Survey Landscapes (TSL; total area of 1.1 million km^2), where the current tiger status was unknown but the area was considered sufficiently large to support at least five tigers. The study highlighted that it would be a major challenge to survey and monitor these TCLs on a regular basis unless new options for mapping and monitoring their state, condition and dynamics were developed and implemented.

Determining the Distribution and Abundance of Tigers

Whilst the ecology of wild tigers has been studied in depth (Sanderson et al. 2006), information on their present status is limited and scattered over a number of publications, with many in the grey literature. In a review conducted in support of this study, 35 papers were identified that provided information relating specifically to assessing tiger populations and their habitat, with these based largely on primary data collection. Approximately half of these studies used capture-recapture methods with a lesser proportion focusing on sign-based indices (such as tracks, visual estimates, scat DNA analysis; 25%), and satellite tracking (7%). Only one paper (Imam et al. 2009) explicitly used remote sensing data to assess tiger habitat suitability in the landscape. The review highlighted the general paucity of information, with only a few local-scale studies, but also the varied and often inconsistent approaches for estimating population distributions and sizes. Furthermore, many population estimates were found to be based on single-studies, typically field-based surveys, with these relying primarily on pugmark census, camera traps, satellite telemetry, scent-matching using dogs, fecal genetics and large scale occupancy surveys. Most studies also relied on capture-recapture techniques conducted over a specified period (Table 1).

From the 1970s to 2005, the pugmark census technique has been widely used and has relied on sign-based abundance estimates. The census is based on intensive monitoring of tigers within areas, identifying individual tigers by visual inspection of the pugmark tracings/plaster casts, measuring pug mark variables (e.g., for toe to pad distance, pad length and width etc.) and subsequently using statistical techniques to identify individuals and numbers (Choudhury 1970; Sharma et al. 2005).

Table 1 Advancement of methods for estimating tiger populations

>1970[a]	1970[b]	1992–2005[c]	2005–2010[d]	2010-(proposed)
Hunting and man-animal conflict records.	Pugmark census	Camera-traps, Prey-predator models, Landscape level habitat analysis.	Large-scale occupancy and Population modelling, faecal genetics and digital pugmarks	Integration of ground-based surveys and satellite-based observations

[a]Rangarajan (2001)
[b]Choudhary (1970), Karanth (1995)
[c]Wikramanayake et al. (1998), Mackenzie et al. (2002)
[d]Sharma et al. (2005), Caroll and Miquelle (2006), Sanderson et al. (2006), Samrat et al. (2009)

Satellite telemetry allows animals with radio collars to be tracked from the ground or air. Animal locations are plotted on base maps from which their home range can be approximated. This method is particularly useful for detecting the movement of individuals and understanding tiger ecology and behaviour (Smith et al. 1998).

Camera trapping has long been used to estimate populations of faunal species (Otis et al. 1978; Balme et al. 2009; Carbone et al. 2001; Harihar et al. 2009) and is currently the most accepted technique for studying tiger populations in the Indian Subcontinent (Karanth and Sunquist 1995; Karanth et al. 2002; Karanth et al. 2006; Royle et al. 2009; Wegge and Storaas 2009). This method works on the *a priori* premise that individual tigers have unique biometrics, such as stripe patterns or tail to body ratios, that can be captured by placing automated camera traps in optimal field locations. The biometric information can then be analyzed within a robust capture-recapture statistical framework to arrive at estimates of numbers. The technique has been used for determining the spatial location and movements of carnivores such as tigers (Karanth et al. 2006; O'Brian et al. 2003; Linkie et al. 2006), leopards (Harihar et al. 2011), pumas (Kelly et al. 2008), jaguars (Davis et al. 2010) and other large predators. New techniques that have been used in conjunction with camera traps include the use of scent-matching dogs for Amur tigers (Kerley and Salkina 2007), fecal genetics to prepare DNA profiles of individual tigers (Samrat et al. 2009) and large-scale occupancy surveys for tigers and their prey (Jhala et al. 2008).

Use of Remote Sensing Data for Tiger Habitat Mapping

To address the mapping of tigers distributions, a handful of studies (e.g., Sanderson et al. 2006; Imam et al. 2009) have used thematic or continuous layers (e.g., representing vegetation, land use and elevation data) derived from remote sensing data as input to Species Distribution Models (SDM) (e.g., occupancy) or population models. Most have focused on using freely available pre-processed optical satellite sensor data, pixel-based image classifications and geometry-based geospatial analysis (Karanth et al. 2009; Ranganathan et al. 2008; Jhala et al. 2008). However, their generation and use has often been compromised by uncertainty and errors (positional and thematic) related to geometric correction accuracy, poor atmospheric and topographic correction (particularly in the mountainous areas that tigers commonly occupy), and inconsistent use of classification and accuracy assessment techniques. There is, however, considerable interest in the use of remote sensing data, including as input to SDMs and population models, and the increased quality, diversity, availability and accessibility of data and derived products is contributing to greater uptake. A summary of information that can be obtained from remotely sensed data and which is relevant to understanding and/or modelling tiger distributions, movement and behaviour is provided in Table 2.

Table 2 Habitat suitability parameters that can be derived from satellite sensor data

Factor	Relevance	Systems (Examples)
Elevation, slope, and aspect	Tigers prefer shallow to moderate slopes and, in winter, southerly facing aspects are often visited.	SRTM, ASTER, Tandem-X, PRISM
Water bodies	Water supply for tigers (primarily for drinking).	Optical, radar
Climate	Long term impacts on tiger distributions, movements and breeding.	TRMM[a], MODIS[b]
Grassland productivity	Associated with fluctuations in ungulate numbers. During periods of low productivity, prey are often found close to water bodies	MODIS, Landsat, Sentinel-2
Biomass	Easier movement for tigers within higher biomass forests because of greater openness of the ground layer.	ALOS PALSAR
Canopy height	Taller trees are typically associated with a more open understorey, which eases movement for tigers	ICESAT GLAS
Canopy cover	Tigers prefer vegetation with a higher percentage canopy cover	Optical
Settlements	Tigers avoid extensively urbanized areas but can occur near villages and isolated houses	SPOT-5, ASTER
Infrastructure	Roads limit the movement of tigers	High resolution data including Worldview and RapidEye
Logging roads	Indicate intrusion into the forest, which can restrict or limit tiger movements and occupation.	
Land cover	Tigers prefer areas of woody vegetation but will venture into grassland areas when hunting or moving between patches.	Optical and/or radar data
Land cover change	Losses and degradation of habitats occurs over time and the replacement habitats are often unsuitable. Regeneration of forests may favour movement of tigers through the landscape.	Time-series of optical and radar data
Fire history	Fire can initially disturb tigers but growth of ground vegetation can increase prey numbers and hence tigers	Fire hotspots mapped from MODIS and fire scars mapped from optical data.

[a]For rainfall and [b]snow cover

The Indo-Bhutan Manas Tiger Conservation Landscape (IBMTCL) and Its Conservation Significance

The IBMTCL lies at the critical juncture of the Indo-Malayan and the Indo-Gangetic bio-geographical pathways and has been lauded for its outstanding natural beauty and unparalleled diversity as a UNESCO World Heritage site (UNESCO 2008). The IBMTCL forms an important connecting link between the Buxa, Nameri and Pakke Tiger Reserves in India and the Hukaung Valley Tiger Reserve in Myanmar. Within

Table 3 PAs supporting tiger populations and methods for their estimation in IBMTCL

Name of protected area	Area (km^2)	Forest type	Previous methods for estimating populations of tigers
Manas Tiger Reserve, India	2837	Subtropical forests interspersed with savannah grasslands	Camera trap studies undertaken since 2009
Royal Manas National Park, Bhutan	1057	Subtropical forests	Camera trap studies undertaken since 2009
Jigme Singye National Park	1730	Temperate and upland broadleaf forests	Single camera-trap study
Phibsoo Wildlife sanctuary	269	Subtropical landscapes	Not available
Khaling Wildlife Sanctuary	335	Subtropical	Not available
Bornadi Wildlife Sanctuary	26	Subtropical	Not available
TOTAL AREA	6254		

this region, a number of Park Areas (PAs) exist, including Jigme Singye National Park (NP), Royal Manas NP, Phibsoo Wildlife Sanctuary (WLS), Khaling WLS in Bhutan and the Manas Tiger Reserve (MTR) in India. Together, these form the single-largest TCL for tigers (*Panthera tigris tigris*) in the world (Dorji and Santiapillai 1989; Sanderson et al. 2006).

The IBMTCL is of historical evolutionary significance as tigers here share the connecting gene pool with the south eastern tiger population and the area represents the entry point of tigers into the Indian sub-continent (Jhala et al. 2011). Administratively, the IBMTCL is comprised of six Protected Areas (PAs; Table 3) but there are several potential landscapes which do not have PA status but nevertheless are able to support tigers (e.g., the Ripu-Chirang Forest complexes in India).

Data Collected

Mammal Surveys

Various organizations, including NGOs such as WWF-India, Aaranyak and the Wildlife Trust of India, assisted the Manas NP Authority in India to lay camera traps between November 2010 and February 2011. This was part of a nation-wide tiger sampling exercise carried out within most of the tiger reserves in India (Jhala et al. 2011). The infrared-triggered camera-traps used within the two ranges of Manas NP (Bansbari and Bhuyanpara) were the Cuddeback (Non Typical Inc. Wisconsin), TrailMaster (Goodson and Associates, Kansas, USA) and Panthera Camera Trap V3 (Panthera, USA). A pair of camera traps was placed within individual 2 × 2 km sized cells and the total area covered by all camera traps approximated 300 km^2 (in India) (Table 4). In a separate but simultaneous exercise in Royal Manas NP, camera trapping was undertaken by the Department of Agriculture, Royal Government of Bhutan (RGoB) in collaboration with the Ugyen Wangchuk Institute for Conservation and Environment (UWICE) and the Bhutan Foundation, specifically to monitor the tiger population (Fig. 1). This study focused only on the camera trapping undertaken in Manas NP.

Table 4 Summary of camera trapping in IBMTCL

Total number of camera traps	102[a]
Sampling occasion	64 days
Sampling effort (number of traps × sampling occasion)	5955
Camera trap polygon area	436.37 km^2
Estimated buffer width (HMMDM)	4.2 km
Effective sampled area	789.2 (±50.98) km^2

[a]78 in or near the border of Manas NP

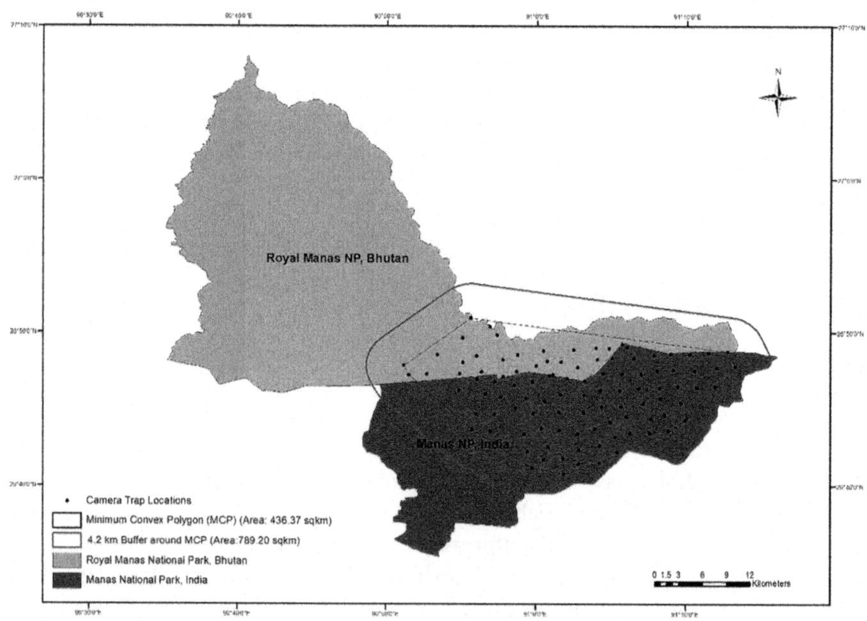

Fig. 1 The location of camera traps placed at IBMTCL between November 2010 and February 2011

Remote Sensing and Other Spatial Data

For the 21 and 30 January and 6 and 7 February 2010, relatively cloud free Landsat Enhanced Thematic Mapper (ETM+) data were downloaded from the Glovis (http://glovis.usgs.gov/) and Land Processes Distributed Active Archive Centre (https://lpdaac.usgs.gov/get_data/data_pool). These data were converted to surface reflectance (%) using the 6S atmospheric correction software. Radiometric normalization between scenes was also undertaken using the approach outlined by Homer et al. (1997). All images were geo-referenced by first establishing ground control points between locations identified from existing maps and the most cloud free Landsat sensor data, warping these images and then generating a mosaic covering the study area. Control points were then established subsequently between the remaining Landsat sensor data and this reference mosaic. The processing of the Landsat sensor data was undertaken using the open source software RSGISLib (www.rsgislib.org; Bunting et al. 2013).

The history of burning with the IBMTCL over the period 2000–2012 was documented by referencing the MODIS product MOD 14 (Thermal Anomalies – Fires and Biomass Burning). This product includes fire occurrence (day/night) and location, the logical criteria used for the fire selection and an energy calculation for each fire. 8-day and monthly day-and-night composite fire occurrence (full resolution) and gridded 10-km and 0.5° summaries (including counts) per fire class (daily/8-day/monthly) are also available. The Level 2 products include various fire related parameters including the occurrence of day and night time thermal anomalies, flagged and grouped into different temperature classes associated with different levels of emitted energy from the fire. These parameters are retrieved daily at 1 km spatial resolution.

An ASTER Global Digital Elevation Model (GDEM) was used (http://asterweb.jpl.nasa.gov/gdem-wist.asp) to establish the topography of the two national parks, with this released at 30 m resolution and available for 1 × 1 tiles (geographic projection). Other data layers were also available to support understanding the movement of large cats, including continuous distance from roads and the locations of anti-poaching camps and settlements. The park boundary, roads and rivers were digitized from topographic sheets and Google Earth images, although the boundary location was revised using the Landsat sensor data available for 2010.

Methods of Data Analysis

Analysis of Camera Trap Data

Each day (24 h) was defined as one sampling occasion (Otis et al. 1978) and this was repeated over a period of 57 days (Otis et al. 1978; Karanth 1995). Individual tigers were identified with reference to their documented stripe pattern, with this undertaken by carefully examining the position and shape of stripes on the flanks, limbs, forequarters and sometimes the tail (Schaller 1967; Karanth and Nichols 1998). Each tiger captured in a photograph was then assigned a unique identification number (e.g. TM1M, TM2F for the first and second tiger of male and female gender). Similar procedures were undertaken for leopard and clouded leopard. Individual capture histories for tigers, leopards and clouded leopards were developed in a standard 'X-matrix format' (Otis et al. 1978) and these were analyzed using models developed for closed populations within the programs CAPTURE (Rexstad and Burnham 1991) and MARK (White 2008). The abundance of all three species for the effective sampled area was estimated by using a buffer around the locations, determined using half the mean of the maximum distance moved (HMMDM) model, and the area associated also with a trapping grid polygon (Karanth and Nichols 1998). A habitat mask was also used to remove non-tiger habitat (e.g., deep water, human settlements; Karanth and Nicols 2002). The density (per 100 km^2) was then estimated by dividing this area by the population size (number of individuals). The density of individual prey species, which were also recorded through the camera traps, was estimated using the program DISTANCE (Thomas et al. 2010). This online software works by using detection functions that model the probability of detecting an animal given its distance from the transect. Further details of both methods can be found in Jhala et al. (2010).

Spatial Analysis of Field Data

The spatial configuration of home ranges of individual cats is an important property of their social systems and has been defined as an area traversed by individuals during normal activities of food gathering, mating and caring for young (Dillon and Kelly 2007). For this study, the GPS locations for each capture event and for identified individuals were extracted and plotted spatially. Where an animal was photographed, the distance from the camera trap was determined and a minimal-bounding polygon was constructed around each of the individual point locations where the same individual was sighted and also for all species of the same type. The resulting spatial datasets were then used to quantify ranging behavior, niche overlap and movement in relation to observed land cover types and landscape dynamics. Home range size is considered an important predictable aspect of an animal's feeding strategy and has been related to food density, metabolic needs and the efficiency of movement. The metric also indicates the degree of overlap as a function of territory and can reflect the variation in population and community characteristics. For example, home-range size is frequently correlated inversely with population density and home-range exclusiveness can indicate significant inter and intra-specific competition (*ibid*). To calculate the ranging patterns of tigers and sympatric carnivores within the habitat, a point density tool was used which calculates a magnitude per unit area from point features falling within a neighborhood around each cell. In addition, a minimum boundary geometry tool was used, which represents a specified minimum bounding geometry enclosing each input feature for each group of input features.

Land Cover Classification

To classify land cover and associated changes between the different years, 150 training areas representing 14 land cover categories (see Table 5) were established, with a further 150 collected to subsequently assess the accuracy of the classification. A hybrid supervised classification of the Landsat ETM+ data (all spectral bands) was then undertaken within Erdas Imagine 9.3.

Table 5 Description of land cover types within the IBMTCL

	LULC type	Description	Suitability for tigers
1	Sub-tropical semi-evergreen forest	Occurs between 500 and 2000 m asl; dominated by a variety of trees and shrub species forming a multi-storey structure. The forests extend along the northern boundary of Manas NP and into Royal Manas NP, which has a wetter moisture regime and lower anthropogenic disturbance.	An ideal habitat for tigers, although supports lower densities of prey and predators
2	Temperate broadleaf forest	Occurs between 2000 and 2500 m asl; dominated by oak and laurel species although pine species are frequent in the inner dry valleys. Typical to Jigme Singye Dorji NP.	Tigers reported in Jigme Singye Wancgchuk NP.
3	Temperate conifer forest	Occurs between 2500 and 3000 m asl. Dominated mainly by conifer species (spruce and fir) and often has understory communities of deciduous broad-leaved and evergreen species (Wangda and Ohsawa 2006). Dominant forest type in the north of Jigme Singye Dorji NP.	Not suitable for tigers, although some reports of tigers present
4	Sub-alpine conifer forest	Occurs up to 4000 m asl. Dominated by cold-adapted and widespread conifer species such as spruce and juniper.	Not suitable for tigers, but other large predators (snow leopards) occur.
5	Alpine scrub	Occurs above 4000 m asl; the limit of tree growth and start of the alpine zone. Scattered shrubs (dwarf junipers and rhododendrons) can occur as high as 4932 m asl.	More suited to snow leopards

(continued)

Table 5 (continued)

	LULC type	Description	Suitability for tigers
6	Closed moist mixed deciduous forest	Occurs up to 700 m asl. Dominated by a single dipterocarp species (*Shorea robusta)* in drier areas. A mix of sub-tropical semi-evergreen tree species also occurs in the moister areas Canopy cover often exceeds 40%. Occurs along the Himalayan foothills and the northern boundary of the MTR.	Suitable for tigers.
7	Open moist mixed Deciduous forest	Occurs up to 700 m. The reduced canopy cover (to <40%) is, in part, the result of selective removal of dipterocarp and other timber species. Dominates much of the MTR buffer.	Not suitable for tigers as highly fragmented and disturbed disturbance.
8	Alluvial short grasslands	Occurs up to 200 m. Soil-climax habitat type dominated by annual grass species that occur over seasonal wetlands. Now confined to patches within Manus NP and highly regulated by fires and annual flooding.	Highly suitable for prey and supports one of the highest densities of tigers in the world (Ahmed et al. 2010). 4.5% of the total habitat in Manas NP.
9	Savannah tall grasslands	Occurs up to 200 m. Soil-climax habitat type dominated by perennial grass species interspersed with colonizing tree species (characteristic of savannahs). Highly regulated by forest fires and flooding with patchy distribution within Manas NP.	Highly suitable for prey with very high densities of tigers (similar to class 8; Ahmed et al. 2010). 10% of the total habitat in Manas NP.
10	Riverine/sandy vegetation	Occur along the fringe of rivers that are characterized by sand and boulders (largely debris carried downstream by Himalayan streams); colonized by lichens and grasses.	Suitable for a range of prey and large cats as adjacent to water sources.
11	Waterbody	All major and minor rivers; typically narrow and fast flowing in the upper reaches but less water in the middle stream area; seasonal flow regulated by rainfall and silt load is variable.	Essential for all wild animals as the primary source of natural water.
12	Cultivation	All non-forest areas that are cultivated (primarily for paddy) during the monsoon season.	Not suitable for tigers.
13	Plantation	Monoculture plantation of tea bushes interspersed with deciduous tree species. Some tea gardens declared as forest reserves. Rubber is the main plantation crop and is typically managed by private growers. Plantation developments often lead to forest encroachment and fragmentation.	Not suitable for tigers.
14	Snow	Only in the higher elevations of Bhutan.	Not applicable

Carnivore Distribution in Relation to Anthropogenic Disturbance

Fires, roads and settlements were considered to be the main anthropogenic disturbances likely to influence the distribution and movement of large cats within the landscape. The point distances tool available in the ArcGIS Geospatial Modelling Environment (GSM; http://www.spatialecology.com/gme and previously known as Hawths tools) was used to calculate the distances from tiger observations to fire locations over the observation period. This tool, which was designed to produce flexible distance matrix outputs, calculates distances between points. The output of distance between each fire location and species capture point was then categorized into 500 m intervals and analyzed using simple regression analysis. The null hypothesis that there was no link between fire and carnivore occurrence was rejected where $p > 0.05$.

Results

Camera Trap Sightings

A total of 14 tigers (8 males and 6 females) were identified from the camera trap photographs (Table 6, Fig. 2). Other top carnivores sighted during the same survey period included 27 leopards (*Panthera pardus*; 11 males, 13 females, 3 unidentified) and 16 clouded leopards (*Neofelis nebulosa*; 4 males, 5 females, 7 unidentified) (Fig. 3). Others species that were photographed included the marbled cat (*Pardofelis marmorata*), Leopard Cat (*Prionailurus bengalensis*), Jungle Cat (*Felis chaus*), Dhole (*Cuon alpinus*), Himalayan Black Bear (*Ursus thibetanus*), Sloth Bear (*Melursus ursinus*), Jackal (*Canis aureus*) and Civet species. Herbivore prey species photo captured included gaur (*Bos gaurus*), wild pig (*Sus scrofa*), sambar (*Rusa unicolor*), hog deer (*Hyelaphus porcinus*) and barking deer (*Muntiacus muntjac*) (Borah et al. 2012).

Table 6 Camera trap sightings of tigers

Number of individual tigers captured	14
Estimated numbers of tigers in the sample area using model Mb	15 (95% CI: 15–29)
Estimated tiger density in sampled area using HMMDM[a]	1.9 (±0.36) tigers 100 km^{-2}
Estimated tiger density using MLSECR[b]	0.75 (±0.21) tigers 100 km^{-2}

[a]Half Mean Maximum Distance Moved
[b]Maximum Likelihood Spatially explicit Capture Recapture based on Half normal method

Fig. 2 Camera trap pictures of individual tigers in the IBMTCL (Nov 2010–Feb 2011; Source: Forest Department, Manas and Borah et al. 2012)

Ranging Pattern and Seasonal Home Range

Within Manas NP, tigers were found to be highly territorial (Fig. 2), with males sharing their territory with only 1 or 2 females. However, one adult male tiger (Tiger 1; and possibly the dominant) shared its territory with almost all the other individuals. The range of one female was significantly greater than that of others (Table 7), which may be attributed to the sub-optimal habitats in her territory. During the observation period, all tigers were moving to the areas of higher elevation to the north (i.e., towards Bhutan). No tigers were observed near the boundary of Manas NP, with most occurring towards the central area.

Leopards and Clouded Leopards shared much of their territory with tigers, although their movement was more restricted to the fringe areas. This also corroborates with studies elsewhere (e.g., Harihar et al. 2011) that sympatric species of carnivores co-exist in sub-optimal habitats (Fig. 4).

Fig. 3 Camera trap pictures of some Leopards and Clouded Leopards in the IBMTCL (Nov 2010–Feb 2011; Source: Forest Department, Manas and Borah et al. 2012)

Table 7 Summary of the dispersal of tigers in Manas NP (Nov 2010–Feb 2011)

ID	Sex	DOFC	DOLC	No of traps captured	Ranging pattern (km^2)	Area of overlap (km^2)
Tiger 1	Male	27 Nov	19 Jan	28	232	
Tiger 2	Female	27 Nov	28 Dec	4	14	Tiger 1 (5.9)
Tiger 3	Female	17 Dec	19 Dec	2	–	
Tiger 4	Female	14 Dec	16 Jan	9	91	Tiger 1 (85.4)
Tiger 5	Female	31 Dec	1 Jan	2	–	
Tiger 6	Male	12 Dec	8 Jan	2	–	
Tiger 7	Male	18 Jan	18 Jan	1	–	
Tiger 8	Female	12 Jan	30 Jan	2	–	
Tiger 9	Male	31 Dec	19 Jan	2	–	

DOFC Date of First Capture, *DOLC* Date of Last Capture

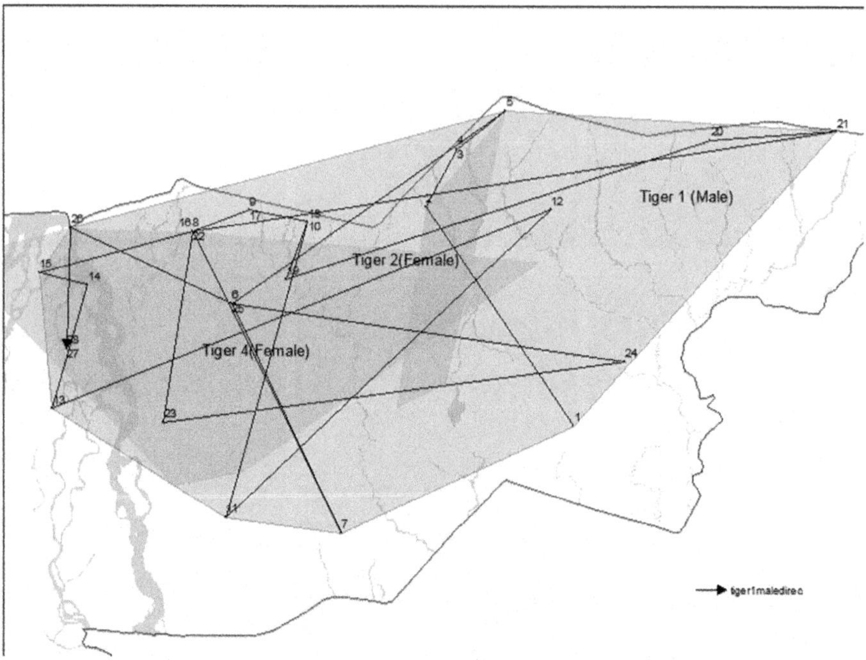

Fig. 4 Ranging pattern of female tigers in relation to the movement of the dominant male v(Tiger 1) in IBMTCL (Camera trap Area = 300 km^2)

Land Cover Classifications and Carnivore Preferences

The classification of land covers for the IBMTCL is illustrated in Fig. 5. The majority of the lowland was occupied by closed moist mixed deciduous forests, with subtropical evergreen forest dominating the lower slopes and progressing into temperate broadleaved and coniferous forests and then subalpine vegetation with increasing elevation. Grasslands were found primarily in the south of the IBMTCL and riverine vegetation and unconsolidated deposits occurred along the wide river channels. Based on standard confusion matrices, the overall accuracy in the classification of land covers was 81.5% (kappa coefficient of 0.78), with the majority of users' or producers' accuracies exceeding 80% but being as low as 40% for some classes (e.g., user's accuracy 60% for cultivation and scrub, 56.7% for closed moist mixed forests). Manas NP supported many of the habitats suitable for tigers including the moist mixed deciduous forests (36.5%), semi-evergreen forest (19.1%), riverine sandy areas (15.1%) and moist alluvial grassland (8.4%). During the observation period, female tigers were most commonly recorded in the closed moist deciduous forests, whereas males were also recorded in sandy riverine areas close to forests. Although considered suitable because of the large numbers of prey, both sexes were not recorded within the tall savanna grasslands but rather remained on the periphery. Leopards and clouded leopards were observed primarily in the riverine areas.

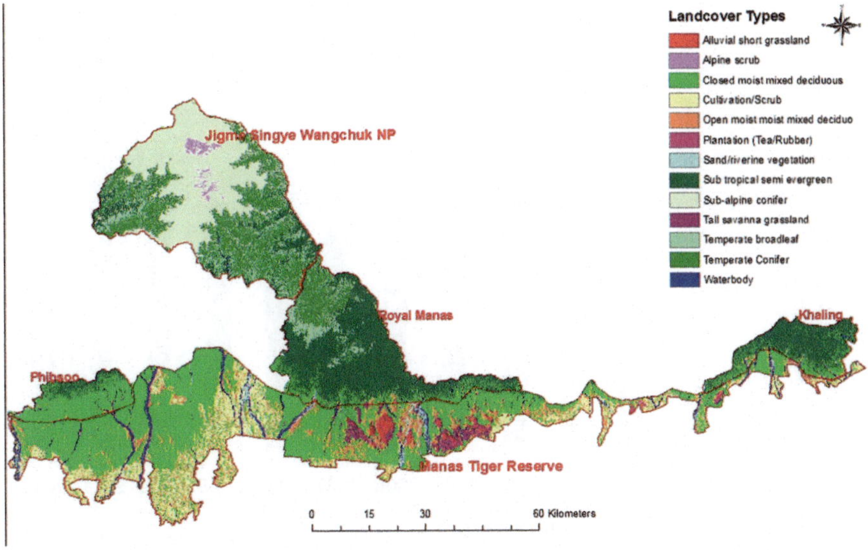

Fig. 5 Land cover classification of the IBMTCL generated from 2010 Landsat TM data

Distributions of Tigers in Relation to Fires and Anthropogenic Disturbance

The Manas NP section of the IBMTCL landscape experienced repeated and extensive burning between 2000 and 2012, with 1749 fire events recorded in the MOD 14 product. The greatest number of fires was in 2011, with 271 recorded (Fig. 6a). In all years, there was considerable intra-annual variability, with the period December to March associated with the highest frequency of fires. The transition period before and after the monsoon season (May to September) (i.e., the first and last 2 months of the dry season) experienced fewer fires. Manas NP is divided into three ranges; the western range (Panbari), central (Bansbari) and eastern (Bhuyanpara), and the density of fires (5.13 km^{-2}) was greatest in the grassland areas of the Bansbari range, with those in the Kuribeel and Pahu field areas repeatedly burnt. The incidence of fires was greatest in the more open moist mixed deciduous forests (Fig. 6b; Table 8), with this attributed largely to human activities. Over the period of the camera trapping, 32 fires were detected in the areas within which camera traps were placed. Reference to the tiger data indicated that the highest number of counts (indicating presence) was at a distance of 5–10 km from fires and often around the margins of burnt areas (Fig. 7). In general, tigers were recorded away from roads and village settlements, with this indicating their sensitivity to human activity, including vehicular movements.

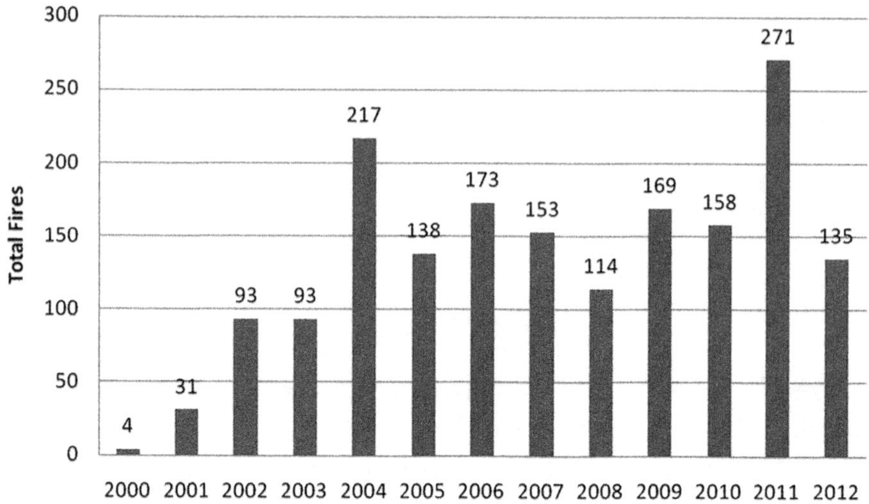

Fig. 6 The frequency of fires in the IBMTCL between 2000 and 2012 as determined from MODIS data

Table 8 Summary of fire distribution in land-cover types in Manas NP, India

Forest type	Area (km^2)	Area (%)	Total Fires	(%)	Fires # km^{-2}
Semi evergreen forest	104.1	19.1	112	6	1.08
Riverine/sandy areas	82.1	15.1	321	18	3.91
Moist mixed deciduous forest	197.9	36.5	569	33	2.88
Moist alluvial grassland	45.6	8.4	434	25	9.51
Degraded/scrub Forest	35.8	6.6	173	10	4.83
Tall savannah grassland	77.3	14.3	140	8	1.81
TOTAL	542.8	100.00	1749		

Discussion

The Role of Earth Observation

The main benefit of using earth observation data was the ability to provide a land cover classification for a similar period as the camera trap surveys. The overall accuracy of the land cover classification for the IBMTCL was relatively high, but there was confusion between a number of classes, largely because of the high seasonal variation in the phenology of vegetation but also fires and flooding. In some areas, dry areas became inundated during the wet season, but as the waters receded, extensive areas of highly productive grassland established. These became senescent as the dry season progressed and many were then subject to extensive burning. Hence, the spectral signatures of the same area of ground varied considerably

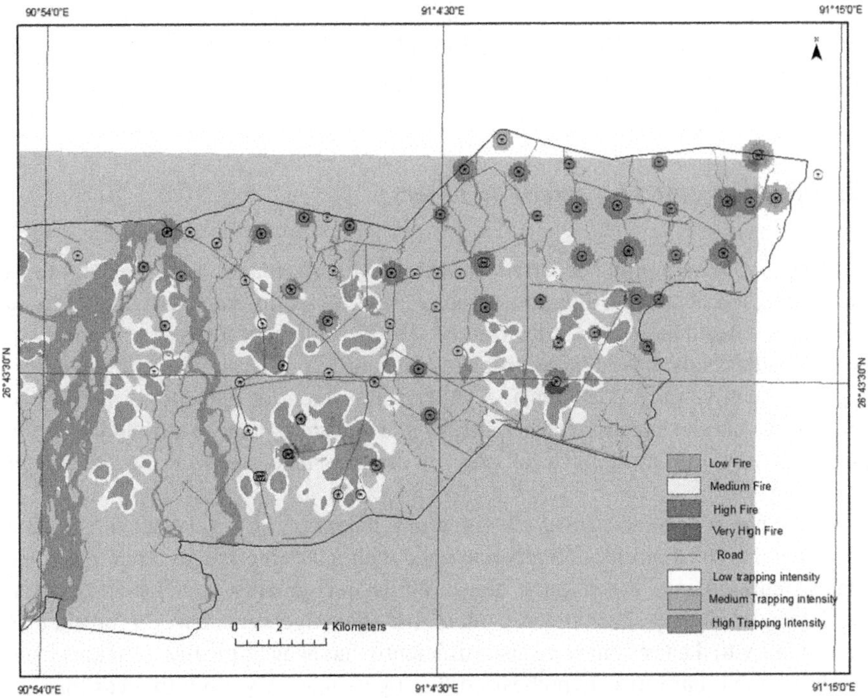

Fig. 7 The distribution of tiger observations in relation to fire density in the Manas NP section of the IBMTCL

(senescent and productive grassland, water and burnt ground). For this reason, the use of dense time-series of optical remote sensing data would provide a better overview of the changing states of vegetated and non-vegetated environments within the IBMTCL. Whilst atmospheric correction of the Landsat sensor data was undertaken, smoke haze and cloud still limited reliable retrieval of surface reflectance for some areas.

A further limitation of the land cover maps was that insufficient information on the structural characteristics of forests that might influence tiger movements was obtained, particularly in relation to the tree canopy density and canopy height. A common approach to estimating canopy cover is to establish relationships between ground based measurements and combinations of reflectance data. However, the lack of ground data and seasonal changes in leaf cover prevented the generation of reliable retrieval algorithms or use of those that are existing. Information on canopy height was also not obtained although there is potential to retrieve this using NASA's Ice, Cloud, and land Elevation (ICESat) satellite Geoscience Laser Altimeter System (GLAS). Future mapping could therefore focus on developing these products and applying classification schemes that consider their inclusion such as the Food and Agricultural Organisation's (FAO) Land Cover Classification System (LCCS; see

Lucas and Mitchell (2017; this book). The MODIS data provided invaluable insights into the frequency and distribution of fires, identifying that the majority occurred in the moist mixed deciduous forests and alluvial grasslands.

Distribution and Movement of Tigers

Although camera trapping occurred over a relatively short period, 52 tiger sightings were obtained. The resulting photographs allowed 14 different individuals to be distinguished and their gender determined. The movement of tigers was variable but several individuals moved large distances (tens of kilometers), with one male having a substantive range (232 km^2). A general movement of tigers towards forested areas in the northern sections (of slightly higher elevation) may be attributable to the comparatively low availability of prey in the grassland areas during the winter months.

Reference to the land cover classification generated from Landsat sensor data indicated that the majority of tigers remained within the forest areas, although some males ventured into the riverine areas. Whilst recognizing that the observation period was relatively short, it was evident that some tigers remained in forest areas in proximity to the fires but were also sufficiently far away (typically 5–10 km) from the last known burn. It is hypothesized that tigers may stay close to areas that are burning (or may be susceptible to burning) in order to take advantage of prey species fleeing from the fire. They may also anticipate or take advantage of the prey species returning to graze on new growth in the periods following the fires. The influence of fires on the distribution of large cats and their prey has also been noted by Sarma et al. (2008) and Takahata et al. (2010) but further studies are need to specifically link occurrence and movements with fire activity. Avoidance of areas occupied or influenced by humans was noticeable, with many individuals located towards the less disturbed central areas where human activity and vehicular movements were minimal and prey was more abundant.

A practical outcome from the study is that information generated on individual movement of tigers can assist future surveillance, especially as the likelihood of tiger movements can be communicated to the forest camps. Training in the use of camera traps and photographic interpretation allows frontline staff to identify individual tigers and contribute to a better understanding of their ranging behaviour in this landscape. Whilst tigers may benefit from fires, they may also be disturbed from those that are human-induced and therefore the establishment of camps in close proximity to fire prone areas may lead to prevention of many and hence less overall disturbance of habitats.

The tigers showed minimal overlap with leopards and clouded leopards and some studies (e.g., Rabinowitz et al. 1987) have suggested that the clouded leopard may alter its range or become less nocturnal when larger competitive species occur. Further research on the movement of tigers and other large cats based on camera trap surveys is nevertheless advocated in order to better determine their interactions and the distribution of their prey. The camera traps also provide a basis for better estimating the abundance and density of large cats and to therefore monitor changes and trends in populations and their behaviour.

When used in combination with camera trapping and other monitoring techniques, earth observation data can play a role in understanding how large carnivores utilize the IBMTCL during different times of the year. These data can also be used to address connectivity issues within and between landscapes. Several new initiatives have included the development of software (e.g., MIST, SMART, M-STrIPES) that facilitate spatial monitoring of tiger populations including through the integration of field observations. Within such systems, there is considerable opportunity to make better use of the large, extensive and publically available archives of satellite sensor data.

Final Conclusions

In 1998, the global tiger population was estimated at 5000–7000 individuals (Seidensticker and McDougal 1993). By 2014, however, the global tiger population was at an historic low of approximately 3500 individuals, having declined by more than 50% of what they were just a few decades earlier (Goodrich et al. 2015). Given this situation, there is an urgent need to prioritize efforts to protect the remaining tiger populations (Stokes 2010; Walston et al. 2010). For the most part, these are found in PAs (Chapron et al. 2008; Rabinowitz 2009; Clements et al. 2010; Walston et al. 2010) but there are considerable opportunities to expand their range, including by connecting areas where they occur. This study has highlighted the potential of using earth observation data in conjunction with ground-based (including camera trap) surveys for better understanding the distribution, movement and preferences of tigers and a recommendation is to develop what is termed a Tiger Observation System (TOS) to maximize these opportunities. Whereas past use of earth observation data has been limited by availability and cost, the release of the Landsat sensor archive and the provision of freely available Sentinel-1 and -2 radar and optical sensor data respectively gives tiger conservation efforts additional impetus to achieve their long term conservation and expansion of populations.

Acknowledgements The authors would like to thank the Honourable Deputy Minister, Bodoland Territorial Areas District Shri Kampa Borgayari and his officials, Shri G.C.Basumatary, Shri A.Swargowari, Assam Forest Department for giving the requisite permissions to carry out the study. Dr. Jimmy Borah and his team from WWF, Dr. Bibhuti Lahkar from Aaranyak, and the Park Authorities of India and Bhutan are thanked for providing the field data. Dr. P.S. Roy, Dr. SPS Kushwaha and Dr. Arijit Roy from Indian Institute of Remote Sensing are also thanked for their technical help and support.

References

Ahmed, M.F., Borah, J., Das, C., Basumatary, A., Sarma, R.N., Gogoi, D.D., Buragohain, S.N., Vasu, N.K., Talukdar, B.K., Jhala, Y.V., Qureshi, Q.: Monitoring Tigers and Prey Animals of Kaziranga National Park, Assam, India. Technical Report, Aaranyak. 50 pp (2010)

Balme, G.A., Hunter, L.T.B., Slotow, R.: Evaluating methods for counting cryptic carnivores. J. Wildl. Manag. **73**, 433–441 (2009)

Blake, S., Hedges, S.: Sinking the flagship: The case of forest elephants in Asia and Africa. Conserv. Biol. **18**, 1191–1202 (2004)

Borah, J., Wangchuk, D., Swargowari, A., Wangchuk, T., Sharma, T., Das, D., Rabha, N., Basumatari, A., Kakati, N., Ahmed, M.F., Sharma, A., Sarmah, A., Dutta, D.K., Lahkar, B., Dorji, T., Brahma, P.K., Ramchiary, L., Tempa, T., Wangdi, Y., Nedup, T., Wangdi, T., Tharchen, L., Dhendup, P., Bhobora, C.R., Pandav, B., Vattakaven, J.: Tigers in Indo-Bhutan Transboundary Manas Conservation Complex. 2012. WWF Technical report (2012)

Bunting, P., Clewley, D., Lucas, R.M., Gillingham, S.: The Remote Sensing and GIS Software Library (RSGISLib). Comput. Geosci. **62**(2–6), 226 (2013)

Carbone, C., Mace, G.M., Roberts, S.C., Macdonald, D.W.: Energetic constraints on the diet of terrestrial carnivores. Nature. **402**, 286–288 (1999)

Carbone, C., Christie, S., Conforti, K., Coulson, T., Franklin, N., Ginsberg, J.R., Griffiths, M., Holden, J., Kawanishi, K., Kinnaird, M., Laidlaw, R., Lynam, A., Macdonald, D.W., Martyr, D., McDougal, C., Nath, L., O'Brien, T., Seidensticker, J., Smith, D.J.L., Sunquist, M., Tilson, R., Shahruddin, W.N.W.: The use of photographic rates to estimate densities of tigers and other cryptic mammals. Anim. Conserv. **4**, 75–79 (2001)

Carroll, C., Miquelle, D.G.: Spatial viability analysis of Amur tiger Panthera tigris altaica in the Russian Far East: the role of protected areas and landscape matrix in population persistence. J. Appl. Ecol. **43**, 1056–1068 (2006)

Chapron, G., Miquelle, D.G., Lambert, A., Goodrich, J.M., Legendre, S., Clobert, J.: The impact on Tigers of poaching versus prey depletion. J. Appl. Ecol. **45**, 1667–1674 (2008)

Choudhury, S.R.: Let us count our Tiger. Cheetal. **14**(2), 41–51 (1970)

Clements, R., Rayan, D.M., Zafir, A.W.A.: Trio under threat: can we secure the future of Rhinos, Elephants and Tigers in Malaysia. Biodivers. Conserv. **19**, 1115–1136 (2010)

Davis, B.W., Li, G., Murphy, W.J.: Supermatrix and species tree methods resolve phylogenetic relationships within the big cats, Panthera (Carnivora: Felidae). Mol. Phylogenet. Evol. **56**(1), 64–76 (2010)

Dillon, A., Kelly, M.J.: Ocelot Leopardus pardalis in Belize: the impact of trap spacing and distance moved on density estimates. Oryx **41**(04), 469–477 (2007)

Dinerstein, E., Loucks, C., Wikramanayake, E., Ginsberg, J., Sanderson, E., Seidensticker, J., Forrest, J., Bryja, G., Heydlauff, A., Klenzendorf, S., Leimgruber, P., Mills, J., O'Brien, T.G., Shrestha, M., Simons, R., Songer, M.: The fate of wild tigers. Bioscience. **57**, 508–514 (2007)

Dorji, D.P., Santiapillai, C.: The status, distribution and conservation of the Tiger *Panthera tigris* in Bhutan. Biol. Conserv. **48**, 311–319 (1989)

Gittleman, J.L., Harvey, P.H.: Carnivore home-range size, metabolic needs and ecology. Behav. Ecol. Sociobiol. **10**, 57–63 (1982)

GOI: Joining the Dots-The report of the Tiger Task Force, Project Tiger, Government of India. 217 pp (2005)

Goodrich, J., Lynam, A., Miquelle, D., Wibisono, H., Kawanishi, K., Pattanavibool, A., Htun, S., Tempa, T., Karki, J., Jhala, Y., Karanth, U.: *Panthera tigris*. The IUCN Red List of Threatened Species 2015: e.T15955A50659951. http://dx.doi.org/10.2305/IUCN.UK.2015-2.RLTS.T15955A50659951.en. Downloaded on 02 June 2017 (2015)

Gratwicke, B., Bennett, E.L., Broad, S., Christie, S., Dutton, A., Gabriel, G., Kirkpatrick, C., Nowell, K.: The world can't have wild tigers and eat them, too. Conserv. Biol. **22**, 222–223 (2008)

Harihar, A., Pandav, B., Goyal, S.P.: Subsampling photographic capture-recapture data of tigers (*Panthera tigris*) to minimize closure violation and improve estimate precision: a case. Popul. Ecol. **51**, 471–479 (2009)

Harihar, A., Pandav, B., Goyal, S.P.: Responses of leopard Panthera pardus to the recovery of a tiger *Panthera tigris* population. J. Appl. Ecol. **48**(3), 806–814 (2011)

Homer, C.G., Ramsey, R.D., Edwards, T.C., Falconer, A.: Landscape cover type mapping and modelling using multi-scene thematic mapper mosaic. Photogramm. Eng. Remote Sens. **63**, 59–67 (1997)

Imam, E., Kushwaha, S.P.S., Singh, A.: Evaluation of suitable tiger habitat in Chandoli National Park, India, using multiple logistic regression. Ecol. Model. **220**, 3621–3629 (2009)

Jhala, Y. V., Gopal, R., Qureshi, Q.: Status of tigers, co-predators and prey in India. National Tiger Conservation Authority, Government of India and the Wildlife Institute of India, New Delhi and Dehra Dun (2008)

Jhala, Y.V., Qureshi, Q., Gopal, R., Sinha, P.R. (eds.): Status of the tigers, co-predators and prey in India, 2010. National Tiger Conservation Authority, Govt of India and Wildlife Institute of India, New Delhi and Dehra Dun (2011)

Jhala, Y.V., Qureshi, Q., Gopal, R.: The status of tigers in India. National Tiger Conservation Authority, Wildlife Institute of India, New Delhi and Dehra Dun (2015)

Karanth, K.U., Sunquist, M.E.: Prey selection by Tiger, Leopard and Dhole in Tropical Forests. J. Animal Ecol. **64**, 439–450 (1995)

Karanth, K.U., Nichols, J.D.: Estimation of tiger densities in India using photographic captures and recaptures. Ecology **79**(8), 2852–2862 (1998)

Karanth, K.U., Nichols, J.D.: Monitoring tigers and their prey: a manual for researchers, managers and conservationists in tropical Asia. Centre for Wildlife Studies, Bangalore (2002)

Karanth, K.K., Nichols, J.D., Hines, J.E., Karanth, K.U., Christensen, N.L.: Patterns and determinants of mammal species occurrence in India. J. Appl. Ecol. (2009)

Karanth, K.U., Nichols, J.D., Kumar, N.S., Hines, J.E.: Assessing tiger population dynamics using photographic capture-recapture sampling. Ecology. **87**, 2925–2937 (2006)

Kelly, M.J., Noss, A.J., Di Bitetti, M.S., Maffei, L., Arispe, R.L., Paviolo, A., De Angelo, C.D., Di Blanco, Y.E.: Estimating puma densities from camera trapping across three study sites: Bolivia, Argentina, and Belize. J. Mammal. **89**(2), 408–418 (2008)

Kerley, L.L., Salkina, G.P.: Using scent-matching dogs to identify individual Amur tigers from scats. J. Wildl. Manag. **71**(4), 1349–1356 (2007)

Kerley, L.L., Goodrich, J.M., Miquelle, D.G., Smirnov, E.N., Quigley, H.B., Hornocker, M.G.: Reproductive parameters of wild female Amur (Siberian) tigers (*Panthera tigris altaica*). J. Mammal. **84**, 288–298 (2003)

Linkie, M., Chapron, G., Martyr, D.J., Holden, J., Leader-Williams, N.: Assessing the viability of tiger subpopulations in a fragmented landscape. J. Appl. Ecol. **43**, 576–586 (2006)

Luo, S.J., Kim, J.H., Johnson, W.E., van der Welt, J., Martenson, J., Yuhki, N., Miquelle, D.G., Uphyrkina, O., Goodrich, J.M., Quigley, H.B., Tilson, R., Brady, G., Martelli, P., Subramaniam, V., McDougal, C., Hean, S., Huang, S.Q., Pan, W.S., Karanth, U.K., Sunquist, M., Smith, J.L.D., O'Brien, S.J.: Phylogeography and genetic ancestry of tigers (Panthera tigris). PLos Biol. **2**, 2275–2293 (2004)

MacKenzie, D.I., Nichols, J.D., Lachman, G.B., Sam, D., Royle, J.A., Langtimm, C.A.: Estimating site occupancy rates when detection probabilities are less than one. Ecology **83**(8), 2248–2255 (2002)

Nowell, K.: Deciding the Tiger's future: Implementation of CITES decision 14.65–14.72, In 58th meeting of the CITES Standing Committee. Geneva (2009)

O'Brien, T.G., Kinnaird, M.F., Wibisono, H.T.: Crouching tigers, hidden prey: Sumatran tiger and prey populations in a tropical forest landscape. Anim. Conserv. **6**, 131–139 (2003)

Otis, D.L., Burnham, K.P., White, G.C., Anderson, D.R.: Statistical-inference from capture data on closed animal populations. Wildl. Monogr. **62**, 7–135 (1978)

Rabinowitz, A.: Stop the bleeding: implementing a strategic tiger conservation protocol. CAT News. **51**, 30–31 (2009)

Rabinowitz, A., Andau, P., Chai, P.P.: The clouded leopard in Malaysian Borneo. Oryx. **21**(02), 107–111 (1987)

Ranganathan, J., Chan, K.M.A., Karanth, K.U., Smith, J.L.D.: Where can tigers persist in the future? A landscape-scale, density-based population model for the Indian subcontinent. Biol. Conserv. **141**(1), 67–77 (2008)

Rangarajan, M.: India's Wildlife history: an introduction. Delhi, Permanent Black (2001)

Rexstad, E., Burnham, K.P.: User's guide for interactive program CAPTURE. Color. Cooperative Fish and Wildlife Research Unit (1991)

Royle, J.A., Karanth, K.U., Gopalaswamy, A.M., Kumar, N.S.: Bayesian inference in camera trapping studies for a class of spatial capture-recapture models. Ecology. **90**, 3233–3244 (2009)

Samrat, M., Karanth, K.U., Kumar, N.S., Gopalaswamy, A.M., Andheria, A., Ramakrishnan, U.: Evaluation of non-invasive genetic sampling methods for estimating tiger population size. Biol. Conserv. **142**(10), 2350–2360 (2009)

Sanderson, E., Forrest, J., Loucks, C., Ginsberg, J., Dinerstein, E., Seidensticker, J., Leimgruber, P., Songer, M., Heydlauff, A., O'Brien, T., Bryja, G., Klenzendorf, S., Wikramanayake, E.: Setting priorities for the conservation and recovery of Wild Tigers: 2005–2015. The technical assessment. WCS, WWF, Smithsonian, and NFWF-STF, New York and Washington, DC, USA (2006)

Sarma, P.K., Lahkar, B.P., Ghosh, S., Rabha, A., Das, J.P., Nath, N.K., Dey, S., Brahma, N.: Land-use and land-cover change and future implication analysis in Manas National Park, India using multi-temporal satellite data. Curr. Sci. **95**, 223–227 (2008)

Schaller, G.B.: The deer and the tiger- a study of wildlife in India. University of Chicago Press, Chicago (1967)

Seidensticker, J., McDougal, C.: Tiger predatory behavior, ecology and conservation. Mamm. Predators. **65**, 105–125 (1993)

Sharma, S., Jhala, Y., Sawarkar, V.B.: Identification of individual tigers (Panthera tigris) from their pugmarks. J. Zool. **267**(1), 9–18 (2005)

Smith, J.L.D., Ahearn, S.C., McDougal, C.: Landscape analysis of tiger distribution and habitat quality in Nepal. Conserv. Biol. **12**, 1338–1346 (1998)

Stokes, E.J.: Improving effectiveness of protection efforts in tiger source sites: Developing a framework for law enforcement monitoring using MIST. Integr. Zool. **5**, 363–377 (2010). doi:10.1111/j.1749-4877.2010.00223.x

Sunquist, M., Karanth, K.U., Sunquist, F.: Ecology, behaviour and resilience of the tiger and its conservation needs. In: Seidensticker, J., Christie, S., Jackson, P. (eds.) Riding the Tiger: Tiger Conservation in Human-Dominated Landscapes, pp. 5–18. Cambridge University Press, London (1999)

Takahata, C., Amin, R., Sarma, P., Banerjee, G., Oliver, W., Fa, J.E.: Remotely-sensed active fire data for protected area management: eight-year patterns in the Manas National Park. India. Environ. Manag. **45**, 414–423 (2010)

Thomas, L., Buckland, S.T., Rexstad, E.A., Laake, J.L., Strindberg, S., Hedley, S.L., Bishop, J.R., Marques, T.A., Burnham, K.P.: Distance software: design and analysis of distance sampling surveys for estimating population size. J. Appl. Ecol. **47**(1), 5–14 (2010)

UNESCO: World Heritage Centre-IUCN: Mission report. Reactive monitoring mission to the Manas Wildlife Sanctuary, India From 12 to 18 February 2008. Available from: http://whc.unesco.org/en/list/338/documents/. (2008)

Walston, J., Robinson, J.G., Bennett, E.L., et al.: Bringing the tiger back from the brink: The six percent solution. PLoS Biol. **8**, e1000485 (2010)

Wangda, P., Ohsawa, M.: Gradational forest change along the climatically dry valley slopes of Bhutan in the midst of humid eastern Himalaya. Plant Ecol. **186**(1), 109–128 (2006)

Wegge, P., Storaas, T.: Sampling tiger ungulate prey by the distance method: lessons learned in Bardia National Park, Nepal. Anim. Conserv. **12**, 78–84 (2009)

White, G.C.: Closed population estimation models and their extensions in program MARK. Environ. Ecol. Stat. **15**(1), 89–99 (2008)

Wikramanayake, E.D., Dinerstein, E., Robinson, J.G., Karanth, U., Rabinowitz, A., Olson, D., Mathew, T., Hedao, P., Conner, M., Hemley, G., Bolze, D.: An ecology-based method for defining priorities for large mammal conservation: the tiger as case study. Conserv. Biol. **12**, 1427–1427 (1998)

Improving the Accuracy of Bird Counts Using Manual and Automated Counts in ImageJ: An Open-Source Image Processing Program

Clive Hurford

Abstract Bird counts estimated 'by eye' are subject to high levels of observer variation. Comparison of bird numbers estimated by experienced ornithologists and through manual counting in Image J indicated that observers typically underestimate by more than 30 %, with the median ranging from −13% to −57%. Image J is an open source software package that provides both manual and automated options for counting birds, whether on the ground or in flight. The manual approach has the least preparation time, and will generate accurate counts of 1000 birds in only c. 20 min. This technique is particularly useful for counting birds in breeding colonies, where the complexity of the background in an image will compromise the accuracy of automated approaches. The manual counter provides markers for counting up to eight species simultaneously and is the preferred option for images containing up to 3000 birds. The automated counter is best suited for estimating bird numbers when large groups occur against relatively plain backgrounds. However, the automated counter will not differentiate between species and will typically underestimate the number of birds in an image, as it is object-based and overlapping individuals will count as one bird. Conversely, birds with strongly contrasting plumage patterns will be overestimated."

Keywords Variation in observer bird count estimates • Object based image analysis (OBIA) • ImageJ • Open-source image processing software • Manual and automated bird counts

---

C. Hurford (✉)
Natural Resources Wales, Cardiff, Wales, UK
e-mail: churford5@gmail.com

© Springer International Publishing AG 2017

R. Díaz-Delgado et al. (eds.), *The Roles of Remote Sensing in Nature Conservation*,
DOI 10.1007/978-3-319-64332-8_13

Introduction

Birdwatchers spend a lot of time counting birds, mostly to collect information for their personal, regional and national archives, but also to contribute to nationally organised surveillance schemes, such as the Wetland Bird Survey in the UK. Count data can also feed back into conservation management (Rowell 2006), perhaps as part of a monitoring project to trigger management change: this type of project aligns well with Habitats Directive (European Commission 1992) obligations to achieve favourable conservation status for internationally rare and declining species.

The Levels of Observer Variation Associated with Bird Counts

Bird counts are often carried out by field observers in sub-optimal counting conditions, i.e. with oblique viewpoints of large numbers of birds and in adverse weather conditions (Fig. 1). Birds in flight can be even more difficult to count as the shape of the flock is constantly changing and there is less time to make an accurate estimate. To illustrate the levels of observer variation associated with bird counts, 50 experienced ornithologists were invited to participate in an exercise to estimate the number of birds in images of bird flocks in flight. Of these, 35 responded positively and contributed flock-size estimates. However, five of the respondents made it known that they did not consider themselves to be experienced bird counters, so their counts were omitted from the results presented here.

Each participant received four images of bird flocks in flight: the images used to demonstrate the automatic and manual counts in this chapter, i.e. Figs. 2, 3, 4 and 5.

Fig. 1 UAVs or light aircraft are the best options for capturing images of large concentrations of birds feeding over low-lying land or water. A vertical image would reveal the personal feeding spaces between these Whooper Swans (Cygnus cygnus) and Greylag Geese (Anser anser): these spaces are less visible in oblique images

Fig. 2 An image of Greater Flamingos in flight over Doñana NP in Andalucia, Spain taken from a Cessna in March 2014. The automatic count carried out in ImageJ generated a total of 871 birds. A manual count undertaken for validation purposes generated a total of 969 birds. Camera model Canon EOS 600D Lens Canon 35 mm ISO 100 F-stop f/9 Exposure time 1/250 s

Fig. 3 Starlings *Sturnus vulgaris* performing aerial displays near their roost on Aberystwyth Pier, Ceredigion, UK. ImageJ generated an automatic count of 9380 birds. A validation exercise added 1737 birds, bringing the estimated total to 11,117. Camera – Canon EOS 7D, Lens – Canon 35 mm, ISO – 400, F-stop – f/6.3, Exposure time – 1/640 s

The participants were asked to spend no more than 30 seconds estimating the number of birds in each image (and not to confer with other observers) before returning their estimates for collation. The 30 second limit replicates the limited time that an observer has to estimate the size of mobile bird flocks in the field. In reality, however, estimating the size of a flock of birds as it flies past presents a more difficult

Fig. 4 A flock of Linnets (*Carduelis cannabina*) and at least one Goldfinch (*Carduelis carduelis*) in flight near West Angle, Pembrokeshire, UK. An automatic count of this image generated a total of 377 birds while a manual count, carried out for validation, gave a total of 458 birds. Camera model – Canon EOS 7D, Lens – Canon 35 mm, ISO – 400, F-stop – f/8, Exposure time – 1/500 s

Fig. 5 A flock of Golden Plovers in flight near Castlemartin, Pembrokeshire, UK. An automatic count of this image carried out in ImageJ generated a total of 832 birds, while a validation exercise, based the on the number of pixels allocated to each individual in the Excel file, generated a total of 848 birds. A manual count generated a total of 842 birds. Camera model – Canon EOS 7D, Lens – Tamron 150–600 mm, Focal length – 150 mm, ISO – 200, F-stop – f/5.6, Exposure time – 1/250 s

challenge as, (a) we often have less than 30 seconds to make an estimate, (b) the birds do not remain static for 30 seconds and (c) the dimensions of the flock are constantly changing.

The charts in Figs. 6, 7, 8 and 9 show the results from the observer variation exercise.

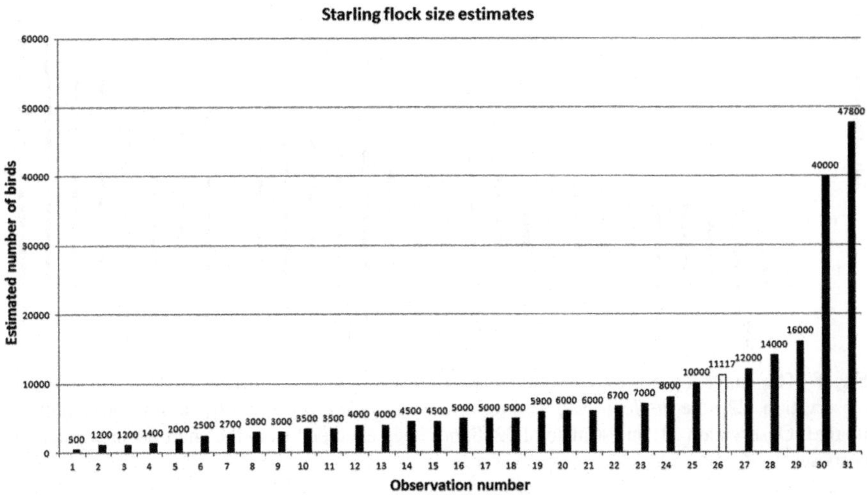

Fig. 6 The range of observer variation recorded for Fig. 3: an image of a Starling murmuration. Observation 26, the *hollow bar*, is the best estimate (11,117) provided by a validated automatic count in ImageJ

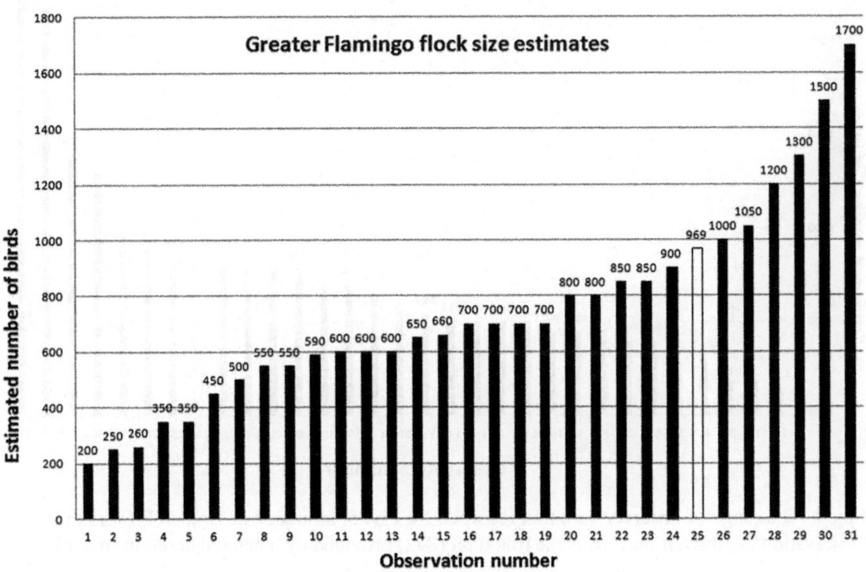

Fig. 7 The range of observer variation recorded for Fig. 2, an image of Greater Flamingos in flight. Observation 24, the *hollow bar*, is the true total (969) provided by a manual count in ImageJ

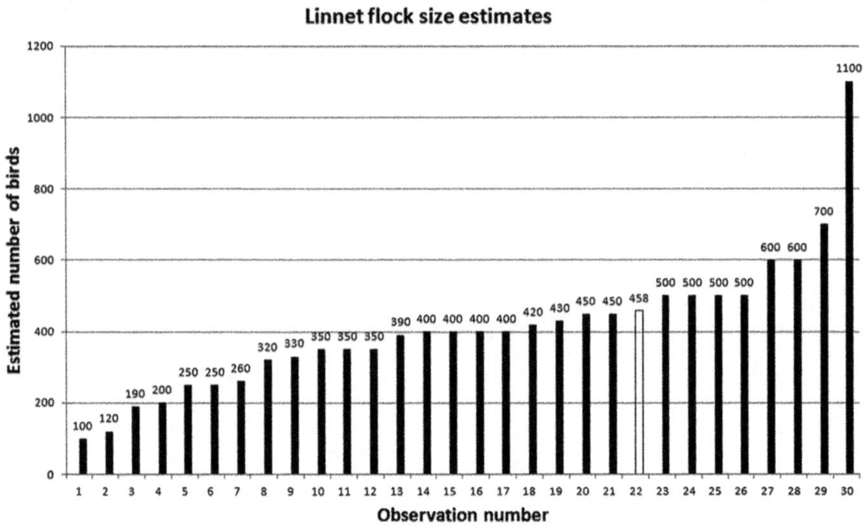

Fig. 8 The range of observer variation recorded for Fig. 4, an image of Linnets in flight. Observation 22, the *hollow bar*, is the true total (458) provided by a manual count in ImageJ. Observation 31, an estimate of 3258, has been excluded from the chart for interpretation purposes

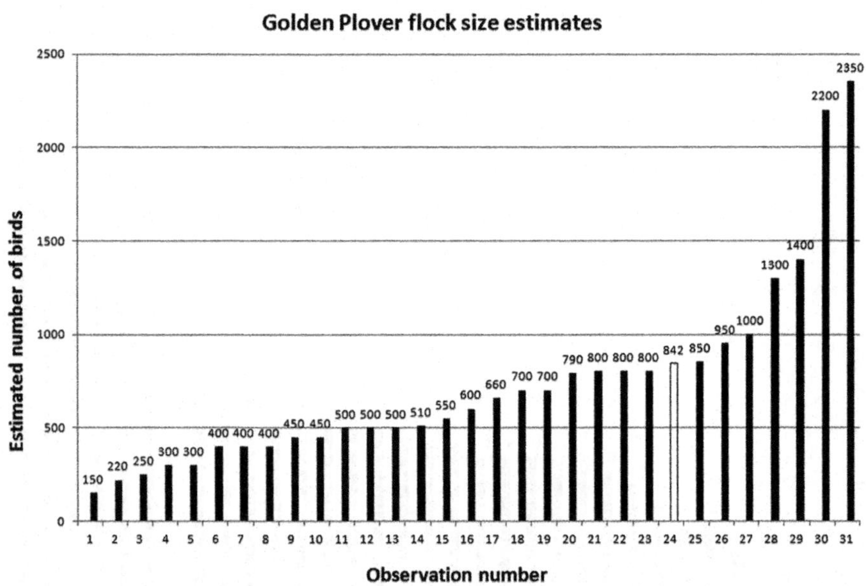

Fig. 9 The range of observer variation recorded for Fig. 5, an image of Golden Plovers in flight. Observation 24, the *hollow bar*, is the true total (842) provided by a manual count in ImageJ

Table 1 The range of variation and the median values recorded for each image used in the observer variation exercise

Species in image	Actual total	Lowest estimate	Highest estimate	Median
Greater Flamingo	969	200	1700	680 (−30%)
Golden Plover	842	100	1100	575 (−32%)
Linnet	458	150	3258	400 (−13%)
Starling	11,117[a]	500	47,800	4750 (−57%)

[a]Best estimate based on a validated automatic count in ImageJ

In all, 30 experienced bird recorders submitted 120 flock estimates, equating to 30 estimates for each of four images. Of these, 95 (79%) were underestimates and 25 (21%) overestimates, suggesting that four out of five counts of bird flocks are likely to be underestimates. These results reflect those recorded by other researchers, e.g. Škorpíková (2006), Grahn (2007).

Because of the uneven balance between the underestimates and the overestimates, they were analysed separately. The mean level of underestimation ranged from −29% in the Linnet image (Fig. 9) to −62% in the Starling image (Fig. 3). The mean level of overestimation ranged from +36% in the Linnet image to +134% in the Starling image. Table 1 shows the range of variation and median value for the estimates in each image.

There is no doubt that these levels of observer variation could mask significant increases or declines in actual bird populations.

Consequently, in recent years, researchers have started to look at ways to use technology to improve the quality and efficiency of bird counts e.g. Chabot and Francis (2016), Groom et al. (2013), Merkel et al. (2016), though these methods have yet to be adopted by mainstream ornithologists. Estimates made in the field remain the main source of bird count data underpinning all of the major bird surveys in the UK. However, given the wide range of observer variation associated with field estimates made by eye (see Results section below), it is clear that any practical and cost efficient method that improves the accuracy of bird counts should be embraced and developed.

The rest of this chapter demonstrates the how 'ImageJ', an open-source image processing program, can be used to improve the accuracy of flock and population size estimates.

ImageJ Software

Originally, ImageJ was developed for medical purposes, and the program is particularly well adapted for counting blood cells in blood samples. However, using suitable images and the appropriate settings, the program is equally well adapted for providing accurate estimates of large aggregations of birds, whether in flight, at breeding grounds or feeding grounds. Essentially, for the purpose of mainstream application, it is straightforward to use and is freely available to download on the internet.

ImageJ can be downloaded from the 'Downloads' page at https://imagej.net, though you may also need to download the 'cell counter' plug-in for manual counts. Alternatively, if you download 'Imagej Fuji', this comes with a wide range of plug-ins already incorporated, including the cell counter. ImageJ Fuji can be downloaded at https://imagej.net/Fiji/Downloads. A user guide is also freely available (Ferreira and Rasband 2012).

Case Study and Image Locations

The species and locations selected for case studies were those illustrated in Figs. 2, 3, 4 and 5, namely:

- Guillemots (*Uria aalge*) at their Elegug Stacks breeding colony in Pembrokeshire, UK;
- Greater Flamingos (*Phoenicopteris roseus*) on their wintering grounds at Doñana National Park in Andalucia, Spain;
- Golden Plovers (*Pluvialis apricaria*) and Linnet (*Carduelis cannabina*) flocks on farmland in the Angle Peninsula in Pembrokeshire, UK; and
- Starlings (*Sturnus vulgaris*) murmurating near a roost on the pier at Aberystwyth in Ceredigion, UK.

These case studies illustrate the potential of ImageJ for providing accurate estimates of bird numbers from images taken in a variety of different situations.

Methods

ImageJ can carry out either manual or automatic counts as appropriate, depending on the number and density of the birds in the image and on the complexity of the background. The sections below detail the key stages in the process, namely:

1. Image collection;
2. Deciding whether to carry out a manual or automatic bird count in ImageJ;
3. The process for carrying out counts in ImageJ;
4. Validating the counts.

Image Collection

Three key components of the image collection process combine to provide suitable images for carrying out bird counts in ImageJ: these are the camera equipment, the camera settings and the method of image capture.

Camera Equipment

A Digital Single Lens Reflex (DSLR) camera and a either a 35 mm or 100 mm lens is the preferred option for collecting images of bird flocks, depending on the size of the flock and how far away it is. If the flock is close, then a 35 mm lens is probably the best, with the 100 mm lens better for distant flocks. A good quality zoom lens, such as 24 mm–70 mm, offers a more versatile option: this would avoid the inconvenience of deciding which lens to use and possibly having to change lenses.

A good quality bridge camera or compact zoom could also suffice, as long as the images are sufficiently clear and detailed. The captions for Figs. 2, 3, 4, 5, 10a, b include the equipment and settings used to take the images.

Camera Settings

The key camera settings centre on the ISO, the f-stop and the shutter speed. The balance of these three settings will determine the sharpness of the image, the exposure of the image and the amount of movement in the image. Ideally, the image will not be too grainy, will be well exposed, i.e. not too bright or too dark, and the birds in the image will be sharply defined and not blurred. The sections below outline the general recommendations for these settings.

ISO Settings

The ISO setting will determine the 'graininess' of the image, where lower ISO settings deliver finer grained, higher quality, images. During summer months, with a reasonable quality camera, you can use ISO 200 in good light conditions. At other times of year, as the light quality declines, ISO 400 should suffice in all but very poor light conditions. Avoid using settings greater than ISO 800, as these will negatively impact on image quality.

The f-stop

The f-stop relates to how wide the aperture is at the time of image capture: this determines the depth of field in the image, which is a factor if the main subject of the image is close to the camera and you want the background to be sharp too. High f-stops of f22 or f32 provide the greatest depth of field. However, large flocks of birds are rarely close enough to need to use these settings. If the birds are all at a fixed distance from the camera, e.g. in photographs of feeding birds taken from a light aircraft or UAV, then an f-stop of f8 should be fine. Similarly, even if the birds are in flight and at various distances from the camera, an f-stop of f11 should ensure all of the birds in the flock are visible.

Fig. 10 (**a**) An image of a Guillemot colony at Elegug Stacks in Pembrokeshire UK. A manual count of the birds in ImageJ revealed that there are 1626 Guillemots, 1 Razorbill (*Alca torda*) and 2 Great Black-backed Gulls (*Larus marinus*) in the image. (**b**) is an inset from Fig. 10a but after a manual count in ImageJ. This shows the marker that has been left on each bird to indicate that the bird has been counted and which species it is (indicated by the marker number – 2 in this case). These markers make it possible to avoid double counting and to avoid losing your place in the image during the count. Camera model – Canon EOS 7D, Lens – Canon 35 mm lens, ISO 800, F-stop – f/10, Exposure time – 1/800 s

Shutter Speed

The shutter speed determines whether there is blurring as a consequence of 'hand-shake' and whether the birds in the flock are sharply defined. Generally, a shutter speed of 1/50th s will be sufficient to eliminate handshake if you are using a 50 mm lens or smaller. However, the risk of handshake increases with the focal length of the lens, so if you are using a 100 mm lens then you would need a shutter speed of 1/100th s to eliminate the risk of handshake, and a shutter speed of 1/500th s to eliminate handshake using a 500 mm lens etc. We must also attempt to freeze the movement of the birds which, in flight, can be more demanding. As a general rule, setting the shutter speed at 1/500th s will be fast enough to still the movement in most flocks of birds: only the Starling murmuration in the case studies below used a higher shutter speed than this, at 1/640th s.

In summary, if you set the camera at ISO 400, with the f-stop at f11 and the shutter speed at 1/500th s, this should suffice to capture well-lit, well-defined images with a good depth of field: images with these characteristics are ideal for carrying out counts in ImageJ.

Capturing the Images

There are two basic alternatives for capturing images of bird flocks:

- Vertical images of feeding or breeding birds taken from light aircraft or Unmanned Aerial Vehicles (UAVs); and
- Oblique images of cliff-based breeding colonies or flocks of birds in flight, typically taken from the ground, but also potentially by UAV.

Capturing Vertical Images of Feeding or Breeding Birds

Until recently, the only practical option for collecting vertical images of birds was by light aircraft, similar to the four-seat Cessna used to capture the flamingo images in Case Studies 3 and 4. This is no longer the case and the potential for using UAVs to capture vertical images of birds is increasing rapidly. Although there are limitations to the area that UAVs can cover during one flight, which is often in the region of 1 square km, this would be sufficient to capture images of many feeding hotspots or breeding colonies. Furthermore, there is always the option of using more than one UAV or carrying out more than one flight.

Vertical images of birds feeding on the ground or in water offer the best chance for accurate counts, because the birds are all the same distance from the camera: making it possible to differentiate different species by size as well as plumage. It also means that we do not need to set a large depth of field on the camera, so an f-stop of f8, or even f5.6, might be sufficient to provide adequate images for counts. Another advantage of vertical images is that, whether breeding or feeding, birds often prefer to have their own personal space: this space can be clearly visible on vertical images but is less obvious on oblique images (see Fig.1).

Capturing Images of Flocks of Birds in Flight

Most of the case studies in this chapter focus on flocks of birds in flight: this is because these present the greatest challenge to a bird counter. Often, however, flying birds also offer the best opportunity of obtaining an accurate count of how many birds are present. This is particularly true of gregarious estuarine waders such as Dunlin and Knot, and of small and sociable ground-feeding birds, such as sparrows, finches, larks and buntings.

As with feeding birds, birds in flight also like their own space: this would be obvious in a 3D image but is much less evident in 2D images, where the birds will overlap. This presents a problem for any automatic counter program that simply counts the number of discrete objects in an image. There are ways to minimise the error caused by overlapping birds in an image and these are covered both in the section on validation below and in the individual case studies. However, life is generally a lot easier if we can minimise the number of overlapping birds in images at the time of capture. We can do this by following a few general rules:

(a) Identify the best vantage point to take the images, for example, as Starlings arrive to roost or as waders arrive to feed on a falling tide;
(b) Try to capture the entire flock in a single image or, if the flock breaks up, then take separate images of the component parts of it: large flocks of Starlings or waders, for example, often aggregate and disaggregate during aerial displays;
(c) If the birds are in a tight flock, try to get an image of the flock as it banks to one side or the other, this often reveals the spaces between the birds and reduces the amount of overlap in the image; and
(d) If birds tend to arrive in sequence, e.g. Cranes or Geese, have a system for recognising when one flock ends and the next one starts – perhaps by taking an image of a tree or a building, something obvious that is in your field of view, to form a break between the end of one flock and the start of the next.

Carrying Out Counts in ImageJ

There are two basic types of count available in ImageJ: a manual count and an automatic count. In general, a manual count is by far the best option if the image contains less than c. 3000 birds, and might be the only option if the image has a noisy or complex background. Alternatively, if an image contains several thousands of birds set against a relatively plain background, e.g. sky, water, mud or sand, then an automatic count may be the more practical option. Both count types offer a more reliable alternative to 'by eye' counts in the field.

Manual Counts

In most cases, except when dealing with exceptionally large flocks of birds, a manual count is the most efficient and accurate option, and this will always be the case if there is a complex background to the image, such as vegetation or rock. For this reason, manual counts are likely to be the only feasible option for counting seabird colonies on cliff-faces, for example. A manual count is also the only option if the birds in the image have a strongly contrasting plumage pattern, such as Avocets (*Recurvirostra avosetta*) or Shelducks (*Tadorna tadorna*) as, in this situation, the automatic counter will typically count each black section (or white section in an inverted image) as a separate object/bird.

Manual counts are the default option because they need virtually no preparation. ImageJ keeps running totals for up to eight species (I have yet to come across a situation where I needed all eight) until the count is complete, at which point there is an option to save totals to an Excel file. Manual counts in ImageJ have a number of advantages over carrying out counts in the field and over automatic counts in ImageJ, including the ability to:

- Zoom into the image and systematically count all of the bird in your own time.
- Pick up the count where you left it if you are disturbed or distracted: this is not an option if you are disturbed while counting birds in the field.
- Stop to rest your eyes, or for a comfort break, and carry on later when you are rested and refreshed.
- Decide what should count as one bird, what is part of a bird, what is several overlapping birds etc.: overlapping birds are counted as 'one' in automatic counts.
- Keep running totals for several species at the same time.
- Commit only one person to the task: it would usually take at least two people to carry out a seabird population count in the field.

As a general guide, a manual count of c. 1000 birds in ImageJ will take 10–15 min: Case Study 1 provides a worked example of how to do this.

Manual Count Validation

There is no need for validation of manual counts in ImageJ as you can export images with markers showing (a) which birds were counted and (b) which birds were allocated to which species (Fig. 10b). If validation is necessary, an independent expert could check the 'marked' image for missed birds or mistaken identification.

Automatic counts

Automatic counts are best suited to situations where, for practical purposes at least, there are simply too many birds in the image to consider counting manually. These situations tend to arise in only a few circumsances in the UK, typically at winter roosts of waders and Starlings. However, the image will need a plain background and validation is essential.

Automatic Count Validation

Automatic counts come with three options for validation:

- A drawing showing the numbered outlines of objects that contributed to the count;
- An Excel file documenting the pixel size of every numbered object in the drawing; and
- The option of carrying out a manual count of all or part of the image.

Validation is an essential source of confidence when applying the automatic counter and the case studies provide examples of each validation option.

Results

The six case studies focus on situations where it would be difficult to carry out an accurate count in the field. In the case of automated counts, the accuracy depends not only on having a suitable image but also on selecting the most appropriate settings for threshold and object size. The case studies below illustrate how to adjust these settings to best effect for different image types.

A Manual Count of Guillemots at a Breeding Colony at Elegug Stacks, Pembrokeshire, UK

Counting seabirds at breeding colonies is fraught with difficulties and frustrations. The number and density of birds being counted is typically high, the birds are constantly coming and going, the birds engage in a lot of territorial disputes, and there are occasional incursions by predators causing unrest or, even worse, dreads when many birds will leave the colony etc.

A big advantage of carrying out photographic counts is that you can wait until most of the birds are present on the colony, take the photo and go home to do the count, safe in the knowledge that the birds are not going to be disturbed or flushed by predators part of the way through the count. Text Box 1 details the process used to carry out a

Text Box 1: The Process and Settings Used for Carrying Out the ImageJ Manual Count of the Guillemots at the Elegug Stacks Colony in Pembrokeshire, UK

Open ImageJ, select File ⇨ Open – and choose the image that you want to analyse

When the image is open in ImageJ select Plugins ⇨ Analyze ⇨ Cell counter

This will open a table showing the cell counter options, click on 'Initialize'

Next, choose your 'Counter Type' for the species that you want to count – some of the markers colours are brighter than others, so you might want to test them before deciding which to use. If there is more than one species in the image, select a different 'Counter Type' for species that you want to count.

After selecting the Counter Type, just left-click on each individual that you want to count and ImageJ will leave a marker on the bird to show that it has been counted and will add one to the total for that marker type, so you do not need to count and cannot lose your place in the image. Note that you can zoom into the image using the 'zoom' option under the 'Image' heading in the main menu and you can move about the image while counting by toggling between the 'hand' scrolling tool and the point selection tool on the main menu.

When you have finished counting, click on 'Results' in the cell counter table and an Excel spreadsheet will pop up showing how many species of each type you have counted, click on 'File'⇨'Save' to save these results in your chosen location

Click on 'Export' and your image will pop up with the markers on it

In the main ImageJ menu select 'File' ⇨'Save as' – then choose the image type from the drop-down menu, e.g. Tiff or Jpeg etc

Finally, choose a file name and location and click on 'Save'

manual count of the Guillemot colony in ImageJ: this process was the same for the manual counts carried out for validation purposes in Case Studies 3, 4, 5 and 6. In the 'automatic count' mode, ImageJ will generate a relatively accurate count of several thousands of birds in seconds, a task that could take many hours, if not days, to carry out manually. However, the accuracy of the count will depend entirely on a) the ability of the program to isolate the birds from their background, b) the threshold settings applied to the image and c) the pixel range settings applied to the image.

To deliver a near-accurate automatic count, ImageJ needs an image that shows clearly defined bird outlines against a clean and high contrast background: each bird needs to be a separate entity, i.e. something that the program can equate to a blood cell. Case Study 2 below outlines the process for carrying out an automatic count in ImageJ and the subsequent case studies illustrate how to adjust the automatic count settings to different situations.

An Automatic Count of a Starling Murmuration at Aberystwyth, Ceredigion, UK

One of the most difficult challenges for any bird counter is to assess the number of Starlings participating in aerial displays (murmurating) before going to roost.

The intricate formations witnessed during Starling murmurations are spellbinding and can involve tens of thousands of birds. Reliably estimating the number of birds involved in these displays by eye is impossible. However, if we can capture an image of a flock of birds against a plain sky background, there is a good chance that it will be suitable for an automatic count in ImageJ. In images with a plain background, the factor most likely to influence the accuracy of the count will be the degree of overlap between individual birds. In images of dense flocks, there will always be some individuals overlapping, and in dense flocks comprising a large number of birds, this could equate to thousands. This will pose a problem for any program that bases its counts on the number of objects in an image, because ten overlapping birds could form a single object and contribute only a single bird to the total. The impact of this problem can be minimised, however, because ImageJ provides two options for validating the counts: a drawing showing the numbered objects that contributed to the count and an Excel file stating the pixel size of every object that contributed to the count. In practice, the drawing is an excellent source of confidence that the objects counted by the program actually equate to the birds in the original image. If this is not the case, then the settings for image threshold and object pixel size will need adjusting. Text Box 2 describes the process involved in carrying out the automatic count of the Starling murmuration shown in Fig. 3, while Fig. 11 shows an inset from the drawing showing the numbered objects that were counted, this indicated that all of the objects counted were birds.

Starling Case Study Results and Validation

The total generated by the automatic count in ImageJ was 9380 birds, but this included a considerable number of overlapping birds. The validation process described below improved the accuracy of this estimate.

Validation

The automatic count details in the Excel table (Table 2) suggested that the area of most individuals in the Starling flock was in the 35–80 pixel range, depending on the position of the bird in the flock (closer or further from the camera) and whether the wings were outstretched or close to the body at the time. However, because there

Text Box 2: The Process and Settings Used for Carrying Out the Automatic Count of a Starling Murmuration (Fig. 3) in ImageJ

Open ImageJ, select 'File' ⇒ 'Open' – and choose the image that you want to analyse

When the image is open select 'Image' ⇒'Type' ⇒'8-bit'

Then, again under the 'Image' menu, select 'Adjust' ⇒'Threshold' – this will open a box containing the Threshold settings

Then select the appropriate Threshold settings which, in the case of the image of the Starling murmuration, were as follows:

Top slider	– 0
Bottom slider	– 119
Default	B&W
Don't tick	'Dark background'

Click on the apply button

Then go back to the main ImageJ toolbar and select 'Analyze' ⇒'Analyze particles' – this opens another box and the following settings were selected:

Size (pixel)	10-infinity
Circularity	0.00–1.00
Show	Outlines
Tick	Display results
Click on OK	

This will generate an Excel spreadsheet with the count details and an image labeled 'Drawing of + file name' this shows the Starling outlines (numbered) that contributed to the count total.

is a drawing showing the numbered outlines of every object counted and a table showing the area (in pixels) for each object, you can cross-reference between the two to confirm this and to check whether the objects with a larger pixel area comprised more than one bird. Cross-referencing between the drawing and Excel table for the Starling image revealed that objects of 100–199 pixels comprised at least two birds. Therefore, as many objects in the spreadsheet had an area of >100 pixels, it was clear that the automatic count of 9380 birds must have been an underestimate, and quite possibly a substantial one.

To improve the accuracy of the Starling estimate, I scrolled through the spreadsheet and for every object with an area of 100–199 pixels I added one bird to the total, and for every object with an area of 200–299 pixels I added two birds etc., etc.

Fig. 11 An enlarged inset from the 'drawing' of Fig. 3. This drawing, generated during an automatic count of a Starling flock in ImageJ, shows the numbered outline of each bird that contributed to the total: the larger objects (e.g. no. 1512) comprise at least two overlapping birds

Table 2 These are the first entries in the Excel results table generated by the automatic count of the Starling flock: the pixel area counts of >100 pixels referred to objects comprising more than one bird

Object/bird number	Pixel area
1	39
2	39
3	65
4	22
5	155
6	46
7	26
8	146
9	179
10	171
11	45
12	67
13	40
14	49
15	77
16	63
17	79
18	54
19	42

This exercise generated an additional 1737 birds, bringing the estimated total in the image to 11,117. This means that the automatic count underestimated the number of birds in the flock by at least 18.5%, though this still compares favourably with the median of −57% recorded by experienced ornithologists. In reality, the revised total of 11,117 birds is also likely to be an underestimate, but a less significant one and probably by fewer than 500 birds.

It is possible, of course, to undertake a full validation of the number of birds in the image by carrying out a manual count, but this would be a demanding and time consuming process that would render the automatic count pointless.

An Automatic Count of Wintering Greater Flamingos (Phoenicopterus roseus) *in Flight over Doñana NP in South-West Spain*

The image used in this case study was taken from a light aircraft (a four-seater Cessna) in March 2014. The image shows a flock of Greater Flamingos in flight over Doñana NP in south-west Spain. Despite the different tones in the background of the image, it was possible to neutralise these using the threshold settings given in the Text Box 3.

Although the image in Fig. 2 shows spacing between many of the flamingos in flight, the estimate generated by the automatic count was 871 birds: an underestimate of c. 10% compared to the manual count of 969. This suggests that c. 10% of the birds in the image were overlapping, though probably more because of the vertical perspective of the image than because the birds were particularly close together.

In practice, a manual count in ImageJ is most appropriate for this type of bird flock, as it would take only c. 20 min to complete and should be c. 99–100% accurate (Figs. 12 and 13).

An Automatic Count of Wintering Greater Flamingos (Phoenicopterus roseus) *Feeding in a Lagoon at Doñana NP in Andalucia, Spain*

The image of Greater Flamingos feeding in a lagoon at Doñana NP (Fig. 14) was taken in from a light aircraft (a four-seat Cessna) in March 2014. Text Box 4 outlines the process and settings used to carry out an automatic count of the feeding flamingos in Fig. 14.

The relatively plain background of standing water and the feeding spaces between individual birds made this image equally suitable for an automatic or manual count. This was reflected in the results from the two counts, with the auto-

Text Box 3: The Process and Settings Used to Carry Out an Automatic Count of a Flock of Wintering Greater Flamingos in Flight over Doñana in South-West Spain (Fig. 2)

Open ImageJ, select File ⇨ Open – and choose the image that you want to analyse

When the image is open select Image' ⇨Type ⇨8-bit

Then, again under the'Image' menu select 'Adjust' ⇨Threshold – this will open a box containing the Threshold settings

Then select the appropriate Threshold settings, which, in the case of the image of the Flamingos in flight, were as follows:

Top slider	– 121
Bottom slider	– 255
Default	– Red
Tick	– Dark background'

Click on the apply button

Then go back to the main ImageJ toolbar and select 'Analyze' ⇨Analyze particles – this opens another box and the following settings were selected:

Size (pixel)	– 60 to infinity
Circularity	– 0.00–1.00
Show	– Outlines
Tick	– Display results

Click on OK

This will generate an Excel spreadsheet with the count details and an image labeled 'Drawing of + file name' this shows the Greater Flamingo outlines (numbered) that contributed to the count total. Save both.

Fig. 12 The drawing generated during the automatic count of the flamingos in flight shows that the counted objects correspond very closely with the distribution of birds in the original image (Fig. 2)

Fig. 13 An enlarged inset from the drawing generated during an automatic count in ImageJ (Fig. 12). This shows the numbered outlines of birds that contributed to the count, including some overlapping birds in the lower right quarter

Fig. 14 An aerial image of wintering Greater Flamingos feeding in a lagoon at Doñana NP. An automatic count of these birds in ImageJ indicated that 215 birds were present in the image, a manual count for validation purposes revealed 214 Greater Flamingos. Camera model – Canon EOS 7D, Lens – Canon 35 mm, ISO 400, F-stop f/8, Exposure time 1/500 s

matic count generating an estimate of 215 birds, compared to the manual count of 214 birds.

In practice, a manual count is the most appropriate for this type of image, as this would be 99%–100% accurate and take no more than 10 min to complete (Fig. 15).

Text Box 4: The Process and Settings Used for Carrying Out an Automatic Count of Wintering Greater Flamingos Feeding in a Laguna at Doñana in South-West Spain (Fig. 14)

Open ImageJ, select 'File' ⇨ 'Open' – and choose the image that you want to analyse

When the image is open select 'Image' ⇨'Type' ⇨'8-bit'

Then, again under the 'Image' menu select 'Adjust' ⇨'Threshold' – this will open a box containing the Threshold settings

Then select the appropriate Threshold settings, which, in the case of the image of the feeding flamingos, were as follows:

Top slider	– 91
Bottom slider	– 233
Default	– Red.

Leave 'Dark background' clear, do not tick

Click on the apply button twice – so that you are seeing black birds on a white background

Then go back to the main ImageJ toolbar and select 'Analyze' ⇨'Analyze particles' – this opens another box and the following settings were selected:

Size (pixel)	300 – infinity
Circularity	0.00–1.00
Show	Outlines
Tick	Display results

Click on OK

This will generate an Excel spreadsheet with the count details and an image labeled 'Drawing of + file name' this shows the Greater Flamingo outlines (numbered) that contributed to the count total.

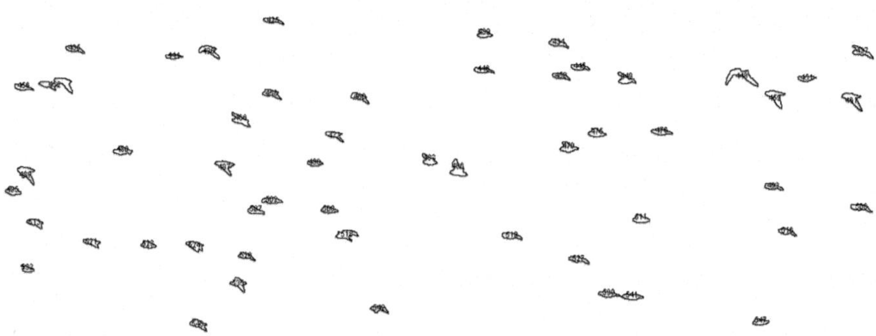

Fig. 15 An enlarged inset from the drawing of Fig. 14: this drawing was generated during an automatic count in ImageJ. This shows the numbered outlines of the Greater Flamingos that contributed to the count total.

An Automatic Count of a Flock of Linnets (Carduelis cannabina) *in Flight over Stubble Fields at Angle, Pembrokeshire, Wales*

In autumn and winter, sizeable flocks of farmland birds present a challenge for bird recorders. These flocks, which often comprise several hundreds of small, ground-feeding birds, are typically difficult to see or count while feeding on the ground. These flocks are dense, very mobile and easily disturbed. Furthermore, the flocks often comprise several different species, typically involving larks, sparrows, finches and buntings. In flight, if there is a mixed flock of several similarly sized species, a manual count is will be the most appropriate option. However, an automatic count is also possible if it is a single-species flock and the birds are not too densely packed.

Text Box 5 outlines the process and settings used to carry out an automatic count of the birds in Fig. 4 in ImageJ. This automatic count generated a total of 377 birds,

Text Box 5: The Process and Settings Used for Carrying Out an Automatic Count of a Flock of Linnets in Flight over Farmland (Fig. 4)
Open ImageJ, select 'File' ⇒ 'Open' – and choose the image that you want to analyse

When the image is open select 'Image' ⇒ 'Type' ⇒ '8-bit'

Then, again under the 'Image' menu select 'Adjust' ⇒ 'Threshold' – this will open a box containing the Threshold settings

Then select the appropriate Threshold settings, which, in the case of the image of the Linnets in flight, were as follows:

Top slider	– 92
Bottom slider	– 255
Default	– Red.
Tick	Dark background'

Click on the apply button twice

Then go back to the main ImageJ toolbar and select 'Analyze' ⇒ 'Analyze particles' – this opens another box and the following settings were selected:

Size (pixel)	500- infinity
Circularity	0.00–1.00
Show	Outlines
Tick	Display results
Click on OK	

This will generate an Excel spreadsheet with the count details and an image labelled 'Drawing of + file name' this shows the outlines (numbered) of the Linnets that contributed to the count total.

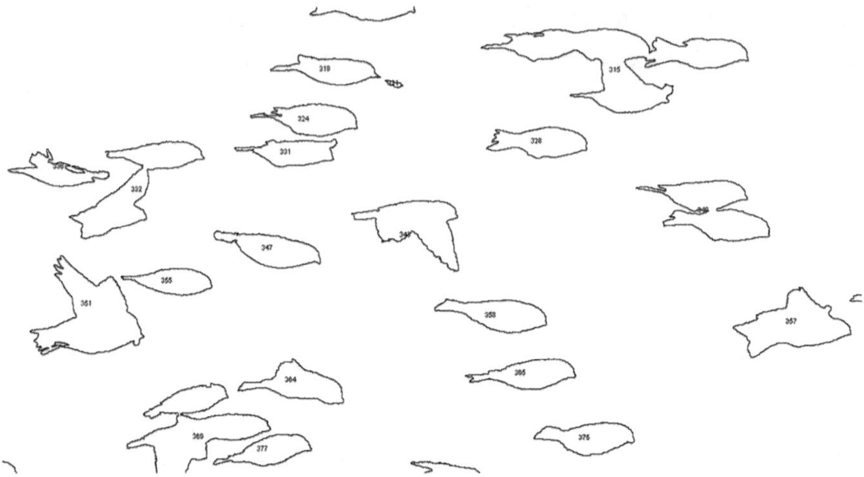

Fig. 16 An enlarged inset from the drawing of the image shown in Fig. 4, generated during an automatic count in ImageJ. This shows the numbered outlines of the Linnets that contributed to the count total. Note that overlapping birds, e.g. the four birds in the top right of the drawing, are counted as one

compared to the manual count total of 458, so the automatic count underestimated the number of birds in the flock by c.18%. This disparity between the two counts was primarily a consequence of overlapping birds counting as a single object during the automatic count process. However, it is also likely that the lower 500 pixel size threshold for the automatic count excluded some of the birds partially represented on the edge of the image. These birds would have contributed to the manual count total.

In the case of larger flocks, it would also be possible to look through the pixel sizes in the Excel results spreadsheet and generate a more accurate estimate of how many birds were present in the image, as illustrated for the Starling murmuration in Case Study 2 (Fig. 16).

An Automatic Count of Golden Plovers (Pluvialis apricaria) *in Flight over Farmland Near Castlemartin, Pembrokeshire, UK*

The Golden Plover is a sociable species that can form large flocks and favours farmland for winter feeding. The birds are typically well spaced when moving between feeding sites, or drifting over a favoured feeding area after being disturbed, though overlapping birds are common in images of birds taking to flight. Figs. 5 and 17 show a large, well-spaced, single-species flock of Golden Plovers set against a plain sky background: this ticks all of the boxes for an accurate automatic count in ImageJ. Text Box 6 outlines the process and settings used to generate an automatic count of the Golden Plovers in Fig. 5.

Text Box 6: The Process and Settings Used to Carry Out an Automatic Count of a Golden Plover Flock in Flight (Fig. 5)

Open ImageJ, select 'File' ⇒ 'Open' – and choose the image that you want to analyse

When the image is open select 'Image' ⇒ 'Type' ⇒ '8-bit'

Then, again under the 'Image' menu select 'Adjust' ⇒ 'Threshold' – this will open a box containing the Threshold settings

Then select the appropriate Threshold settings, which, in the case of the image of the Golden Plovers in flight, were as follows:

Top slider	– 122
Bottom slider	– 255
Default	Red.
Tick	Dark background'

Click on the apply button twice

Then go back to the main ImageJ toolbar and select 'Analyze' ⇒ 'Analyze particles' – this opens another box and the following settings were selected:

Size (pixel)	50 to infinity
Circularity	0.00–1.00
Show	Outlines
Tick	Display results
Click on OK	

This will generate an Excel spreadsheet with the count details and an image labelled 'Drawing of + file name' this shows the outlines (numbered) of the Golden Plovers that contributed to the count total.

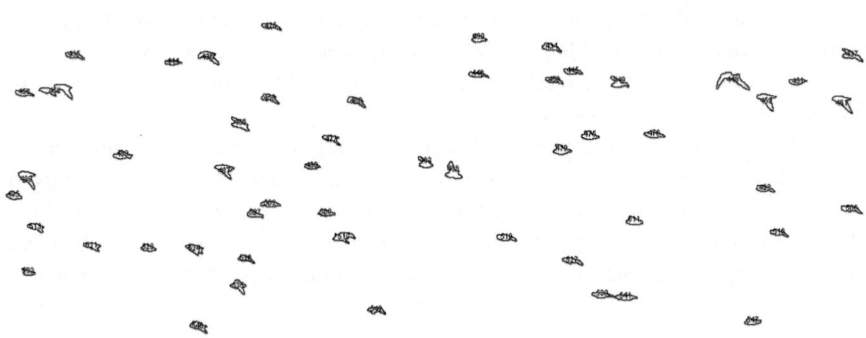

Fig. 17 An inset from the 'drawing' of Fig. 5: this shows the numbered outline of each Golden Plover contributing to the count total

The total generated by the automatic count was 832: this was revised to 848 birds after using the pixel sizes in the Excel file to validate the count. A manual count revealed 842 birds in the image. Therefore, the automatic count of 832 was an underestimate of almost 1% and the validation of the automatic count of 848 was an overestimate of c.0.7%.

Discussion

The results of the exercises described in this chapter suggest that the range of observer variation associated with 'by eye' bird counts is considerably greater than any real change we would tolerate without some form of political or conservation management response. The results of these exercises should be a concern for any ornithologist or researcher using 'by eye' count estimates for scientific purposes. In contrast, we can expect the results of careful manual counts in ImageJ to be 99–100% accurate, with the accuracy of automatic counts also likely to exceed 90% accuracy, particularly after being revised through validation.

The results from the case studies suggest that technology has a significant role to play in improving the accuracy of bird counts, particularly on breeding grounds, in wintering areas and at staging sites during migration periods. This is particularly true where counts involve gregarious species which, in western Europe, include Common Scoters (*Melanitta nigra*), Cranes (*Grus grus*), Whooper Swans (*Cygnus cygnus*), White Storks (*Ciconia ciconia*), Greater Flamingos, Guillemots, Golden Plovers, Lapwings, Dunlins (*Calidris alpina*), Knots (*Calidris canutus*), Starlings, and many other species of geese, wader, wildfowl, lark, pipit, finch and bunting.

There are a small number of examples in the literature of researchers experimenting with remote counts using segmentation analysis methods (e.g. Chabot and Francis 2016, Merkel et al. 2016, Delord et al. 2015, and Groom et al. 2013) but the recommendations have not yet penetrated mainstream thinking in ornithological circles. Some of these segmentation exercises have been carried out in eCognition software (Nussbaum and Menz 2008), but this software is both specialised and expensive and would not be readily available or accessible to most birdwatchers, in direct contrast to ImageJ software.

Given suitable images to process, ImageJ can carry out relatively accurate automatic counts of large aggregations of birds. These automatic counts take seconds to complete and come with three options for validation, (a) a numbered drawing of the objects contributing to the count, (b) an Excel file listing the pixel sizes of each object in the drawing and (c) partial or complete manual counts of the birds in the image.

The Opportunities for Counting Birds in the Future

The true potential of ImageJ will be realised when used in conjunction with images collected by light aircraft or drones, particularly in relation to inaccessible, or difficult to see, feeding or breeding grounds (see Chapt. 15). Furthermore, thermal imaging cameras mounted on drones will offer the option of carrying out low disturbance night surveys of ground-nesting bird colonies, again providing images suitable for processing in ImageJ.

The manual counter in ImageJ is a deceptively powerful tool and there is no doubt that it has the potential to revolutionise how we collect and process bird count data.

Conclusions

Bird surveillance schemes are among the most heavily subscribed citizen science projects in the UK. Several of these are long-running national schemes such as the Bird Atlas 2007–2011 (Balmer et al. 2013a, b), the Wetland Birds Survey (e.g. Frost et al. 2016) and the Seabird Monitoring Programme (e.g. JNCC 2015). The BTO coordinates the Wetland Bird Survey (WeBS), which has been operational since 1947, while the Seabird Monitoring Survey is organised by JNCC and has been operational since 1986.

All of these rely heavily on untrained volunteers to collect data on large numbers of birds, including wildfowl, waders, seabirds and farmland birds, yet there is nothing in the literature to suggest that the data provided by the volunteers is subject to any form of validation. There is also nothing to suggest that photographic counts have contributed to the data collection process, despite the longevity of the schemes.

The results of the observer variation exercises in this chapter suggest not only that some validation of 'by eye' counts is essential, but also that the ornithological community would benefit from embracing new technologies to improve both the accuracy and precision of bird counts in the future.

Acknowledgements Thanks are due to Dan Guest, Chris Millican and Jonas Grahn for helping with the thankless task of copy-editing and for providing invaluable comments on the content. I am also grateful to the 35 experienced ornithologists who participated in the observer variation exercise by contributing flock size estimates of the images presented in this chapter: as agreed, these contributors will remain nameless. Finally I would like to thank Ricardo Diaz-Delgado, Manuel Mañez and their colleagues at Estación Biológica de Doñana for giving me the opportunity to accompany their wetland bird counters, both on the ground and in the air: an exercise that rekindled my interest in the subject area and inspired me to write this chapter.

References

Balmer, D.E., Gillings, S., Caffrey, B.J., Swann, R.L., Fuller, R.J.: Setting the scene. In: Balmer, D.E., Gillings, S., Caffrey, B.J., Swann, R.L., Downie, I.S., Fuller, R.J. (eds.) Bird Atlas 2007–11: The Breeding and Wintering Birds of Britain and Ireland, pp. 17–31. BTO Books, Thetford (2013a)

Balmer, D., et al.: Bird Atlas 2007-11: The Breeding and Wintering Birds of Britain and Ireland. British Trust for Ornithology (2013b)

Chabot, D., Francis, C.M.: Computer-automated bird detection and counts in high-resolution aerial images: a review. J. Field Ornithol. **87**(4), 343–359 (2016)

Delord, K., Roudaut, G., Guinet, C., Barbraud, C., Bertrand, S., Weimerskirch, H.: Kite aerial photography: a low-cost method for monitoring seabird colonies. Eco. Inform. **86**(2), 173–179 (2015)

European Commission: Council Directive 92/43/EEC of 21.5.1992 on the conservation of wild habitats and of wild fauna and flora. Official Journal of the European Communities: No L 206: 22.7.1992 (1992)

Ferreira, T., Rasband, W.: ImageJ User Guide IJ1.46r. (2012). Available to download at: https://imagej.nih.gov/ij/docs/guide/user-guide.pdf

Frost, T.M., et al.: *Waterbirds in the UK 2014/15: The Wetland Bird Survey*. BTO, RSPB and JNCC, in association with WWT. British Trust for Ornithology, Thetford (2016)

Grahn, J.: Räknar vi rätt? Om de övervintrande gräsänderna i Umeå och rapporter på Svalan. In Fäglar i Västerbotten, nr 2, 44–45 (2007)

Groom, G., Stjernholm, M., Due Nielsen, R., Fleetwood, A.: Ib Krag Petersen. Remote sensing image data and automated analysis to describe marine bird distributions and abundances. Ecol. Inform. **14**, 2–8 (2013)

JNCC: Seabird population trends and causes of change: 1986–2014 report (2015)

Merkel, F.R., Johansen, K.L., Kristensen, A.J.: Use of time-lapse photography and digital image analysis to estimate breeding success of a cliff-nesting seabirds. J. Field Ornithol. **87**(1), 84–95 (2016)

Nussbaum, S., Menz, G.: eCognition image analysis. In: Nussbaum, S., Menz, G. (eds.) Object-Based Image Analysis and Treaty Verification: New Approaches in Remote Sensing. Springer, Dordrect (2008)

Rowell, T.A.: The relationship between monitoring and management. In: Hurford, C., Schneider, M. (eds.) Monitoring nature conservation in cultural habitats: A practical guide and case studies. Springer, Dordrecht (2006)

Škorpíková, V.: Velká hejna – velké problémy (Big flocks – big problems). In: Crex – Zpravodaj Jihomoravské Pobočky ČSO 26 (2006): 152–153 (2006)

Using UAVs to Map Aquatic Bird Colonies

Ricardo Díaz-Delgado, Manuel Mañez, Antonio Martínez, David Canal, Miguel Ferrer, and David Aragonés

Abstract In this chapter, we present the results of several flight campaigns carried out in 2015 and 2016 using multirotor Unmanned Airborne Vehicles (UAVs) over Slender-billed Gull (*Chroicocephalus genei*) colonies in the Doñana Nature Space, south west Spain. The images were taken at different times during the breeding season. The requirements for the flight campaigns were to acquire sufficient visible and nadir pictures at 5 cm pixel resolution and to cover the entire nesting colony with maximum overlap. Although we carried out the flights under clear skies, low wind speed was not always possible, causing a few blurred pictures. After georeferencing and mosaicking the set of raw pictures, we adopted photo-interpretation as the first technique to identify and delineate birds, either lying, standing or flying. A nest position was assigned when the clear pattern of a lying birds was recognised. We then selected a set of breeding individuals (nests) to train a supervised classification in semi-automatic nest delineation. We applied two different algorithms and tested their accuracy in identifying gulls with an independent set of manually delineated individuals. We chose the best method according to the accuracy results and applied it to the whole colony. We found major issues for nest identification and delineation for nests under tree and shrub canopies. The different campaigns and flight characteristics were useful to improve bird identification accuracy. As a result,

R. Díaz-Delgado (✉)
Estación Biológica de Doñana (EBD), Consejo Superior de Investigaciones Científicas (CSIC), Sevilla, Spain
e-mail: rdiaz@ebd.csic.es

M. Mañez • A. Martínez
Natural Processes Monitoring Team, ICTS-RBD, Estación Biológica de Doñana, CSIC, Avda. Americo Vespucio, 26, 41092 Sevilla, Spain
e-mail: mmanez@ebd.csic.es; antonio.martinez@ebd.csic.es

D. Canal • M. Ferrer
Estación Biológica de Doñana, CSIC, Avda. Americo Vespucio, 26, 41092 Sevilla, Spain
e-mail: davidcanal@ebd.csic.es; mferrer@ebd.csic.es

D. Aragonés
Laboratory of Remote Sensing and GIS, Estación Biológica de Doñana, CSIC, Avda. Americo Vespucio, 26, 41092 Sevilla, Spain
e-mail: davidcanal@ebd.csic.es

© Springer International Publishing AG 2017

R. Díaz-Delgado et al. (eds.), *The Roles of Remote Sensing in Nature Conservation*,
DOI 10.1007/978-3-319-64332-8_14

we provided estimates of the number of breeding pairs per year to managers and cross-checked these with estimates from the ground monitoring and colony sampling. As an added value, the spatial coordinates of nests can be used for spatial analysis and investigate nest aggregation, density and distribution in order to reveal spatial relationships with environmental factors such as distance to colony edges, distance to colony centroid, distance to predators, etc.

Keywords UAVs • Slender-billed Gull • Orthomosaic • Bird delineation • Photointerpretation • Supervised classification • Colony monitoring

Overview

- UAVs are increasingly helpful and accessible for nature conservation and management;
- Picturing bird colonies with UAVs has proved to be a very efficient procedure for estimating colony size with low or no disturbances;
- Ground and manned aerial surveys of bird colonies are time-consuming and have different constraints in retrieving accurate colony size estimates;
- Most of the studies have identified birds and other animals by visual photo-interpretation;
- Automatic supervised classification provides acceptable results for delineating ground nesting birds unless under a canopy;
- Some added-values arise from the use of UAV images in bird colony surveys, such as the spatial distribution of nests, their attributes and digital surface models of the colony;

This study compares automatic *versus* manual approaches to delineating lying and standing gulls on UAV images of a breeding colony at Doñana Natural Space.

Introduction

In this chapter, we evaluate the use of Unmanned Aerial Vehicles (UAVs) for automated mapping of Slender-billed gulls (*Chroicocephalus genei*) inside the Doñana Natural Space, southwest Spain. Particular focus was on monitoring bird colony size (number of breeding birds or lying individuals) and productivity (clutch size, total number of chicks). Ground validation of colony size estimates was achieved through concurrent visual surveys.

Unlike previous studies using UAVs to retrieve bird colony information, we used two different automatic methods to estimate colony size in addition to visual photo-interpretation. The selected automatic classification procedures were: (a) Support Vector Machine algorithm and (b) Random Forests machine learning methodology.

Both algorithms have proven to be very efficient in delineating and targeting features in digital image analysis (Foody and Mathur 2004; Pal 2005). These approaches can be used for any other colonial species to provide quick status assessment and colony size estimates. Automatic techniques based on pattern recognition and image classification have been previously applied for bird detection in flocks (Abd-Elrahman 2005; Grenzdörffer 2013) showing acceptable results.

Additionally, visual assessment allowed us to identify standing birds from lying birds and, for a few cases, count eggs lying exposed on the ground. We were also able to identify other species breeding in the colony, such as the Black-headed Gull (*Chroicocephalus ridibundus*) and Yellow-legged Gull (*Larus michahellis*), the latter being an active chick predator of the other two species.

Background

Two of the most critical factors in assessing the conservation status of a bird population are the number of breeding pairs and annual breeding success. Usually, managers and scientists estimate changing population sizes by counting nesting pairs and fledging chicks throughout the breeding season and from year to year. Additional research or monitoring activities are however needed to obtain ancillary information over the course of the breeding season, such as clutch size and the number of born and fledging chicks per nesting colony. This information requires very intensive sampling but generally involves entering the breeding colonies. This usually leads to significant disturbances (Anderson and Gaston 2013), to which Slender-billed Gulls are particularly sensitive. (Oro and Tavecchia 2008).

As an alternative to ground sampling, manned aerial surveys have been used widely to reduce disturbances in deriving counts while also providing synoptic information (Frederick et al. 1996; McEvoy et al. 2016). This is the case in Doñana Natural Space, where aquatic bird populations are estimated on a monthly basis by manned flights over wetlands (Fig. 1). In addition to these observations, photos are usually taken during the flight for after-flight assessment. However, these images are not usually acquired with adequate planimetric planning (zenithal view, appropriate overlapping and pixel size) or illumination conditions, making them an unsuitable source for mapping – shadowed and overlapping individuals are usually the main constraints.

In recent years, many studies have shown the usefulness of UAVs for deriving counts of colony-nesting birds (Chabot et al. 2015; Hodgson et al. 2016; Sardà-Palomera et al. 2012). The recent increase in the availability and ease of use of UAVs has also motivated scientists to use them for ecological studies (Anderson and Gaston 2013). Professional UAV campaigns may resolve these kinds of issues since proper mission planning can provide accurate and spatially explicit information. UAVs have also proven efficient in overflying bird colonies and providing valuable scientific information (Jones et al. 2006; Sardà-Palomera et al. 2012). However, in the context of bird surveys, the use of UAVs has not been sufficiently

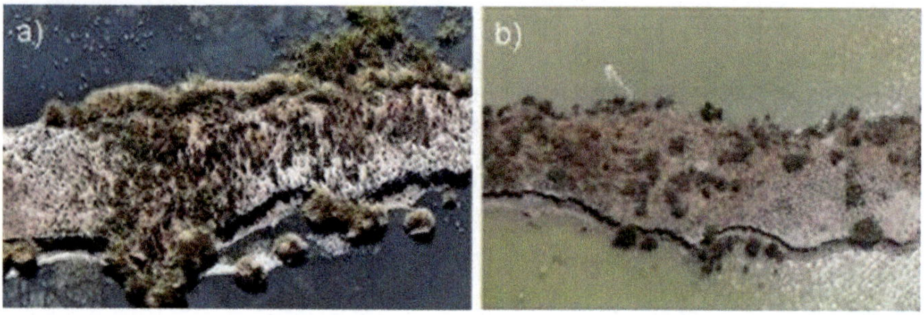

Fig. 1 Examples of (**a**) inappropriate and (**b**) adequate pictures, taken during manned aerial surveys over the Doñana Slender-billed Gull colonies in different years. Notice the differences in illumination in the left picture and the effects of shadowing and overlapping (Source: H. Garrido. Doñana Natural Processes Monitoring Team)

evaluated in terms of their cost efficiency or their precision in retrieving population data of sufficient accuracy (Hodgson et al. 2016). Several studies have indicated this technology can be used to support bird observations and sampling but few have proposed automated counting of either individuals or nests.

Targeted Species: The Slender-billed Gull

The Slender-billed Gull is a medium size *Laridae* (42–44 cm long and 220–350 g weight) species that breeds widely at isolated, scattered localities, from Senegal to Mauritania, and from the south and east of the Iberian Peninsula, through the Mediterranean, Black Sea, Minor Asia and the Middle East to east Kazakhstan, Afghanistan, Pakistan and north-west India (del-Hoyo et al. 1996). Least Concern is their current IUCN Red List category (BirdLife International 2016).

The Spanish population is very small and sparsely distributed and therefore is labelled as "vulnerable" by the Red Book on Spanish Birds (Madroño et al. 2004).

It usually breeds in marshes and saltpans with a typical laying period between late April and late May (exceptionally till July). The nests, formed of scat and feathers, are hardly visible as they are built on the ground or on low halophytic vegetation. It is a very social species that forms colonies together with other gulls, terns and shorebirds. However, the species is very sensitive to human disturbances and shows a distinctive breeding biology. Colonies are typically very dense and chicks cluster into nurseries just after birth. Such characteristics make the estimation of breeding pairs from the ground very challenging, often leading to overlooked individuals and chicks (Oro and Tavecchia 2008). Ideally, the colony should be visited at the end of the incubation period only once, which should be enough as the species shows a tight synchronicity during the egg-laying period (Oro and Tavecchia 2008).

Study Area: The Doñana Colonies

Since 2002, the Slender-billed Gull colonies in Doñana Natural Space have usually been inside the Veta la Palma fisheries area, where the birds lay their eggs on the islands inside the fish ponds (Fig. 2). This area, transformed to aquaculture ponds from original marsh, combines conservation with fish farming. The availability of islands in the middle of flooded pools offers protection against terrestrial predators, and the abundant food supply makes it a suitable habitat for nesting. However, in the case of island colonies, such as the one in *Veta la Palma*, the disturbance times associated with visiting the colony to ring chicks and count nests are increased due to the time needed to reach the colony: extended disturbances of this kind should be avoided. A much better estimation is gained by counting chicks while enclosing them. However, this, again, is not a recommended practice.

The species also breeds in the marshes of the Doñana Natural Space and on the *Salinas de Sanlúcar* saltpans located on the other side of the Guadalquivir River (Fig. 2).

The number of Slender-billed Gull breeding pairs has fluctuated in the last 15 years from 216 in 2008 to 766 in 2014, with an annual average of 500 pairs (Fig. 3). In the natural marshes of Doñana, the species also suffers from intense wild boar

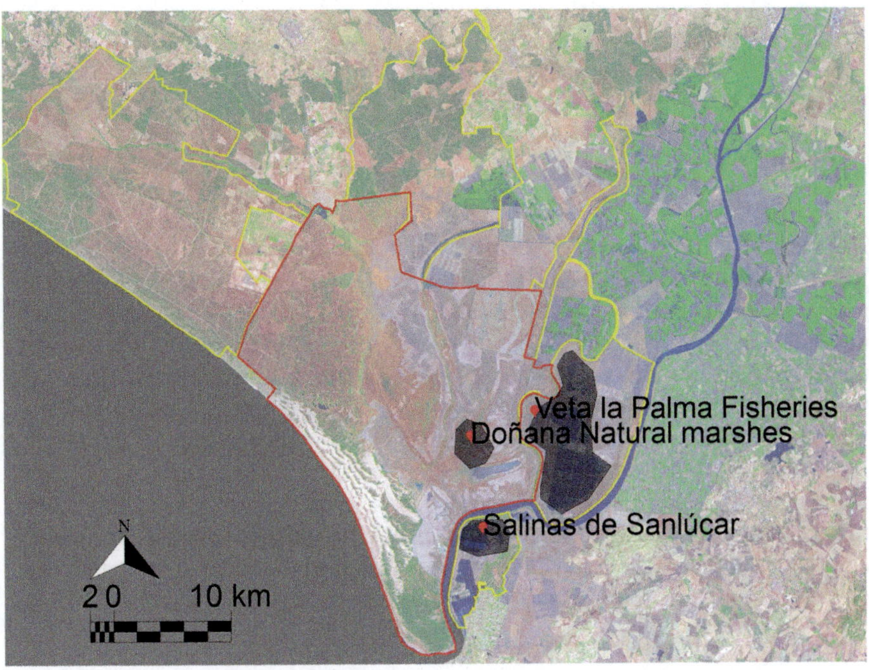

Fig. 2 Location of the Slender-billed Gull colonies in Doñana Natural Space, composed by the National Park (*red line*) and the Natural Park (*yellow line*)

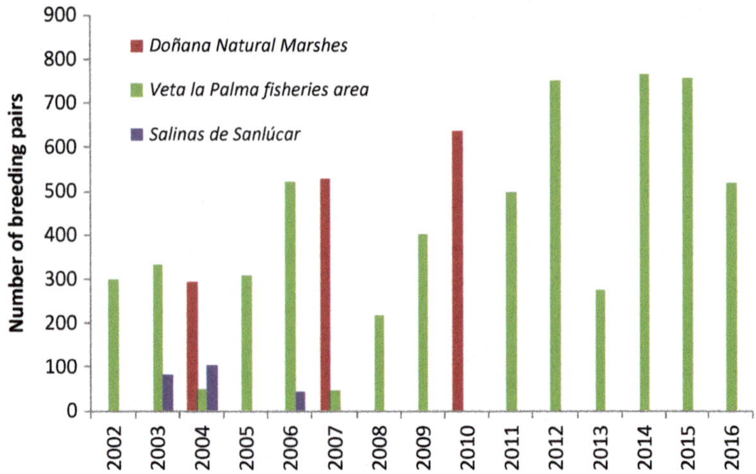

Fig. 3 Slender-billed Gull colony size in the last 15 years for every nesting location in the Doñana area (Mañez et al. 2015)

predation and the National Park administration fenced the area in 2004 in order to reduce the impacts of these ground-based predators as well as trampling by cattle. Black Kites (*Milvus migrans*) also predate on the eggs and chicks.

Methods

UAVs Flight Campaigns

The first flight campaigns were carried out in 2015 as part of the research project RECUPERA2020 funded by the European Regional Development Fund. The main goal of the project was to transfer research knowledge on the topic of environmental sustainability to regional farmers. The specific goal of this study was to assess the use of UAV technology for monitoring environmental integrity. The UAV system provided by the Center for Advanced Aerospace Technologies (Fada-Catec) was an electric multirotor equiped with a camera Sony Alpha-5100, with 24 Megapixels and a focal distance of 22 mm. Table 1 shows the characteristics of every flight campaign. Flight height and speed were selected following Vas et al. (2015). Missions were planned to collect as many pictures as possible instead of having just one frame for the whole colony. Flights were designed to retrieve a minimum pixel size of 6 cm in order to be able to identify gulls in the captured pictures (Table 1).

Table 1 Characteristics of the flight campaigns

Flight #	Date (DD/MM/YYYY)-Time (HH:MM)	Flight Height (m)	Pixel size (mm)	N of images
1	18/05/2015-11:26	80	8	60
2	18/05/2015-12:00	50	5	90
3	17/06/2015-11:26	50	5	90
4	30/05/2016-14:30	48	49	11
5	06/06/2016-12:09	51	58	70

In 2016, two more flight campaigns were carried out with a MD4-1000 electric multicopter from Microdrones GmbH equipped with a Olympus EPM2, 16 Mpix and a focal distance of 17 mm. The colony traditionally occupied one island, but in 2016 a new island located 1 km away was occupied to breed and hence it was surveyed.

During every flight, the photograph overlap was maximized to 80% longitudinal and 60% lateral. In every flight, the take-off and landings were carried out 500 m away from the colony where the UAVs base station was established. There were no signs of disturbance observed at the Slender-billed gull colonies during the flights, with the exception of flight #5, where a flock of Spoonbill was flushed by the UAV and a black kite attacked the colony.

Geometric Processing

Images were mosaicked with different photogrammetrical software. Orthomosaics and Digital Surface Models (DSMs) were derived from the processing. For the 2015 campaigns, there was no preparatory field work on the site and no Ground Control Points (GCPs) were located in the island. Therefore, the orthomosaics were georeferenced using 40 common GCPs identified on the available Bing Maps image. The geometric correction used a splines model a Root Mean Square (RMS) error lower than 2 pixels.

Before the start of the breeding season in 2016, we accessed the island and delineated, with centimetric precision, the perimeter with a differential GPS Leica 1200 (horizontal position error < 30 cm). In addition, eight large stones found on the island were geolocated to provide more features to be used as GCPs for future UAVs flights. After the flight, we used 80% of the GCPs to produce the orthomosaic and 20% as Check Points to assess the geometric correction. The final 2016 orthomosaics also had RMS error lower than 2 pixels.

Methods for Identifying Birds From UAV Imagery

Photointerpretation

Photointerpretation of the orthomosaics was based on simple eye identification of birds, either lying, standing or flying. A point vector file was created while on-screen digitizing at 1:50 scale. Clutches, whenever visible (temporarily uncovered by the parent), were also located in the image. Nest position was assigned by the same observer when the clear pattern of a lying bird was recognized (lower shadowed area surrounding the body as shown in Fig. 4). Labels were assigned to every delineated bird.

Supervised Classification

Supervised classification relies on the definition of homogeneous training areas for the different thematic classes present in the image, and an algorithm assigns every pixel a probability of belonging to one of these classes (Richards 2013). The algorithms use mainly the spectral information from the pixels of the training areas to estimate the similarity of any pixel to them. We used two different algorithms:

Support Vector Machines: They are originally binary classifiers and separate the classes with a decision surface that maximizes the margin between the classes. SVM provide good classification results from complex and noisy data (Camp-Valls and Bruzzone 2009).

Fig. 4 A photographic example of the visual identification of lying Slender-billed gulls (*purple circle*), standing individuals (*red circle*) and eggs (*blue circle*). *Yellow circles* highlight Black-headed gulls. The *black square* shows an example of a flying gull

Random Forests. A random forest is a collection of classification trees trained on a subsample of the training data (Chen 2007). Random Forests are currently one of the top performing algorithms for data classification and regression. They are widely used because of their ability to classify large amounts of data with high accuracy.

In both cases, we used chicks and adult gulls from the orthomosaic as training areas for the class "bird" by selecting body pixels as in Grenzdörffer (2013). The rest of the training classes represented open water, vegetation and bare soil. Finally, where the number of pixels associated with a bird was less than a certain threshold, these were removed from the classification. We carried out an accuracy assessment against the photo-interpretation results for the whole orthomosaic. Swimming birds were excluded.

Colony Monitoring

Ground-based monitoring surveys comprised periodic visits to the colony during the breeding season and the estimation of pairs using binoculars and telescopes from the closest position to the island to avoid disturbance. In a separate one-off visit during the breeding season, the monitoring team entered the colony to estimate the number of nests and to ring the chicks brought offshore. The date on which the colony was accessed was determined and informed by the continuous observations of the breeding birds.

Results

Visual Identification of Birds

Visual identification of the birds took about three days of work for each orthomosaic (Fig. 5). An experienced ornithologist used one working day to identify birds and a second day to label them according to their activity (flying, lying, standing up, etc.). Finally, on the third day, the supervisor assessed the identification and labelled the individuals. As *C. genei* nests are not visible, we assigned nests to lying birds. Table 2 shows the results for UAV flight #2. That was the flight-date when most of the colony was lying: we selected this flight as the reference for 2015. A total of 915 individuals of Slender-billed Gull were identified together with 149 individuals of Black-headed Gull. Another 27 birds were geolocated in the orthomosaic, corresponding to several individuals of White-headed Duck (*Oxyura leucocephala*) and Yellow-legged Gull (*Larus michahellis*). In addition, 97 visible eggs from 47 nests (up to 4 eggs per nest) were identified and labelled. UAV flight #3 took place late in the season and, by this time, many chicks were already grouped in nurseries. Therefore, for this flight and in addition to adults, we also identified chicks that were either alive or dead (Table 3).

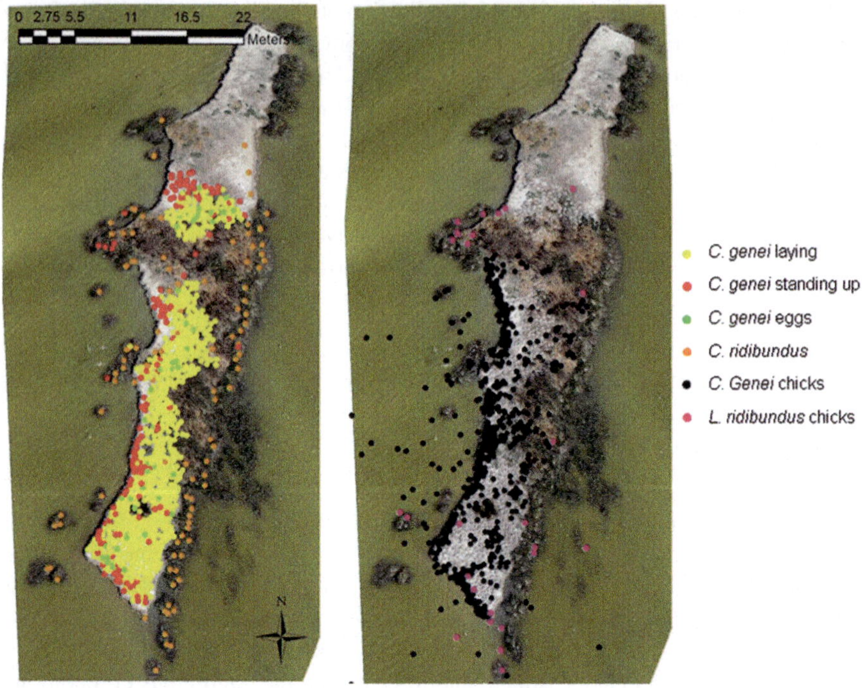

Fig. 5 The location of the different features visually identified and geolocated on the orthomosaics of UAV flights #2 and #3 from 18th May and 17th June 2015

Table 2 Number of individuals identified by photointerpretation on the orthomosaic of UAV flight #2 on 18th May 2015

Species	Attitude	Number
Slender-billed Gull	Laying	718
	Standing	144
	Clutches	47
	Swimming	6
Black-headed Gull	Laying	74
	Swimming	39
	Standing	33
	Flying	3
Other species	All attitudes	27
Total		1091

According to photo-interpretation, 718 Slender-billed Gulls were identified as laying and subsequently producing 47 clutches, which compares with an estimate of 757 laying birds from a preliminary visual count.

Table 3 Number of individuals identified by photointerpretation on the orthomosaic of UAV flight #3 on 6th June 2015

Species	Age	Number
Slender-billed Gull	Adult	418
	Alive chicks	575
	Dead chicks	63
Black-headed Gull	Adult	47
	Chick	28
Other species	All	9

Table 4 Overall agreement (OA), commission and omission errors for bird automatic classification for the different UAV flights and classifiers.

Flight Number	Classifier	Overall Agreement (%)	Commission Error (%)	Omission Error (%)
Flight #2	SVM	82	7	0.7
	Random Forests	98	5.5	1
Flight #3	SVM	85	10	1.3
	Random Forests	96	6	2
Flight #4	SVM	89	3	0.5
Flight #5	SVM	87	5	0.9

Image Automatic Classification

The process of delineating training classes and applying classification procedures took less than 2 hours. Random forests were slightly lower than SVM according to the Percent of Overall Agreement (OA). Both classifiers performed well and provided an accurate classification of birds (Table 4). Patches associated with the class "bird" were grouped and individuals were identified where these were discrete and their size exceeded 24 cm^2. No automatic discrimination was possible between *C. ridibundus* and *C. genei*. Neither eggs nor clutches were classified with the automatic classification. A few individuals were overlapping, which prevented an automatic single individual classification. Most of the delineated polygons corresponded to an actual bird shape. However, several resulting polygons just delineated a part of the bird's body. Chicks, when present, were also considered as birds in the accuracy assessment.

Colony Monitoring

The estimates of colony size and total chick numbers from the ground surveys and preliminary visual estimates were also used to assess the accuracy of those estimated by photo-interpretation and for the total number of lying gulls (OA between 85 and 95%). Ground surveys and visual estimates were, in all cases, higher than the automatic counts on images from UAVs.

Discussion

In this study, we were able to provide quick and accurate estimates of colony size (lying birds) and productivity (number of chicks). One of the most challenging tasks in estimating breeding pairs in a bird colony from the ground is the limited horizontal visibility (Sardà-Palomera et al. 2012). Zenithal views from UAVs definitely contribute to improving colony size estimates. However, birds are often fully or partially obscured by trees or shrubs, with causes difficulty in their detection and counting through photointerpretation. This occurs frequently for Ardeids since nests are often settled in the middle of reed beds or stands of other tall graminoids (Prosper and Hafner 1996). This constraint has also been found relevant in ungulate surveys with UAVs (Chrétien et al. 2016). Given these kind of difficulties, only a complete capture of chicks can eventually provide accurate estimates of productivity, though these disturbances might also cause chick abandonment (Oro and Tavecchia 2008).

No disturbances were observed for gulls during the UAV flights, suggesting that our methodology was appropriate for colony monitoring, which agrees with the results from previous studies (McEvoy et al. 2016; Ratcliffe et al. 2015; Vas et al. 2015). In addition, we employed only three persons to supervise a single whole mission, two of them for flight operations and one for a ground survey of the colony.

Two big challenges arise when working with photographs taken from UAVs . On the one hand, images are not always acquired in the best conditions. Firstly, wind gusts can yield blurred images making it difficult to delineate the target features although it might be avoided by setting a high shutter speed on the camera. Moving birds also appeared distorted. Secondly, the mosaicking process can randomly select the image without the bird that was present in the overlapping discarded image, as reported by Bakó et al. (2014). We experienced both situations in our study, which increased the omission error.

Although visual photo-interpretation ensures high accuracy in bird delineation, automatic classification takes less time and produces satisfactory results (Abd-Elrahman 2005; Grenzdörffer 2013). However, we had to use a size criterion to discard small classified patches. Other approaches using pattern recognition such as Object Based Image Analysis (OBIA) may be able to provide better results (see chapter by Hurford).

Finally, UAV flights provide added-value to the classical colony survey methods by producing, as the main output, a map including the location of the target species. Thus, for instance, one can investigate the relationship between nest density and clutch as a possible factor driving breeding success. On the other hand, 3D points derived from stereocorrelation may also enhance the automatic bird delineation by adding the subtle differences in height between birds and the ground (Anderson and Gaston 2013).

Conclusions

- We could provide accurate estimates of Slender-billed Gull colony size (lying birds) and productivity (number of chicks) by using UAVs flights.
- Flights did not cause any disturbance to the gull colony.
- Visual photo-interpretation always performed better than automatic supervised classification.
- Although automatic classification produced satisfactory estimates, minimum class patch size had to be defined to discard features other than birds.
- Monitoring bird colonies with UAVs in comparison to ground surveys resulted in faster procedures and in sufficiently accurate colony size and productivity estimates.

Practical Lessons for Nature Conservation

- Estimating the number of breeding pairs and/or the number of chicks in bird colonies is an essential monitoring activity to inform on the conservation of protected species.
- Ground surveys usually require costly procedures and often generate disturbances to the colonies.
- Nowadays, UAVs flights can be easily carried out to map any interesting feature for nature conservation
- Image geometric processing is also easily and successfully achieved by using both commercial and free software.
- Visual identification of ground nesting birds in the orthomosaics is an affordable task for any user and can be cross-checked by several users.
- Automatic classification is also available in open source software. This provides quick and reliable estimates but does not easily distinguish between bird species.

Acknowledgments RECUPERA 2020 project partially funded Miguel Ferrer for this study. We are grateful to the Doñana Natural Processes Monitoring Team, especially José Luis Arroyo and Fernando Ibáñez, and also Luis García, Héctor Garrido, José Luis del Valle, Rubén Rodríguez, and Alfredo Chico. The authors want also to thank to Consejería de Medio Ambiente de la Junta de Andalucía for the funding of the Long-Term Monitoring Program of Doñana Natural Space and the annual census of Doñana breeding birds. Thanks to the owners of *Veta La Palma* for their cooperation and help in accessing the colonies.

References

Abd-Elrahman, A.: Development of pattern recognition algorithm for automatic bird detection from unmanned aerial vehicle imagery. Surv. Land Inf. Sci. **65**, 37–45 (2005)

Anderson, K., Gaston, K.J.: Lightweight unmanned aerial vehicles will revolutionize spatial ecology. Front. Ecol. Environ. **11**, 138–146 (2013). doi:10.1890/120150

Bakó, G., Tolnai, M., Takács, Á.: Introduction and testing of a monitoring and colony-mapping method for waterbird populations that uses high-speed and ultra-detailed aerial remote sensing. Sensors. **14**, 12828–12846 (2014). doi:10.3390/s140712828

BirdLife International: Species factsheet: Larus genei, http://datazone.birdlife.org/species/factsheet/22694428

Camp-Valls, G., Bruzzone, L.: Kernel Methods for Remote Sensing Data Analysis. Wiley, New York (2009)

Chabot, D., Craik, S.R., Bird, D.M.: Population census of a large common tern colony with a small unmanned aircraft. Plos One. **10**, e0122588 (2015). doi:10.1371/journal.pone.0122588

Chen, C.H.: Image Processing for Remote Sensing. CRC Press, Boca Raton (2007)

Chrétien, L.-P., Théau, J., Ménard, P.: Visible and thermal infrared remote sensing for the detection of white-tailed deer using an unmanned aerial system. Wildl. Soc. Bull. **40**, 181–191 (2016). doi:10.1002/wsb.629

Del-Hoyo, J., Elliott, A., Sargatal, J.: Handbook of the birds of the world, vol. 3, Lynx Edicions edn. Hoatzin to Auks, Barcelona (1996)

Foody, G.M., Mathur, A.: A relative evaluation of multiclass image classification by support vector machines. IEEE Trans. Geosci. Remote Sens. **42**, 1335–1343 (2004). doi:10.1109/TGRS.2004.827257

Frederick, P.C., Towles, T., Sawicki, R.J., Bancroft, G.T.: Comparison of aerial and ground techniques for discovery and census of wading bird (Ciconiiformes) nesting colonies. The Condor. **98**, 837–841 (1996). doi:10.2307/1369865

Grenzdörffer, G.J.: UAS-based automatic bird count of a common gull colony. ISPRS – Int. Arch. Photogramm. Remote Sens. Spat. Inf. Sci. **1**, 169–174 (2013). doi:10.5194/isprsarchives-XL-1-W2-169-2013

Hodgson, J.C., Baylis, S.M., Mott, R., Herrod, A., Clarke, R.H.: Precision wildlife monitoring using unmanned aerial vehicles. Sci. Rep. **6**, 22574 (2016). doi:10.1038/srep22574

Jones, G.P., Pearlstine, L.G., Percival, H.F.: An assessment of small unmanned aerial vehicles for wildlife research. Wildl. Soc. Bull.. 1973-2006. 34, 750–758 (2006)

Madroño, A., González, G.G., Atienza, J.C.: Libro rojo de las aves de España. Organismo Autónomo Parques Nacionales (2004)

Mañez, M., Arroyo, J.L., Chico, A., del Valle, J.L., García, L., Garrido, H., Martínez, A., Rodríguez, R.: Seguimiento de Aves Acuáticas. Espacio Natural de Doñana. Reproducción 2016. Estación Biológica de Doñana, CSIC, Sevilla, España (2015)

McEvoy, J.F., Hall, G.P., McDonald, P.G.: Evaluation of unmanned aerial vehicle shape, flight path and camera type for waterfowl surveys: disturbance effects and species recognition. Peer J. **4**, e1831 (2016). doi:10.7717/peerj.1831

Oro, D., Tavecchia, G.: Gaviota picofina. In: Gaviotas cabecinegra, picofina, de Audouin y tridáctila, y gavión atlántico en España. Población en 2007 y método de censo. pp. 21–43. SEO/BirdLife, Madrid, España (2008)

Pal, M.: Random forest classifier for remote sensing classification. Int. J. Remote Sens. **26**, 217–222 (2005). doi:10.1080/01431160412331269698

Prosper, J., Hafner, H.: Breeding aspects of the colonial Ardeidae in the Albufera de Valencia, Spain: population changes, phenology, and reproductive success of the three most abundant species. Colon. Waterbirds. **19**, 98–107 (1996). doi:10.2307/1521952

Ratcliffe, N., Guihen, D., Robst, J., Crofts, S., Stanworth, A., Enderlein, P.: A protocol for the aerial survey of penguin colonies using UAVs. J. Unmanned Veh. Syst. **3**, 95–101 (2015). doi:10.1139/juvs-2015-0006

Richards, J.A.: Remote Sensing Digital Image Analysis: An Introduction. Springer Science & Business Media, Berlin, (2013)

Sardà-Palomera, F., Bota, G., Viñolo, C., Pallarés, O., Sazatornil, V., Brotons, L., Gomáriz, S., Sardà, F.: Fine-scale bird monitoring from light unmanned aircraft systems. Ibis. **154**, 177–183 (2012)

Vas, E., Lescroël, A., Duriez, O., Boguszewski, G., Grémillet, D.: Approaching birds with drones: first experiments and ethical guidelines. Biol. Lett. **11**, 20140754 (2015). doi:10.1098/rsbl.2014.0754

Part IV
Looking Ahead Through Current Research Projects and Expected Advances

Integrated Land Cover and Change Classifications

Richard Lucas and Anthea Mitchell

Abstract For nature conservation, regular provision of consistent, timely and useable classifications of land covers and change is highly beneficial but is rarely achieved. This chapter outlines the concepts behind the Earth Observation Data for Ecosystem Monitoring (EODESM) system, which facilitates the description and classification of any site worldwide according to the Food and Agriculture Organisations (FAO) Land Cover Classification System (LCCS; Version 2) and with reference to environmental variables retrieved from earth observation. Changes in land cover, as well as causes and consequences, are described through the accumulation of evidence and the system recognises these to be numerous, highly variable and specific to different elements of the landscapes. Hence, they can be captured by considering information provided by a range of sensors operating in different modes and over different temporal frequencies and scales. The EODESM system is available at no cost and its ease of use makes it well suited to supporting nature conservation.

Keywords Land cover • Land cover change • Environmental variables • Earth observation • Classification • Ecopotential

Introduction

Imagine you are driving through a landscape and you are able to select any area of ground and go back in time, seeing all its transitions and freezing the frame as and when you liked. Was it covered in snow last winter, and was this deep or just a light covering; or when did the spring leaves start to appear and then fall? Or, you want to know whether the road you are now driving along is flooded or clear given there had been intense rainfall in the mountains the night before? What kind of landscape might this be in 50 years time and what might determine how it got there?

Within decades or less, the ability to routinely look back in time, assess current situations and perhaps predict the future will most likely be a reality, particularly

R. Lucas (✉) • A. Mitchell
Centre for Ecosystem Science, School of Biological, Earth and Environmental Sciences,
The University of New South Wales, High Street, Kensington (Sydney), NSW 2052, Australia
e-mail: richard.lucas@unsw.edu.au; anthea.mitchell@unsw.edu.au

© Springer International Publishing AG 2017

R. Díaz-Delgado et al. (eds.), *The Roles of Remote Sensing in Nature Conservation*,
DOI 10.1007/978-3-319-64332-8_15

given progress towards high resolution digital and multi-spectral temporal images (the equivalent of videos) from space. In the next few years, people will be able to look back in time to see how whole landscapes have changed over their lifetimes. Linked with equivalent advances in ground-based observations, this will provide an unprecedented view of our planet and the opportunity to tell stories of what we have done and how things might change over our lifetimes.

The ability to observe the events and processes that have shaped our landscape over the past 32 years (at least since 1985) has, to some extent, already been achieved through time-lapses of Landsat sensor imagery provided by the Google Earth Engine. Within this system, we can observe erupting volcanoes, surging glaciers, large floods, shifting coastlines, clearing of forests and expanding cities. We have a record of man's impact on the planet in the recent past and some of it makes for uneasy viewing.

The public release of the Landsat archive allowed us to have this unique perspective and many scientists have subsequently provided detailed temporal classifications of land cover. These have included forest losses and gains (Hansen et al. 2013), tree canopy density (Hansen et al. 2013; Sexton et al. 2013), hydro-periods (Pekel et al. 2016), open water (Feng et al. 2016), bare ground (Hansen et al. 2013), impervious surfaces (Langanke et al. 2013) and ecosystem extent and dynamics (e.g., mangroves; Giri et al. 2011). Using coarser spatial resolution (typically 0.25–1 km) sensors, such as the NOAA AVHRR and MODIS, an extensive historical archive of other features of the Earth's surface, including snow cover (Hall and Riggs 2016) and land and sea surface temperature (Merchant et al. 2008), has been generated over past decades, giving us a unique insight into recent global change.

The amounts of data that have currently been acquired and will be provided in the future are vast. However, our ability to handle large amounts of, what is often termed big data, is being addressed through cloud and other high performance computing, with these providing substantive storage and processing capability. Furthermore, image data can be downloaded and distributed to users rapidly and, in some cases, in real or near real time (as in the case of the Planet Lab's CubeSat data). This new capability provides opportunities to understand the changes that have happened, both over past decades and more recently, and to monitor and plan into the future. If used effectively, these systems can be used to prevent or reverse some of the damage that is being or has been inflicted on the planet and to conserve what is remaining.

In this chapter, we describe the Earth Observation Data for Ecosystem Monitoring (EODESM) system, which uses retrieved environmental variables and specified classifications from earth observation data to characterise and map land covers. Changes are identified by considering evidence obtained from earth observation data and from other sources. The system provides insight into the causes and consequence of change and redistribution of physical elements (e.g., water, sediments and carbon). The system can also be used to recommend where and how to restore or protect ecosystems. The approach we describe is easy to understand, simple to operate and revise, and provides a wealth of information that can be used for a wide range of purposes, including for the conservation of nature.

Recognizing User Needs

Individuals, groups or organisations charged with managing, conserving, protecting and/or restoring environments desire both historical, recent and, often, real time spatial information on landscapes. Whilst the satellite and aircraft images themselves provide a pictorial (and also digital) overview, the information extracted or derived from these is often far more useful, particularly if this is: consistent over time and within and between areas of interest; includes environmental variables (e.g., biomass, soil moisture, salinity and water flows) or thematic classifications of land cover, and changes in these; considers historical contexts, present situations and future prospects; is provided at scales that are appropriate to the questions being asked; and accurately reflects the state and dynamics of landscapes over varying time frames. Accessibility of information is also critical, whether provided as products (e.g., tree cover density) or as software or processes that allow the users to extract the required information by themselves based on their own requirements or those of others. These requirements have been considered during the design and development phases of the EODESM system.

The EODESM System

The EODESM System was developed through the EU Horizon 2020 Project, ECOPOTENTIAL and was designed to provide consistent classifications of land covers and change at multiple scales. The EODESM System was a later iteration of the Earth Observation for Dynamic Habitat Monitoring (EODHaM; Lucas et al. 2014), which was conceptualised through the FP7 Biodiversity Multi-SOurce Monitoring System (BIOSOS) project.

Both the EODHaM and EODESM system use the Food and Agricultural Organisation's (FAO's) Land Cover Classification System (LCCS; Version 2; Di Gregorio 2005) taxonomy to classify land covers within protected areas and their immediate surrounds. However, for classification, the earlier EODHaM system applied a rule-based classification to very high resolution (VHR) Worldview-2 (acquired in the pre- and peak-vegetation flush periods) and (if available) airborne LIDAR to extract the components of the LCCS classes. These included life form (i.e. shrubs, trees, grasses, forbs, lichens or mosses), leaf type (broadleaved, needleleaved or aphyllous), phenology (e.g., evergreen or deciduous), water movement (standing or flowing) and sediment loads in water (turbid or clear). These extracted components were then combined to generate a string of codes (e.g., A3.A10.B2.C1. D1.E1), which were translated subsequently and automatically to descriptive text (in this case, trees of closed canopy (>70–60%) that are tall (14–30 m), continuous, broadleaved and evergreen). The classifications of each of the layers within the EODHaM system were conducted by defining and adjusting thresholds of spectral bands or indices, including the Normalised Difference Vegetation Index (NDVI),

Plant Senescence Reflectance Index (PSRI), Water Band Index (WBI) and (where available) Canopy Height Models (CHMs) derived from LIDAR. Whilst providing highly detailed classifications, the main limitation was the consistency in the use of spectrally-based rules as these had to be adjusted regularly to allow for differences in atmospheric, illumination and environmental (e.g. phenological) conditions prevalent at the time of the Worldview-2 overpasses. For this reason, a new concept was developed for the EODESM system.

Our Unchanging World

Whilst there are significant changes in land cover arising from both natural and human-induced events and processes, the basic building blocks of landscapes (e.g., foliage, wood, rocks, water in various states) generally do not change and neither do the quantitative measures that are used to describe these (e.g. biomass, canopy cover, amounts of dead or senescent material, species type, temperature, water flows). The measures that satellite sensors record are largely consistent, with these including spectral reflectance (%), radar backscatter (e.g., γ°), surface heights and dimensions (m) and temperature (°C). Therefore, regardless of what happens in the future, descriptors of the building blocks of our environment will largely be the same, as will the data and measures obtained from satellite, airborne and ground-based systems. The challenge is to define the best algorithms and combinations of data to describe these building blocks in a way that is consistent, reliable over time, accurate and understandable. In effect, what is needed is a system with longevity that will allow classifications of landscapes in, for example, 2100 to be compared to those of the 1970s, when the Landsat sensors first acquired spectral data, and even before then.

The FAO LCCS-2 is a taxonomy that is fundamentally well suited for providing consistent classifications of land covers in the long term as many of its inputs are derived from well-defined and established environmental descriptors and variables. For example, for natural and semi-natural vegetation, key descriptors are life form, canopy cover, the vertical and horizontal distribution of plant material, leaf type and phenology, all of which can be derived from earth observation data acquired in different or similar modes. For this reason, rather than focusing on providing the best classification algorithm, the EODESM system places emphasis on retrieving continuous environmental variables as well as generating thematic classifications (e.g., of life form or leaf type), which are combined subsequently to form the LCCS-2 classes. For purposes of nature conservation, an additional and essential descriptor is plant species or genus type (that is not considered in the LCCS classification but is derived independently), which can be mapped remotely although is often restricted to those that are spectrally distinct. An overview of the main layers that are required as direct input to the LCCS-2 scheme are outlined in Table 1, with these relating to essential variables associated with the broad categories of agriculture, biodiversity and ecosystems, human settlements, bare surfaces and water/renewable energy/climate.

Table 1 Variables retrieved from earth observation data and used as direct input to the LCCS classification

Theme	Description	FAO LCCS-2 categories
Agriculture	Crop area	Cultivated area and spatial size
	Crop management and agricultural practices	Crop combinations, sequences, cultural irrigation, cultural practices (time factors) and water seasonality.
	Crop phenology	Evergreen and deciduous
Biodiversity and ecosystems	Phenology (Species traits)	Evergreen, deciduous, leaf type
	Vegetation structure	Vegetation height and cover (all layers)
	Fragmentation	Spatial distribution
Human settlements	Urbanization	Built up or not built up
		Linear/non-linear structures and density
		Urban vegetation
Bare surfaces	Extent and type	Bare surface macro-pattern and materials
Water/renewable energy/climate	Snow and ice cover, glaciers, ice caps and sheets	Water state (water, ice or snow)
	Tidal (min, max, sea surface elevation)	Daily variations in water
	Hydro-period	Hydro-periods, waterlogged
	Water discharge and lakes	Standing or flowing water
	Water quality and suspended particulates	Water sediment loads

Knowledge of the state and dynamics of environments requires additional information on variables that are not relevant or appropriate for land cover classification (Table 2). These relate to the primary uses and components of the landscape, namely agriculture and forestry (e.g., crop and timber yields), vegetation (e.g., biomass, leaf area index), human settlements (populations), bare surfaces (e.g., soil moisture content) and water (e.g., pH, nutrient content, snow grain size or moisture content). As such, information on the magnitudes and changes in these variables can be included as attributes of the land cover classification and inform on current states and past changes. Furthermore, many of these variables, as well as those used as direct input to the LCCS-2 classification, can be modelled, which gives the capacity to generate predicted land cover maps (and associated variables).

Whilst the concept of using environmental variables as the basis for classification and description of land covers is logical, an issue is the practicality of obtaining these. It is unrealistic to expect nature conservation practitioners to generate this information themselves in order to produce land cover maps and so these need to be made available or capacity provided to generate these. Fortunately, because of the past and current efforts of a large number of engineers and scientists, environmental variables are now being routinely retrieved from satellite and airborne data and made freely available. Notable examples include those generated at the global level from Landsat sensor 30 m data, including tree canopy cover (2000 and 2010; Hansen et al. 2013; Sexton et al. (2013)), bare ground proportions (2010; Hansen et al. 2013),

Table 2 Examples of retrieved from EO data that provide additional descriptions of land cover

Variable	Variable
AGRICULTURE	CLIMATE (continued)
Crop type	Leaf Area Index (LAI) (Land)
BIODIVERSITY	Ocean colour (Ocean surface)
Net primary productivity (ecosystem function)	Permafrost (Land)
Population structure by age class and species	Phytoplankton (Ocean surface)
OTHERS	Precipitation (Atmosphere surface)
Elevation, Orography	Sea ice (Ocean surface)
Land surface temperature	Sea level (Ocean surface).
Ocean bathymetry	Sea state (Ocean surface)
Wave, height, direction, period	Sea-surface temperature (Ocean surface)
CLIMATE	Soil moisture (Land)
Above ground biomass (Land)	Surface current (Ocean surface)
Albedo (Land)	Wind speed and direction (Atmosphere surface)
FAPAR[a] (Land)	

[a]Fraction of absorbed photosynthetically active radiation

hydro-period (1987–2015; Pekel et al. 2016), and MODIS 500 m derived data (e.g. 8-day snow cover from 2000; Hall and Riggs 2016). For Europe, 20 m resolution maps of tree cover density (2012), forest leaf type (2012), permanent water (2006–2012) and impervious surfaces (2011–2012) have been generated through the Copernicus project from optical satellite sensor data from 2001 and 2011 (Langanke et al. 2013). At local levels, more detailed retrieval has occurred using VHR resolution and LIDAR as well as spaceborne optical and radar sensors, including ocean wind speeds (Rana et al. 2016), soil moisture (Pasolli et al. 2015) and snow moisture content (Nagler and Rott 2000). In each case, specialist algorithms for retrieving environmental variables have been developed through years or even decades of research and the resulting datasets are often well suited to support the classification and attribution of land cover classes and change according to the LCCS-2 taxonomy. The algorithms used for the generation of environmental variables are also being made available with associated software and these can be used for self-generation of the required data layers, though calibration and/or validation is essential in some cases. The outputs from the EODESM system can also be used to describe additional variables (Table 3), with these relating to, for example, disturbance regimes.

Classification of Land Covers

The FAO LCCS-2 taxonomy used in the EODESM system (Fig. 1) is hierarchical and allows for the progressive classification of a comprehensive range of land covers from earth observation data with these corresponding to those observed at

Table 3 Variables that can be derived from the EODESM system

Theme	Description
Biodiversity	Disturbance regime (Ecosystem function)
	Ecosystem composition by functional type (Ecosystem structure)
	Ecosystem extent and fragmentation (Ecosystem structure)
	Habitat structure (Ecosystem structure)
	Primary and secondary productivity (Ecosystem function)
	Population structure by age/size class (Species populations)
	Species distribution (Species populations)
	Species interactions (Community composition)
Climate and Water	Fire disturbance (Land)
	Land cover, including vegetation type (Land)
	River discharge (Land)
	Water use (Land)
Ocean	Mangrove, saltmarsh and sea grass area (Biology and Ecosystems)
Urban	Land use and land cover in relation to urban development and change
Health	Famine early warning, short term forecasting of communicative diseases

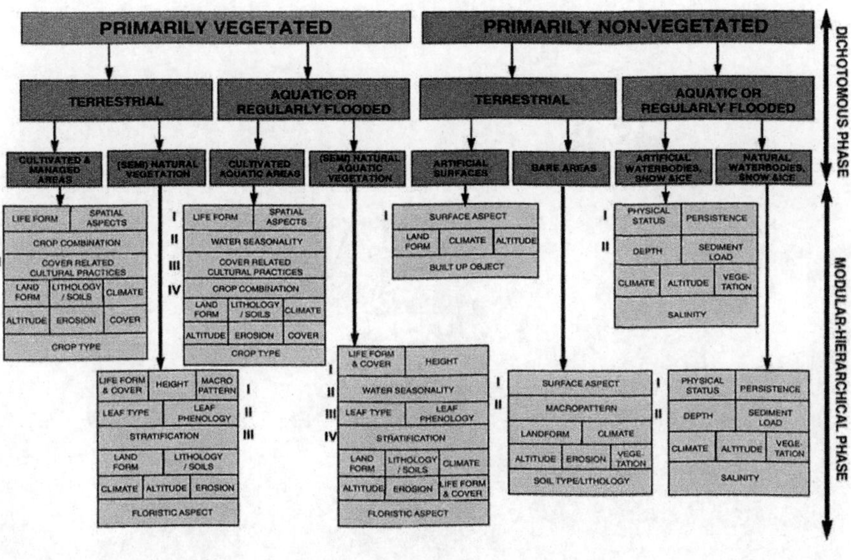

Fig. 1 The FAO Land Cover Classification System (LCCS) Taxonomy

ground level (Kosmidou et al. 2013; Tomaselli et al. 2013). The LCCS system has been used as the basis for EO-based classifications in many studies but the typical approach has been to establish training areas for the 'end classes' of the taxonomy (such as broadleaved evergreen forests; see Yang et al. (2017) for a review of the LCCS and other commonly used taxonomies). The EODESM takes a different view in that it follows the sequences of classifications through the hierarchy using derived products from EO data, with these including environmental variables but also other ancillary spatial information such as cadastral and urban maps, process models (e.g., hydrological) and knowledge. The EODESM system accepts 30 primary inputs (e.g., relating to crop sequences, leaf type, cadastral information), with thematic layers requiring specific class codes (e.g., 1 for woody, 2 for herbaceous vegetation). Continuous layers (e.g., canopy cover and hydro-period) are automatically translated to pre-set thematic classes within the EODESM system. Once entered, the system automatically translates each input to LCCS component codes, which are then combined subsequently to generate a class description. Each class is then coloured according to a standardized scheme, as illustrated in Fig. 2. The advantage of the classification approach is that it is relevant and applicable to any site globally and can be applied independent of scale. The accuracies of both classification and change maps are assessed by referencing ground-based classifications, generated using the LCCS taxonomy (e.g., by exploiting mobile applications), or measures of uncertainty associated with retrieved environmental variables.

Fig. 2 EODESM classification of land covers in the Camargue, southern France. Over 200 classes are represented with each associated with a detailed description according to the LCCS taxonomy. These broadly relate to water (*blue*), bare ground (*brown*), urban areas (*grey*), agriculture (*light greens*) and natural vegetation (*darker greens*)

Classification of Change

Within any landscape, changes are often the result of specific events that are either natural (e.g., fires, floods or storms) or the result of human activities (e.g., deforestation, mine excavation and cultivation). However, changes may be the result of longer-term processes, which again are also natural (e.g., vegetation growth, increased tidal inundation) or human induced (e.g., urban expansion, agricultural homogenisation). Climatic fluctuation may lead to changes in the frequency and intensity of events or alterations of long-term processes (e.g., mangrove extent because of progressive rises in sea level). Changes across a landscape also occur at different times, rates and frequencies and across different scales. For this reason, detection and classification from earth observation data has proved difficult as the acquisition dates and frequencies often do not match those associated with events and processes occurring at the ground level. Indeed, many studies focusing on change detection have typically dealt with only one type of change, with notable examples being deforestation monitoring and flood mapping, and little or no consideration is given to changes occurring within adjacent or proximal classes or at different times and rates. Furthermore, change is often detected on the basis of the differences in only one or a few remotely sensed variables, whether they are spectral reflectance or radar backscatter, indices or retrieved environmental variables.

Within the EODESM system, events and processes are detected when changes in the components of LCCS classes are observed. As an example, the annual period of inundation within a wetland may decline from 292 to 182 days, with this corresponding to a reduction in annual hydro-period from B1 (>9 months) to B8 (4–6 months). However, there may be additional evidence that supports the interpretation that such a change might be the result of drying of the landscape. This might include a change from flowing to standing water and/or turbid to clear water over a similar time frame, which are both thematic categories, but also in environmental variables such as an increase in salinity or algal amounts. By referencing this additional information, the probability of this change being attributed to long-term drying is increased. A further example is given in Table 4, which illustrates a change in both life form and canopy cover (Case A; associated with selective logging) and water state (snow to water) and flow rates (standing to flowing) (Case B; snow melt

Table 4 Examples of change detected by comparisons of LCCS component codes and environmental variables

	Thematic	Continuous	LCCS code (Period 1; P1)	Component code (P1)	Component code (P2)	Biophysical change
A	Life form (Codes of 1–9)	Canopy Cover) (0–100%)	A12.A4.A10. B4.C1.D1.E2	A4 (Trees)	A3 (Shrubs)	Cover of 80% to >40%
B	Water State (1,2,3)	Water movement ($m^3\ s^{-1}$)	B28.A1.B4. C1.D2	A2 (Snow)	A1 (Water)	Velocity of 0 to >20 $m^3\ s^{-1}$

and increased river discharge). The accumulation of evidence to support the interpretation of the change event or process provides the user with information to facilitate a response and manage change effectively.

Whilst the LCCS-2 provides a means to describe land covers, few taxonomies are available for describing and documenting change. However, a review of transition events and processes conducted in support of the EODESM change detection modules identified 80 change categories, with these associated primarily with natural vegetation, agriculture, urban areas, water and bare ground (Table 5). In each case, possible transitions from one LCCS component class to another that are relevant for each of the 80 change categories have been documented, as have changes

Table 5 Main categories of change considered within the EODESM system

Natural vegetation	Agriculture	Urban	Water	Bare ground
Deforestation	Herbicide spraying	Road abandonment	Flooding	Lava flows
Degradation	Burning	Greening	Inundation	Sedimentation
Selective logging	Cutting	Browning	Drying event	Erosion
Defoliation	Grazing	Planning	Long term drying	Dune change
Thinning	Growth	Urban densification	Snow accumulation	
Dieback	Stubble formation	Urban renewal	Snow loss	
Growth	Agri. expansion	Waste dumps or extraction	SnowFall	
Thickening	Agri. water supply	Communication installation and abandonment.	SnowMelt	
Encroachment	Agri. time factor	Rail conversion	Waterlogging	
Abandonment	Tillage	Rail construction	Water OutBurst	
Hedgerow loss	Pasture degradation	Urban expansion	Dam creation	
	Pasture replanting	Road conversion	Land drainage	
	Crop change	Road construction	Freezing	
	Crop growth	Road improvement	Thawing	
	Crop sequence change	Industrialisation	Glacial flow	
	Agri. homogenisation	Infilling/levelling	Sea level rise	
	Agri. division		Water pollution	
	Plantation establishment		Tidal loss	
	Plantation growth			
	Grass fertilization			
	Orchard planting			
	Slurry or sediment spreading			
	Liming			

in retrieved environmental variables (including spectral indices). By considering these transitions, the EODESM system allows for automated detection of these changes based on evidence and can highlight those change events or processes that are adverse or beneficial, although opinion varies depending upon the nature of the environment being affected. For example, the establishment and growth of pine plantations may be beneficial in terms of biomass accumulation and carbon sequestration and storage but may have adverse impacts on the abundance and diversity of faunal species.

The automated detection and description of changes over varying periods of time and based on the accumulation of evidence often results in a large number of events and processes being identified over the period of a time-series. However, more targeted detection and description of change may be achieved by identifying breaks (e.g., using the BFAST algorithm; Verbesselt et al. 2010) or longer-term trends in the time series of, for example, Normalized Difference Vegetation Index (NDVI) data obtained from Landsat or Sentinel-2A/B data. Where an event is identified, imagery acquired just prior to and following the date of change can be accessed. A LCCS-2 class is then assigned and changes in the components of this class are reviewed. When used in combination with time-series of retrieved environmental variables, a better assessment of the change event can be provided. In the case of longer-term processes, the transitions in component classes over the change period (e.g., from trees to grasslands, to shrubs and back to trees in the case of regeneration following deforestation) can be used to track the nature of change in land cover. Changes in environmental variables can similarly be tracked. A particular advantage of this approach is that changes can be automatically highlighted depending upon their severity or benefit.

Causes and Consequences of Change

Often when we detect a change, there are clear drivers and consequences of this. For example, dieback of trees may occur because of a prolonged flooding event, with this evidenced primarily by a decrease in canopy cover. There are only a few likely causes of the flooding, with these including those that are natural (e.g., increased rainfall over an extended period or an intense rainfall event) or human-induced (the creation of a dam and reservoir). The immediate consequence of the flooding is the loss of foliage cover followed by full or partial mortality of all or some of the trees. Follow-on consequences that would be considered negative include the loss of terrestrial elements of biodiversity and carbon in vegetation, with these occurring over variable periods, whilst positive benefits might include an increase in aquatic biodiversity and long-term storage of carbon. The consequences may be relevant to the specific area of ground that is affected or experienced in areas that are proximal or even far removed. For all areas (or objects within a scene), the causes and consequences of change can often be pre-determined and hence mapped alongside the change. The causes and, more often, the consequences of change also relate to the movement of materials within a landscape. For example, a deforestation event within a catchment can result in the loss of carbon to the atmosphere, reduced uptake of carbon dioxide (CO_2), the increased movement of water through the catchment and the transfer of sediment down slope, into water courses and ultimately to coastal regions. This movement of material can be modelled but can also often be observed within earth observation data or quantified within derived products (e.g., temporal vegetation biomass maps reflecting the accumulation of carbon).

The EODESM system has been designed to associate a change (described through the accumulation of evidence) with a number of causes and consequences (including movements of materials), which the most likely determined through consideration of evidence. As such, the system provides a range of users (e.g., scientists, nature conservation managers, politicians) with knowledge that can be used to make informed decisions on many aspects of the landscape relating to, for example, emergency response to adverse events, land management over varying time frames, the impacts of past and current policies and planning future landscapes.

Concluding Remarks

Using the vast archives of historical earth observation data and new concepts, such as those developed through the EODESM system, we can already place ourselves within a landscape, both currently and at specific points over the past 30 or so years, and describe the key elements relating to vegetation, water, bare areas and artificial and cultivated environments. This capacity has been enhanced considerably through the recent provision of near daily data from multiple sensors on board satellites including the RapidEye, Planetscope and Sentinel-1/2 and viewing platforms such as Google Earth, Google Earth Engine and Planet Lab's Explorer. Through knowledge of past landscapes, we can now better understand the reasons for their composition today and plan for future landscapes that balance human use of the land with the requirements of its flora and fauna, with this ultimately leading to societal and economical benefits. This capacity is set to increase significantly with advances in computing technology and engineering. This will also raise our understanding of the environment and how it functions and adapts in response to change. The ability to observe changes now and back in time and into the future is therefore becoming possible and the path is open for this to occur routinely and on demand.

These new advances create significant opportunities for nature conservation as events and processes within the landscape can now be observed in near real time and historically. Much of the perceived complexity in obtaining and pre-processing imagery, and extracting information that is of practical use, has been overcome by the provision of analysis ready datasets and classification and change detection systems such as EODESM. Many of the algorithms used for the retrieval of environmental variables and classifications of landscape, as well as local to global products, are becoming openly and freely available and transparent, with these generated by scientists with decades of experience in earth observation. As illustration, the software used in the development and implementation of the EODESM system is open source and freely available and is based primarily on python scripts and the RSGISLib (Bunting et al. 2013; Clewley et al. 2014). Numerous options are becoming available for routinely evaluating the accuracy and reliability of these products, giving confidence to many users. For these reasons, earth observation datasets can now be better used to transform the way that our environment is managed and conserved. No longer are procedures and products remaining within the realms of the scientific community; they now can transition into being used to support nature conservation in a more practical sense. For this, we give credit to many individuals, groups and organisations (e.g., space agencies, governments and businesses such as Google) for facilitating free and public distribution.

Acknowledgements The authors would like to thank all participants of the H2020 ECOPOTENTIAL and FP7 BIOSOS funded projects. This project has received funding from the European Union's Horizon 2020 research and innovation programme under grant agreement No. 641762.

References

Bunting, P., Clewley, D., Lucas, R.M., Gillingham, S.: The remote Sensing and GIS Software Library (RSGISLib). Comput. Geosci. **62**, 206–226 (2013)

Clewley, D., Bunting, B., Shepherd, J., Gillingham, S., Flood, N., Dymond, J., Lucas, R., Armston, J., Moghaddam, M.: A python-based open source system for geographic object-based image analysis (GEOBIA) utilising raster attribute tables. Remote Sens. **6**(7), 6111–6135 (2014)

Di Gregorio, A.: Land Cover Classification System (LCCS). Classification concepts and user manual for software version 2. FAO Environment and Natural Resources Series, **8**, 190 pp (2005)

Feng, M., Sexton, J.O., Channan, S., Townshend, J.R.: A global, high-resolution (30 m) inland water body dataset for 2000: first results of a topographic-spectral classification algorithm. Int. J. Digit. Earth. **9**(2), 113–133 (2016). doi:10.1080/17538947.2015.1026420

Giri, C., Ochieng, E., Tieszen, L.L., Zhu, Z., Singh, A., Loveland, T., Masek, J.L., Duke, N.: Status and distribution of mangrove forests of the world using earth observation satellite data. Glob. Ecol. Biogeogr. **20**, 154–159 (2011)

Hall, D.K., Riggs, G.A.: MODIS/Terra Snow Cover 8-Day L3 Global 500 m Grid, version 6. Boulder, Colorado, USA. NASA National Snow and Ice Data Center Distributed Active Archive Center. doi:http://dx.doi.org/10.5067/MODIS/MOD10A2.006 (2016)

Hansen, M.C., Potapov, P.V., Moore, R., Hancher, M., Turubanova, S.A., Tyukavina, A., Thau, D., Stehman, S.V., Goetz, S.J., Loveland, T.R., Kommareddy, A., Egorov, A., Chini, L., Justice, C.O., Townshend, J.R.G.: High-resolution global maps of 21st-century forest cover change. Science. **342**(6160), 850–853 (2013)

Kosmidou, V., Petrou, Z., Bunce, R.G., Mucher, C., Jongman, R.H., Bogers, M.M., Lucas, R.M., Tomaselli, V., Blonda, P., Pado-Schioppa, E., Manakos, I., Petrou, M.: Harmonization of the land cover classification system (LCCS2) with the General Habitat Categories (GHC) classification system. Ecol. Indic. **36**, 290–300 (2013)

Langanke, T., Büttner, G., Dufourmont, H., Iasillo, D., Probeck, M., Rosengren, M., Sousa, A., Strobl, P., Weichselbaum, J.: GIO land (GMES/Copernicus initial operations land) High Resolution Layers (HRLs) – summary of product specifications. ESA GIO Land Team. 16p (2013)

Lucas, R., Blonda, P., Bunting, P., Jones, G., Inglada, J., Arias, M., Kosmidou, V., Petrou, Z., Manakos, I., Adamo, M., Charnock, R., Tarantino, C., Mucher, C.A., Jongman, R., Kramer, H., Arvor, D., Honrado, J., Mairota, P.: The Earth Observation Data for Habitat Monitoring (EODHaM) System, JAG. Int. J. Appl. Earth Observ. Geoinf. Spec. Issue Earth Obser. **37**, 17–28 (2014)

Merchant, C.J., Le Borgne, P., Marsouin, A., Roquet, H.: Optimal estimation of sea surface temperature from split-window observations. Remote Sens. Environ. **112**(5), 2469–2484 (2008)

Nagler, T., Rott, H.: Retrieval of wet snow by means of multitemporal SAR data. IEEE Trans. Geosci. Remote Sens. **38**(2), 754–765 (2000)

Pasolli, L., Notarnicola, C., Bertoldi, G., Bruzzone, L., Remelgado, R., Greifeneder, F., Niedrist, G., Della Chiesa, S., Tappeiner, U., Zebisch, M.: Estimation of soil moisture in mountain areas using SVR technique applied to multiscale active radar images at C band. IEEE J. Sel. Topics Appl. Earth Observ. Remote Sens. **8**(1), 262–283 (2015)

Pekel, J.-F., Cottam, A., Gorelick, N., Belward, A.S.: High-resolution mapping of global surface water and its long-term changes. Nature **540**(7633), 418–422 (2016)

Rana, F.M., Adamo, M., Pasquariello, G., De Carolis, G., Morelli, S.: LG-Mod: a modified local gradient (LG) method to retrieve SAR sea surface wind directions in marine coastal areas. J. Sens. **2016**(8), 1–7 (2016)

Sexton, J.O., Song, X.-P., Feng, M., Noojipady, P., Anand, A., Huang, C., Kim, D.-H., Collins, K.M., Channan, S., DiMiceli, C., Townshend, J.R.G.: Global, 30-m resolution continuous fields of tree cover: landsat-based rescaling of MODIS vegetation continuous fields with lidar-based estimates of error. Int. J. Digit. Earth. **6**(5), 427–448 (2013)

Tomaselli, V., Panayotis, D., Marangi, C., Kallimanis, A., Adamo, M., Tarantino, C., Panitsa, M., Terzi, M., Veronico, G., Lovergine, F., Nagendra, H., Lucas, R., Mairota, P., Mucher, S., Blonda, P.: Translating land cover/land use classifications to habitat taxonomies for landscape monitoring: a Mediterranean assessment. Landsc. Ecol. **28**(5), 905–930 (2013)

Verbesselt, J., Hyndman, R., Zeileis, A., Culvenor, D.: Phenological change detection while accounting for abrupt and gradual trends in satellite image time series. Remote Sens. Environ. **114**, 2970–2980 (2010)

Yang, H., Li, S., Chen, J., Zhang, X., Xu, S.: The standardization and harmonization of land cover classification systems towards harmonized datasets: a review. Int. J. Geo-Inf. **6**, 154–170 (2017)

Expected Advances in a Rapidly Developing Work Area

Richard Lucas, Ricardo Díaz-Delgado, and Clive Hurford

Abstract Rapid changes in the global environment, including those associated with climatic fluctuation, are necessitating new approaches to nature conservation, which are being facilitated and partly driven by the introduction and advancement of earth observation technologies. These include ground, airborne and spaceborne platforms and sensors as well as advanced computing hardware and software. A key development in recent years has been the provision of free and open earth observation data and derived datasets as well as methods used for their processing and analysis. Many organisations (e.g., space agencies, governments) are increasingly recognising the need to provide relevant information to a wide range of users, including those charged with nature conservation, but there is still a need to ensure that requirements are conveyed and adequately addressed. Furthermore, practitioners should ensure that they obtain the capacity, knowledge and skills necessary to ensure correct and informed use of these data, particularly in relation to management of protected and also unprotected areas. Systems that effectively integrate data from a wide range of sources also need to be developed, particularly for monitoring.

Keywords Environmental change • Nature conservation • Earth observation • New technologies • Field measurements • Classification • Monitoring

R. Lucas (✉)
Centre for Ecosystem Science, School of Biological, Earth and Environmental Sciences, The University of New South Wales, HIgh Street, Kensington (Sydney) NSW 2052, Australia
e-mail: richard.lucas@unsw.edu.au

R. Díaz-Delgado
Estación Biológica de Doñana (EBD), Consejo Superior de Investigaciones Científicas (CSIC), Sevilla, Spain
e-mail: rdiaz@ebd.csic.es

C. Hurford
Natural Resources Wales, Cardiff, Wales, UK
e-mail: churford5@gmail.com

© Springer International Publishing AG 2017

R. Díaz-Delgado et al. (eds.), *The Roles of Remote Sensing in Nature Conservation*,
DOI 10.1007/978-3-319-64332-8_16

Global Driving Forces Over Protected Areas

Compared to today, nature conservation could have been regarded as relatively straightforward, although that was never really the case. The main focus historically had been on the establishment of protected areas where nature would thrive and sustain itself whilst those areas outside, although still supporting varying types of habitats and wildlife, would primarily be available for exploitation by humans. Whilst conservation reserves are vital, and have played an important role in species, habitat and ecosystem protection, this view has had to change, largely because of a human population that has doubled in the past 40 years from 3 to over 7 billion, through expansion at an average rate of 80 million people per year, and which is anticipated to increase to over 9 billion by 2038. A major consequence of this increase has been and continues to be a rapid rise in levels of greenhouse gases in the atmosphere, primarily because of fossil fuel consumption since the industrial revolution but also losses of natural vegetation, including carbon rich forests. The resulting net increase in global temperatures has contributed to the loss of ice sheets and glaciers, greater frequencies and intensities of fires across the globe and the melting of permafrost, with these further exacerbating changes in climate through positive feedbacks. The result now is that protected areas are no longer protected, as the majority of ecosystems are now vulnerable to the global phenomenon of climate change as well as other changes such as increased atmospheric nitrogen depositions.

The sheer magnitude of environmental change across the world has only been fully realized because of regular observations by satellites, which really only started from the 1960s onwards. These early sensors provided insights and warnings of what was to come, including the initial signs of deforestation and associated burning in the Amazon (from NOAA AVHRR data; Setzer and Pereira 1991), the existence of the ozone hole (Farman et al. 1985) and the losses of ice in the Arctic (Strove et al. 2007). As the value of Earth observations became greater, more and more sensors were launched, with these providing different views of the Earth's surface and sub-surface and progressively documenting the changes that have been occurring over the past 60 years. The development of algorithms for describing and quantifying features of the landscapes has also resulted in new insights into the functioning of natural systems, with examples being retrieval of glacial velocities (Rignot and Kanagaratnam 2006), the three-dimensional structure and biomass of vegetation (Lefsky 2010; Saatchi et al. 2011) and the frequencies and durations of water inundation (Pekel et al. 2016; Díaz-Delgado et al. 2016). By providing new information of the environment from these earth observation data, we have been able to gain a better understanding of how our world has changed and the causes and consequences. Such knowledge is vital if we are to address nature conservation over the coming years and decades.

Where to From Here? The Nature Conservation Perspective

For nature conservation, a key message is to embrace and adopt the new technologies that are becoming available and to ensure that decisions are based on a sound and up-to-date understanding of changes occurring from local to global scales and the interactions between these. This is particularly important as local-scale changes may be the direct or indirect result of, for example, national policies (e.g., relating to agriculture or forestry), regional processes (e.g., urbanization and people migration or wars) or global climate change. Earth observation data provides an essential framework that can be used to address many of the changes that are occurring or may take place in the future.

As highlighted in this book, a major issue is which data to use and how. The number of Earth observing sensors is substantial and increasing, as are the products arising from these; and this can be overwhelming. However, major steps have been taken to ensure that many of the previous limitations to using these data are overcome. There is no longer reliance on just one or two satellite scenes; entire archives are publically available and the software and algorithms needed to process these data are increasingly being provided at no cost. Open source GIS, image processing and statistical software for desktop usage are also now available, with notable examples being QGIS, Image-J and the R statistical package. The release of satellite archives has further allowed changes to be observed and placed in context, particularly as time-lapses of change superbly convey the processes of change whether it be substantive deforestation in tropical regions, loss of water bodies (such as the Aral Sea) or retreating glaciers in high mountain regions. There has also been an increased drive to provide analysis ready satellite data and also products that are being made available across a range of scales and temporal frequencies. However, despite these efforts, there are still the age-old limitations of time, finances, resources, knowledge and skills.

A common assumption is that conservation practitioners have the time, manpower and other resources (e.g., computing) to make maximum use of Earth observation data and so why are these not yet being used to their full potential? Many organizations simply do not have the time, staff or the skills to undertake the essential search, download and processing of satellite data or products. To address this, some focus needs to be placed on establishing and reviewing the nature of the overall savings or other benefits associated with investing in Earth observation data and to then balance these against a 'business as usual' scenario; and the following might be useful to consider.

(a) Better targeting of field campaigns, surveys or compliances, including of fauna surveys, that link to mapped habitats and their different states and dynamics.
(b) Establishing reference datasets (e.g., regional or national airborne LiDAR coverages or phenology) that can be used to determine whether changes within protected areas and surrounds are out of the ordinary.
(c) Collating and augmenting information that can be used to gauge the impact of management activities, assist with conservation planning and enforcement, and ensure the integrity of protected areas.

When Earth observation data are obtained, there is often a lack of confidence in what individuals or groups produce from these and how this stands up when reviewed, particularly if they are not experts in the field. This often arises from a lack of training as many conservation managers have greater expertise in fields other than Earth observation, such as ecology and water management. Similarly, Earth observation scientists may not have expertise in these areas. Understanding the theory and concepts behind Earth observation takes time and effort. In order to address this, the following approach may be useful.

(a) Identify and nurture staff in the organization who show a keen interest in Earth observation for nature conservation.
(b) Identify the key skills that exist or are needed.
(c) Seek or provide funding that allows individuals or groups to be trained, with clear objectives and outcomes that benefit protected areas and ensure long-term engagement of staff and continuity in effort. These might include, for example, pilot training for drones, basic remote sensing image processing and GIS analyses, or computer programming in common languages.
(d) Collaborating (ideally long-term) with individuals or institutions with existing skills, expertise and knowledge in the processing and analysis of Earth observation data.

A large number of short to long-term training opportunities are available, with these including those that are online, provided as short courses or conducted through formal tuition (e.g., Masters or Ph.D. programmes). Linking a series of training courses with qualifications or accreditations is often beneficial and motivating for staff, including volunteers, and allows them to advance careers and increase capacity to achieve nature conservation goals.

Where to From Here? A Remote Sensing Perspective

At the time of writing, we have already entered a new era in Earth observation with the launch of the Sentinel-1/2 satellites providing free radar and optical data at spatial resolutions as high as 10 m and with observations at least every 5 days across multiple wavelength regions. The Landsat-8 sensor data are also available at no cost to the user. In both cases, analysis ready data (i.e., orthorectified and corrected to surface reflectance or with radar calibrations and topographic corrections) are being delivered as well as pre-processing and analysis software. Argentina's and the USA's forthcoming SAOCOM and NiSAR, will be providing L-band Synthetic Aperture Radar (SAR) data at no cost, with this complementing the freely available annual 25 m mosaics of Japanese L-band SAR data, archives of which extend back to the mid 1990s. NASA's Ice, Cloud and land Elevation Satellite (ICESat) Geoscience Laser Altimeter System (GLAS; 2003–2009) is to be superseded by the Global Ecosystem Dynamics Investigation (GEDI) LiDAR, with both providing the capacity to retrieve information on the vertical distribution of plant material with canopy volumes. Open source software, including Sen2Cor and ARCSI (Bunting 2017; this book), are increasingly being provided to support refined correction of satellite sensor data, allowing compensation for differences in topographic illumination and cloud and cloud shadow detection and removal.

Whilst vast amounts of satellite sensor data are being provided, a major issue that is widely acknowledged by Earth observation scientists is where and how to store and process these data. It is unfeasible that users individually download these data to their own computing systems. For this reason, there has been an increasing move towards the use of data cubes whereby algorithms for pre-processing and analysing Earth observation data are submitted to centralized storage facilities, with these allowing targeted access to and analysis of the full time-series of satellite sensor (e.g., Landsat) data. Using such a facility, the Australian Geoscience Data Cube (AGDC) Water Observations from Space (WOfS) has generated maps of water inundation for Australia between 1987 and 2014 using all available Landsat sensor data (Mueller et al. 2016). Other products that are being generated include intertidal mudflats and mangrove dynamics over similar time frames. Data from optical sensors can also be interrogated to retrieve data profiles (e.g., of NDVI) extending back in time and informing on both rapid and gradual changes across the Australian landscape. Similar work has been produced using the Google Earth Engine, including a decadal survey of global tree cover extent, loss and gain (Hansen et al. 2013). The capacity to store these data and conduct analysis has been achieved by rapid advancements in computing storage and parallel processing technologies (e.g., within high performance computing environments). Access to these facilitates has also widened with the provision of web-based services such as the Amazon Cloud and Microsoft Analytic Platform System (APS). Whilst users are charged for these systems and some knowledge of data processing is needed, these are providing the capacity for a greater number of users to access and process data and generate useable products. Many of these products are also being made publicly available, which presents the opportunity for these to be used in support of nature conservation.

One of the major advances in Earth observation that is likely to significantly benefit nature conservation is the development of CubeSat technologies (e.g., PlanetLabs and Surrey Satellite Technology Ltd.), with these providing small, micro and nano satellites (defined on the basis of their weight and size) that currently observe the Earth's surface on a daily basis in different wavelength combinations (e.g., red, green, blue (RGB), red edge and/or near infrared). These data can be used in combination with the more moderate resolution sensors (e.g., Sentinel and Landsat) to determine, in more detail, the nature and extent of changes. As examples, these allow seasonal variations in vegetation phenology to be captured or the patterns of inundation to be mapped at high (typically <3 m) spatial resolution. Perhaps of greatest importance to nature conservation is that these sensors provide early evidence of changes (e.g., deforestation, ploughing of species-rich pastures, pollution events), giving time for enforcement and/or management teams to prevent further damage including loss of habitats and associated species. The increased future use of targeted (e.g., red edge) channels on these satellites will also allow better discrimination of, for example, plant species, particularly given the capacity to analyze seasonal trajectories. Increasingly, a greater number and diversity of spectral bands (e.g., coastal, red edge and short wave infrared) are being provided by other spaceborne VHR sensors, with the Worldview-2 and 3 sensors being notable examples.

Alongside the release of satellite data archives, algorithms and software for processing dense time series of remote sensing data has also been developed, with these including those that identify breakpoints and trends associated with change events or processes (e.g., the TIMESAT program for analyzing time-series of satellite sensor data (Jönsson and Eklundh 2004) or Breaks For Additive Season and Trend or BFAST (Verbesselt et al. 2010)). Others use past data to establish typical trends, deviations from which can indicate a change in land cover (e.g., Zhu et al. 2012). The benefit of these algorithms is that they can be used to better describe and target analysis of change, which allows greater responsiveness to and management of adverse events and processes.

Airborne observations have proved to be highly beneficial, if not essential, in the interpretation of data acquired by spaceborne sensors and derived products and as standalone sources of information, and key amongst these has been airborne LiDAR. Whilst discrete return LiDAR have been available for over two decades, many LiDAR today are operating using full waveform technology (Lim et al. 2003) and there is a drive towards the use of multi-spectral LiDAR (Morsdorf et al. 2009). These sensors now have the potential to provide highly detailed information on the three-dimensional structure of vegetation but also facilitate the discrimination of plant species as well as component materials (leaves and wood) and the generation of Digital Surface Models (DSMs). Time-series of these data are also being exploited to better understand the dynamics of plant communities, with notable successes being in savanna woodlands (Cho et al. 2012). Digital Terrain Models (DTMs) generated from these data also provide new opportunities to understand ecological processes, including loss of sediments through erosion and water flows.

Drone technology has also revolutionized the capacity to obtain very high (centimeter) resolution information that supports the interpretation, calibration and validation of products from spaceborne but also airborne sensors (Colomina and Molina 2014). The data from these sensors fill a significant gap in terms of scaling ground-based measurements to larger areas, with these including distributions of plant species (through classification of RGB or multi-spectral data) or canopy height maps from point clouds derived through stereo imaging (photogrammetry). Nowadays, drones (either with fixed or rotatory wings) are available at affordable prices, and can support a range of sensor types (e.g., RGB and multispectral cameras). As the weight of hyperspectral and LiDAR sensors decreases, these are also being deployed on drones allowing, for example, better discrimination of plant species and interrogation of land covers in multiple dimensions. Accordingly, many applications addressed by spaceborne or airborne images can also exploit drones, with a major advantage being the possibility to obtain images à la carte. Other new applications beyond pixel mapping are emerging in relation to fauna detection, surveying and censoring (e.g. counting nests in bird colonies; Díaz-Delgado's et al. (2017 - this book)).

Other ground-based sensors are also highly beneficial to nature conservation, notable amongst which are Terrestrial Laser Scanners (TLS; Mass et al. 2008; Telling et al. 2017), phenocams (Brown et al. 2016) and, as mentioned in earlier chapters, camera traps. TLS provide millimeter resolution point clouds of surfaces (e.g., the terrain) and structures (e.g., vegetation), albeit for areas often covering less than 1 ha. For vegetation, they can be used to locate and extract the dimensions of individual plants, from which biophysical measures such as component (branch and trunk) biomass and height can be obtained. They can provide a permanent record of the distribution of plants and plant material and also change when taken in a time-series, with this made possible through high precision in georeferencing. These datasets can therefore provide viable substitutes for or can complement permanent plots, which have traditionally been recorded through more intensive ground-based location and measurement of plants. Phenocams are also well suited to monitoring changes in leaf flush and fall and other metrics describing leaf phenology, and provide useful interpretation of Earth observation data. Finally, we need to mention the use of mobile devices that allow collection (including in near real time) of information on the environment, such as land cover and habitat types, biophysical variables and biodiversity distributions. This capability provides a new opportunity to link observations from airborne and spaceborne sensors to those on the ground.

Accessing and using these technologies can be a significant obstacle to uptake and the best advice is to decide in advance what is needed to support nature conservation and then to select the tools that are best suited for the specific job. This will inevitably involve extra effort in terms of research capabilities and the processing that is needed, as well as increased training requirements (as mentioned earlier). However, the development of a longer-term strategy for planning the use of these data and technologies should result in their progressive adoption and integration and ultimately lead to more effective and efficient management of protected areas and surrounds.

Realistic Expectations

Whilst the potential contribution from Earth observation and related technologies is enormous, there is a need to ensure that expectations are not raised to unrealistic levels. Earth observations cannot provide the answers to many of the questions posed by conservation practitioners for a multitude of reasons. As examples, many plant species, particularly if small, rare, or occuring in mosaics or the lower strata, cannot be observed let alone discriminated from Earth observation data. The timing and frequency of observations is also often sub-optimal and the georeferencing inadequate to allow exact matching with ground data. Nevertheless, Earth observation data provide information that can be used in parallel with ground data, including the extent of different land cover and habitat types and environmental measures of relevance to conservation (e.g., phenology and water supply). These data also provide information on how dominant species or genera are responding to pressures and threats and allow targeting of effort, particularly where unexpected changes are detected. Of importance is that these data provide a historical record of the states and dynamics of habitats over areas that are far larger than those that can be surveyed from the ground alone. So, the overall message is to find ways to integrate both approaches, where appropriate, to maximize and optimize the provision of information in support of nature conservation.

A common approach to demonstrate and convey the benefits of advances in Earth observation for nature conservation has been to undertake joint studies between scientists and conservation practitioners, focusing on one or only a few sites. Whilst successes have been obtained, a better approach would be to develop Earth observation techniques that are more widely applicable beyond these sites or support national or international efforts. An example of this is the EODESM system (Lucas and Mitchell (2017), this book), which provides scalable classifications across protected areas using the consistent and globally recognized FAO LCCS taxonomy and, at the same time, attributes classes with universally recognized environmental variables. However, when encouraging the use of a specific taxonomy, it is important to convey or demonstrate also how this can be translated to other taxonomies that may be more familiar or relevant. Examples might be the Ramsar Classification System for Wetlands, which can be reproduced from the LCCS but requires additional information on context (e.g., marine, estuarine, tundra, alpine). Recognition of taxonomies used commonly by ecologists is also important when developing or designing approaches to classification from Earth observation data.

A further criticism of Earth observation products is that these are not fit for purpose or are of insufficient accuracy. As an example, grassland maps may not consider soil acidity measures in their taxonomy. The accuracy of grassland maps may also be stated as exceeding a value that is often considered acceptable (80% correct 80% of the time), but that of grassland habitats of conservation importance may be far lower. Hence, there is a need for remote sensing scientists to consider the characteristics of these more specific and critical habitats and how these might be better discriminated, described and mapped; and also provide realistic measures of accuracy. Perhaps more importantly, there needs to be a dialogue between remote sensing and conservation practitioners in terms of the taxonomies used for classification to ensure that these meet the needs on the ground, with this occurring before the commencement of projects. Communication of ideas and methods and collaborations between all parties is essential to ensure that funding and resources are effectively and efficiently utilized.

When more detailed classifications are generated for sites, acknowledgement of both the actual and potential presence of plant species is needed. As an example, the training of a classification of Australian open woodland tree species from airborne hyperspectral data relied on a set of tree species defined through ground observations (Lucas et al. 2008). However, when conducting an independent ground-based validation of the classifications in other areas, rare tree species that had not been previously observed in the area were found. These were therefore not mapped as their spectral characteristics and statistics were not integrated in the training dataset. This is also the case where plant species are associated with habitat types or structures but many may have been lost from sites as a consequence of historic management practices (e.g. neglect, over-grazing, burning or catastrophic management). In these cases, species may be mapped where they no longer exist. These two cases demonstrate the need for timely and comprehensive ground observations to inform on the classifications of Earth observation data.

Concluding Remarks: What Has Remote Sensing Done for Us?

Over the past 60 years or so, Earth observation data have progressively infiltrated many aspects of society, economy and environment, with this achieved through step-change and often parallel developments in Global Positioning Systems (GPS), communications and navigation technologies, Google Earth and other web-based delivery platforms, Geographic Information Systems and services, mobile technologies and devices, drone capability, hyperspectral and LiDAR remote sensing, new sensor designs, constructions and operations, cloud and high performance computing, and the provision of free and open analysis ready data to name a few. Often without our knowledge, these data have played, and will continue to play, a major role in all of our activities, whether it be weather forecasting, assessing the impact of climate change, monitoring deforestation, controlling wildfires or responding to adverse events such as flooding, oil spills or storms. Remote sensing has already made an enormous but often understated contribution to how we view, manage and respond to our environment. Indeed, without Earth observations, we might not have been aware of the scale of the changes that are occurring across the world and instead we would be at a point where we no longer have choices in how we can respond. However, there is no room for complacency and it is now more important than ever to ensure that these technologies make a significant contribution to ensuring that plant and animal species continue to form and inhabit our environment for the benefit of today's and tomorrow's populations.

References

Brown, T.B., Hultine, K.R., Steltzer, H., Denny, E.G., Denslow, M.W., Granddos, J., Henderseon, S., Moore, D., Nagai, S., SanClements, M., Sánchez-Azofeifa, A., Sonnentag, O., Tazik, D., Richardson, A.D.: Using phenocams to monitor our changing Earth: toward a global phenocam network. Front. Ecol. Environ. **14**, 84–93 (2016)

Bunting, P.: Pre-processing of remotely sensed imagery. In: Díaz-Delgado, R., Lucas, R., Hurford, C. (eds.), The Roles of Remote Sensing in Nature Conservation. A Practical Guide and Case Studies. Springer, Cham (2017)

Cho, M.A., Mathieu, R., Asner, G.P., Naidoo, L., van Aardt, J., Ramoel, A., Debba, P., Wessels, K., Main, R., Smit, I., Erasmus, B.: Mapping tree species composition in South African savannas using an integrated airborne spectral and lidar system. Remote Sens. Environ. **125**, 214–226 (2012)

Colomina, I., Molina, P.: Unmanned aerial systems for photogrammetry and remote sensing. A review. ISPRS J. Photogrammetry Remote Sens. **92**, 79–97 (2014)
Díaz-Delgado, R., Aragonés, D., Afán, I., Bustamante, J.: Long-Term Monitoring of the Flooding Regime and Hydroperiod of Doñana Marshes with Landsat Time Series (1974–2014). Remote Sens. **8**, 775 (2016)
Farman, J.C., Gardiner, B.G., Shanklin, J.D.: Large losses of total ozone in Antarctica reveal seasonal ClO_x/NO_x interaction. Nature. **315**, 207–210 (1985)
Hansen, M.C., Potapov, P.V., Moore, R., Hancher, M., Turubanova, S.A., Tyukavina, A., Thau, D., Stehman, S.V., Goetz, S.J., Loveland, T.R., Kommareddy, A., Egorov, A., Chini, L., Justice, C.O., Townshend, J.R.G.: High-resolution global maps of 21st-century forest cover change. Science. **342**, 850–853 (2013). doi:10.1126/science.1244693
Jönsson, P., Eklundh, L.: TIMESAT—a program for analyzing time-series of satellite sensor data. Comput. Geosci. **30**, 833–845 (2004). doi:10.1016/j.cageo.2004.05.006
Lefsky, M.A.: A global forest canopy height map from the moderate resolution imaging spectroradiometer and the geo - science laser altimeter system. Geophys. Res. Lett. **37**, L15401 (2010)
Lim, K., Treitz, P., Wulder, M., St-Onge, B., Flood, M.: Lidar remote sensing of forest structure. Prog. Phys. Geogr. **27**, 88–106 (2003)
Lucas, R., Mitchell, A.: Integrated land cover and change classifications. In: Díaz-Delgado, R., Lucas, R., Hurford, C. (eds.), The Roles of Remote Sensing in Nature Conservation. A Practical Guide and Case Studies. Springer, Cham (2017)
Lucas, R.M., Bunting, P., Paterson, M., Chisholm, M.: Classification of Australian forest communities using aerial photography, CASI and HyMap data. Remote Sens. Environ. **112**, 2088–2103 (2008)
Mass, H.-G., Bienert, A., Scheller, S., Keane, E.: Automatic forest inventory parameter determination from terrestrial laser scanner data. Int. J. Remote Sens. **29**, 1579–1593 (2008)
Morsdorf, F., Nichol, C., Malthus, T., Woodhouse, I.H.: Assessing forest structural and physiological information content of multi-spectral LiDAR waveforms by radiative transfer modelling. Remote Sens. Environ. **113**, 2152–2163 (2009)
Mueller, N., Lewis, A., Roberts, D., Ring, S., Melrose, R., Sixsmith, J., Lymburner, L., McIntyre, A., Tan, P., Curnow, S., Ip, A.: Water observations from space: Mapping surface water from 25 years of Landsat imagery across Australia. Remote Sens. Environ. **174**, 341–352 (2016)
Pekel, J., Cottam, A., Gorelick, N., Belward, A.S.: High-resolution mapping of global surface water and its long-term changes. Nature. **540**, 418–422 (2016)
Rignot, E., Kanagaratnam, P.: Changes in the velocity structure of the Greenland Ice Sheet. Science. **311**, 986–990 (2006)
Saatchi, S.S., Harris, N.L., Brown, S., Lefsky, M., Mitchard, E.T.A., Salas, W., Zutta, B.R., Buerman, W., Lewis, S.L., Hagen, S., Petrova, S., White, L., Silman, M., Morel, A.: Benchmark map of forest carbon stocks in tropical regions across three continents. Proc. Natl. Acad. Sci. U. S. A. **108**(24), 9899–9904 (2011)
Setzer, A.W., Pereira, M.C.: Amazonia biomass burning in 1987 and an estimate of their tropospheric emissions. Ambio. **20**, 19–22 (1991)
Strove, J., Holland, M.M., Meier, M., Scambos, T., Serreze, M.: Arctic sea ice decline: faster than forecast. Geophysical Research Letters. **34**(9), L09501 (2007)
Telling, J., Lyda, A., Hartzell, P., Glennie, C.: Review of earth science research using terrestrial laser scanning. Earth Sci. Rev. **169**, 35–68 (2017)
Verbesselt, J., Hyndman, R., Zeileis, A., Culvenor, D.: Phenological change detection while accounting for abrupt and gradual trends in satellite image time series. Remote Sens. Environ. **114**, 2970–2980 (2010)
Zhu, Z., Woodcock, C.E., Olofsson, P.: Monitoring of forest disturbance using all available Landsat imagery. Remote Sens. Environ. **122**, 75–91 (2012)

Printed by Printforce, the Netherlands